Molecular and Cellular Biology

Molecular and Cellular Biology

Edited by Gloria Doran

SYRAWOOD
PUBLISHING HOUSE

New York

Published by Syrawood Publishing House,
750 Third Avenue, 9th Floor,
New York, NY 10017, USA
www.syrawoodpublishinghouse.com

Molecular and Cellular Biology
Edited by Gloria Doran

International Standard Book Number: 978-1-68286-570-5 (Hardback)

Cataloging-in-Publication Data

Molecular and cellular biology / edited by Gloria Doran.
        p. cm.
Includes bibliographical references and index.
ISBN 978-1-68286-570-5
1. Molecular biology. 2. Cytology. 3. Biology. I. Doran, Gloria.
QH506 .M65 2018
572.8--dc23

TABLE OF CONTENTS

PREFACE

Every book is a source of knowledge and this one is no exception. The idea that led to the conceptualization of this book was the fact that the world is advancing rapidly; which makes it crucial to document the progress in every field. I am aware that a lot of data is already available, yet, there is a lot more to learn. Hence, I accepted the responsibility of editing this book and contributing my knowledge to the community.

Cellular biology focuses on the different functions and structures of cells. Molecular biology studies biological activity at a molecular level. Some of the techniques used in molecular biology are molecular cloning, gel electrophoresis, polymerase chain reaction, microarrays, etc. The recent discoveries in the field of molecular and cellular biology have made major contributions to medical research and drug development. This book is a compilation of chapters that discuss the most vital concepts and emerging trends in the field of molecular and cellular biology. A number of latest researches have been included to keep the readers up-to-date with the global concepts in this area of study.

While editing this book, I had multiple visions for it. Then I finally narrowed down to make every chapter a sole standing text explaining a particular topic, so that they can be used independently. However, the umbrella subject sinews them into a common theme. This makes the book a unique platform of knowledge.

I would like to give the major credit of this book to the experts from every corner of the world, who took the time to share their expertise with us. Also, I owe the completion of this book to the never-ending support of my family, who supported me throughout the project.

Editor

Transgene Expression under the Adenoviral Major Late Promoter, Tripartite Leader Sequence and E1 Genes in Absence and Presence of Adenovirus Infection

Mohamed El-Mogy[1] & Yousef Haj-Ahmad[1]

[1] Department of Biological Sciences, Brock University, ON, Canada

Correspondence: Mohamed El-Mogy, Department of Biological Sciences, Brock University, 500 Glenridge Ave, St. Catharines, ON L2S 3A1, Canada. E-mail: melmogy@hotmail.com

The research is financed by the Egyptian Government and the Egyptian Cultural & Educational Bureau in Canada

Abstract

Active gene expression driven by the adenoviral major late promoter (MLP) and the tripartite leader sequence (TPL) influenced their utilization to drive vaccines expression. It is unclear if the complete TPL sequence is required for this active expression or if it can be further boosted in presence of the adenoviral E1 genes. In this study, we investigated the effect of TPL exones and E1 genes, out of the viral genome or under adenoviral infection, on green fluorescence protein (GFP) expression driven by the MLP. Gradual increase in the transcript copies with the addition of TPL exons indicated an additive effect especially under infection. The presence of E1 genes significantly enhanced transgene transcription. Infected cells showed higher GFP transcription and translation than non-infected cells. These results indicate higher exogenous gene expression driven by the MLP can be achieved in the presence of adenoviral components other than the TPL and E1 genes.

Keywords: transgene expression, adenovirus, major late promoter, tripartite leader sequence, adenovirus E1 genes

1. Introduction

The adenoviral major late promoter (MLP) is known to be one of the most active viral promoters in mammalian cells (Markrides, 1999; Papadakis, Nicklin, Baker, & White, 2004). The late phase of adenovirus infection is characterized by the production of an abundant amount of late proteins required to form and assemble the new viral capsids. Active expression in this phase is attributed to the activity of the MLP and the presence of the tripartite leader sequence (TPL). The expression of all the late viral proteins is driven by the MLP which has its full activity during the late phase of the viral infection and is transactivated by the adenoviral E1A proteins (Parks & Shenk, 1997; Ziff & Evans, 1978).

TPL is a 5' untranslated sequence present in all of the late, but none of the early, viral mRNA. Adenovirus serotype 5 (Ad5) leader sequence is 201 bp formed by the splicing of three exons during the post-transcriptional modifications. TPL facilitates mRNA transport and accumulation in the cytoplasm and is responsible for the selective translation of the late viral proteins in preference to the cellular proteins (Zhang, Feigenblum, & Schneider, 1994).

Adenovirus E1B-55K and E4-orf6 play the main role in active transport of TPL-containing mRNA from the nucleus to the cytoplasm (Bridge & Ketner, 1990; Leppard & Shenk, 1989) and also inhibits protein synthesis by reducing phosphorylation of the translation factor E1F2α during the late phase of viral infection (Spurgeon & Ornelles, 2009). The viral transcription sites in the nucleus contain a complex of E1B-55K and E4-orf6 (Gonzalez & Flint, 2002). Evidence suggests that viral mRNA interacts with this complex through the ability of E1B-55K to bind RNA (Horridge & Leppard, 1998) and facilitate its transport to the cytoplasm using the E4-orf6 proteins nuclear localization and transport signals (Dobbelstein, Roth, Kimberly, Levine, & Shenk, 1997). Cellular mRNA transport is blocked by the same complex (Flint & Gonzalez, 2003). Translation of any TPL-attached mRNA is eIF-4F-independent (Dolph, Racaniello, Villamarin, Palladino, & Schneider, 1988). The relaxed secondary structure of TPL facilitates its function in translation initiation even when eIF-4F is inhibited (Dolph, Huang, & Schneider, 1990).

In order to assess the effectiveness of the adenoviral TPL exones and E1 genes to achieve high transgene expression levels, driven from the MLP and independent from the adenoviral infection, we looked at the effect of these components on gene expression in non-infected as well as infected cells. This helped to understand the functional activity and the sufficiency of these components to maintain high expression levels outside of their viral genome context and infected environment.

2. Materials and Methods

2.1 Plasmid Constructs

Five plasmids were constructed based on the pUC19 backbone and all contain the MLP as a common promoter and TPL with all (pMTGA), partial (pMT1GA & pMT1,2GA) or no exons (pMGA). One construct contains the complete TPL and the adenoviral ITRs-E1 (pE1). All of these plasmids contain the GFP expression cassette with differences in the regulatory sequences. The maps of the different GFP expression cassettes are shown in Figure 1.

2.2 Cell Lines and Maintenance

The used Chinese hamster ovary (CHO) cells were the subclone K1 (ATCC CCL-61) derived from the parental cell line initiated by Puck, Cieciura, and Robinson (1958). Cells were maintained as a monolayer in Petri cell culture dishes and cultured in advanced Dulbecco's Modified Eagle Medium (Advanced D-MEM: Invitrogen Corp., Gibco), containing 5% (v/v) fetal bovine serum (FBS, PAA Laboratories Inc.), 1% (v/v) penicillin/streptomycin (Invitrogen Corp., Gibco) and 1% (v/v) glutamine (Invitrogen Corp., Gibco). Growing cells were incubated in a water-jacketed incubator (Fisher Scientific, Pittsburgh PA) at 37°C with 96% relative humidity and 5% CO_2.

2.3 Lipofectamine 2000 Transfection

Lipofectamine 2000 (Invitrogen) was used to transfect plasmids DNA into mammalian cells. The confluency of the monolayer was ensured to be at least 70% at the transfection time. The culture medium was replaced prior the transfection with 2 mL of antibiotic-free medium. The transfection mix for each well (of a 6-well plate) was prepared in 500 µL by mixing plasmid DNA and Lipofectamine 2000. First, 5 µg plasmid DNA was diluted in a total volume of 250 µL using Opti-MEM I Reduced Serum Medium (Invitrogen Corp., Gibco). Similarly, 5 µL Lipofectamine 2000 was diluted in 250 µL total volume by using Opti-MEM. Both the diluted DNA and Lipofectamine 2000 were incubated at room temperature for 5 minutes then mixed together and incubated at room temperature for additional 20 minutes. The 500 µL transfection mixture was added dropwise onto the well and shaken to distribute the mixture evenly. Finally, the plate was incubated at 37°C for 6 hours before changing the medium with the regular, antibiotic-containing, medium.

2.4 Adenovirus and Its Infection

Adenovirus dl309 was used in this study. The viral titre was determined by plaque assay according to the method described by Cromeans, Lu, Erdman, Humphrey, and Hill (2008). Viral infection of mammalian cells was carried out in 6-well plates by using a volume of the viral stock equivalent for the required multiplicity of infection (MOI). This viral volume was mixed with PBS++ (0.01% $CaCl_2.2H_2O$ and 0.01% $MgCl_2.6H_2O$ dissolved in PBS) in a total volume of 500 µL/well and then added to the cell monolayer (after aspirating the medium). The 6-well plate was then incubated for 1 hour at 37 °C with 96% relative humidity and 5% CO_2, with swirling the plate every 15 minutes. After that, 2 mL of the culture medium were added to each well and the plate was returned to the incubator.

2.5 RNA/DNA Isolation

DNA and RNA were isolated, all from the same sample, using the RNA/DNA/Protein Purification Kit (Norgen Biotek Corp.), according to the manufacturer's instructions.

2.6 DNase Treatment of RNA

The digestion of residual DNA in the isolated RNA samples was performed by using TURBO DNase (Ambion). Each 50 µL of sample was digested in a 100 µL reaction mixture containing the provided buffer and four units of TURBO DNase, with incubation at 37°C for 30 minutes.

2.7 RNA Cleaning

All cleaned RNA samples were carried out by using the RNA CleanUp and Concentration Kit (Norgen Biotek Corp.), according to the manufacturer's instructions.

2.8 Reverse Transcription

Two to five hundred nanograms of total RNA was used in reverse transcription (RT) reactions. RNA was mixed with 0.5 µL of 100 mM oligo(dT)18 primer (Sigma), and completed to a final volume of 5 µL using RNase/DNase-free water (Ambion). This mixture was heated up for 5 minutes at 70°C, then chilled to 4°C. During the cooling step, 15 µL of the RT reaction solution is added to the mixture. The added RT reaction solution contains 4 µL of 5X First Strand Buffer (250 mM Tris-HCl pH 8.3, 375 mM KCl and 15 mM MgCl$_2$), 2 µL of 0.1 M Dithiothreitol (DTT), 1 µL of 10 mM dNTPs, 0.1 µL Superscript III reverse transcriptase (Invitrogen) and 7.9 µL RNase/DNase-free water (Ambion). The reaction was continued by an incubation step at 25°C for 5 minutes, followed with 90 minutes incubation at 42°C and 15 minutes incubation at 70°C before finally holding the reaction at 4°C.

2.9 Quantitative PCR

Quantitative PCR (qPCR) was performed on a known concentration of template DNA or complementary DNA (cDNA), using the Bio-Rad iCycler thermal cycler. Specific primers within the GFP gene were used (GFP-F: 5' ATCCTGATCGAGCTGAATGG 3' and GFP-R: 5' TGCCATCCTCGATGTTGTG 3') with an amplicon size of 484 bp. The reaction mixture contained 10 µL of 2X SYBR GREEN master mix (Bio-Rad), and 1.2 µL of each primer (5 mM stock). The total volume of the reaction was completed with dH$_2$O to 20 µL. A 15 minutes heating at 95°C was used to activate the hotstart enzyme. Forty amplification cycles were performed as follow: 15 seconds at 95°C, 30 second at 59°C and 1 minute at 72°C. The reaction was kept at 57°C for 1 minute before starting a melting curve analysis by a 0.5°C increment every 10 sec over 80 rounds. A standard curve of known plasmid concentration (10 fg to 1 ng) was used to determine the initial concentration of plasmid in each sample.

2.10 GFP Fluorescence Intensity Quantification

The fluorescence intensity of green fluorescence protein (GFP) was quantified directly from mammalian cells by measuring the relative fluorescence units (RFU) using the BioTek Synergy HT Multi-Mode Microplate Reader. Transfected cells were washed twice with PBS, lifted from the plate and counted. Fifty thousand cells per well were then transferred to a black rounded-bottom 96-well plate (Costar) in a total volume of 200 µL of PBS. The RFU was then measured at an excitation wavelength of 485 nm and an emission wavelength of 528 nm, using non-transfected cells as a blank.

3. Results

Gene expression from the different constructed plasmids was investigated in the absence and presence of adenovirus dl309. An initial experiment was carried out to optimize the viral MOI in CHO cells by using pE1 and then we used the appropriate MOI in the following gene expression evaluation from the different plasmid constructs.

3.1 Gene Expression from pE1 in Cells Infected at Different MOIs

Two sets of monolayer CHO cells were prepared, each set contained five groups of triplicate wells each. First, all wells were transfected with pE1, using Lipofectamine 2000. The transfection medium was changed 6 hours after transfection and cells were allowed to recover from the transfection for another 6 hours. Then, infection with adenovirus dl309 at MOIs of 0, 1, 5 and 10 PFU/cell was carried out. Two negative control groups were prepared; the first is CHO cells transfected with pUC19 and used as a control for transfected cells, while the second is pUC19 transfected and infected with the virus. Samples were collected over time (0, 12, 36 hours, 4.5, 9.5 and 14.5 days post-transfection) from the two sets where one sets was used to isolate DNA and RNA from the attached cellular monolayer, both from same sample, and the second set was used for GFP fluorescence intensity quantification. Three biological replicates were carried out and the statistical analysis of all the data obtained was performed by two way ANOVA using Tukey's test, at a significance level of <0.05.

Cells remained attached in monolayers with MOIs of 0 and 1 PFU/cell, while cell detachment was observed with MOIs of 5 and 10 PFU/cell. The amount of detached cells, on day 5 post-transfection compared to the MOI of 0 PFU/cell, was estimated to be 40-50% with MOI 5 PFU/cell and 80-90% with MOI 10 PFU/cell using a light microscope. Cell detachment seems to be proportional to the MOI. On the other hand, equal amounts of DNA were used in a qPCR reaction and the amount of DNA per CHO cell (3.1 pg/cell) (Gregory, 2012) was used to determine the plasmid copy number per cell, within the different collected samples (Figure 2). A significant (at

P<0.05) increase in plasmid copy number was observed at 12 hours post-transfection, then it decreases significantly on 36 hours post-transfection but is still significantly higher than the baseline. Insignificant increase from the baseline was reached on day 4.5 and afterwards. The same trend was observed for plasmid stability with all of the MOIs.

In addition, RT was performed on equal amounts of the DNase-treated and cleaned RNA samples. Equal volumes of the RT product were used in a qPCR reaction to quantify GFP mRNA copy numbers per cells, based on a total RNA content of 8.17 ± 1.17 pg per each CHO cell (Figure 3). Insignificant (at $P < 0.05$) increment in mRNA copies was seen when using MOI of 1 PFU/cell. MOIs of 5 and 10 PFU/cell showed lower mRNA copies than at MOI of 1 PFU/cell. The GFP intensity was determined in equal cell counts from each sample by measuring the RFU (Figure 4). A significant (at $P < 0.05$) increase in fluorescence intensity was obtained over 36 hours to 5.5 days post-transfection with all of the MOIs used and reached the baseline on day 9.5 and thereafter. Infection at an MOI of 1 PFU has higher expressed GFP than all the other used MOIs (0, 5 and 10 PFU/cell).

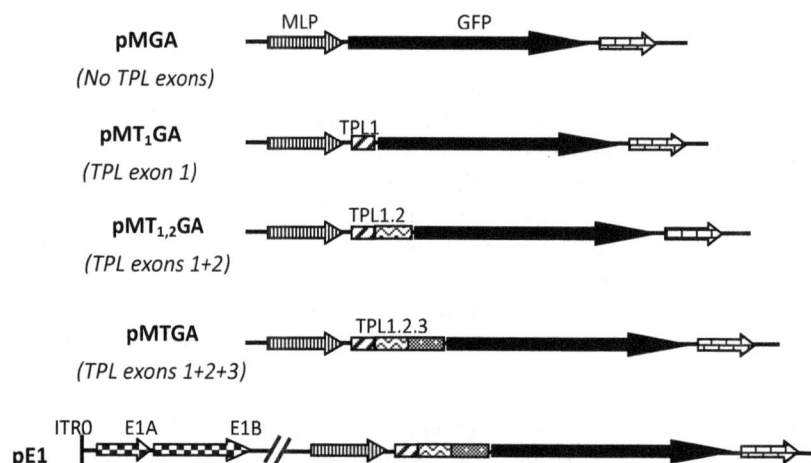

Figure 1. Schematic diagrams of the GFP cassettes and their regulatory sequences in the different plasmids. All of these sequences were constructed into the pUC19 backbone and used in all experiments in the circular plasmid form

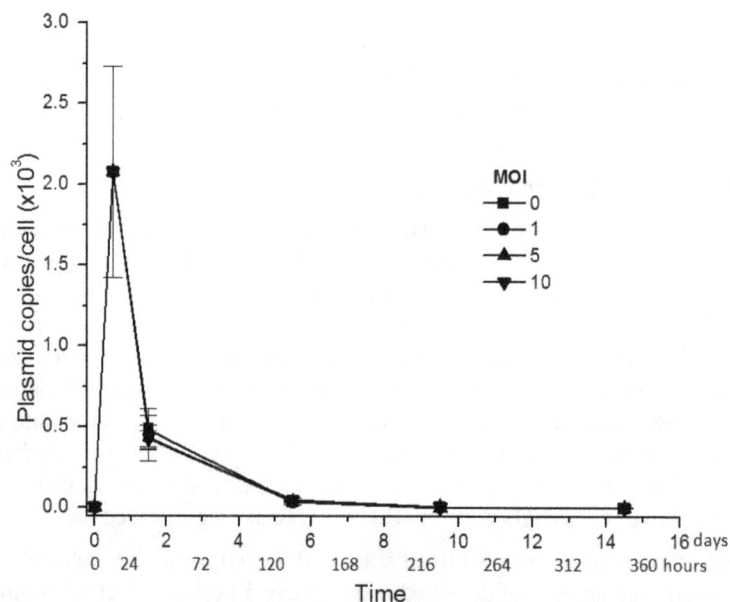

Figure 2. Copy numbers of pE1 over 14.5 days post-transfection in CHO cells, with infection using different MOIs. Copy numbers were obtained by qPCR using a standard curve of known plasmid DNA concentration. qPCR was performed on equal amounts of DNA isolated from collected samples. The MOIs used and their characteristic symbols are shown on the Figure

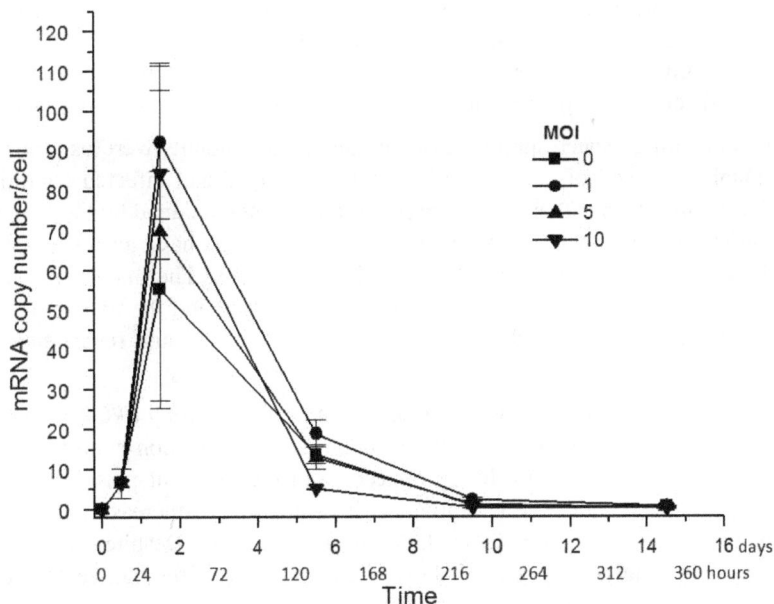

Figure 3. GFP mRNA transcripts from pE1 over 14.5 days post-transfection in CHO cells, with infection using different MOIs. Copy numbers were obtained by qPCR using a standard curve of known plasmid DNA concentration. qPCR was performed on equal volumes of RT product from equal amount of RNA isolated from collected samples. The MOIs used and their characteristic symbols are shown on the Figure

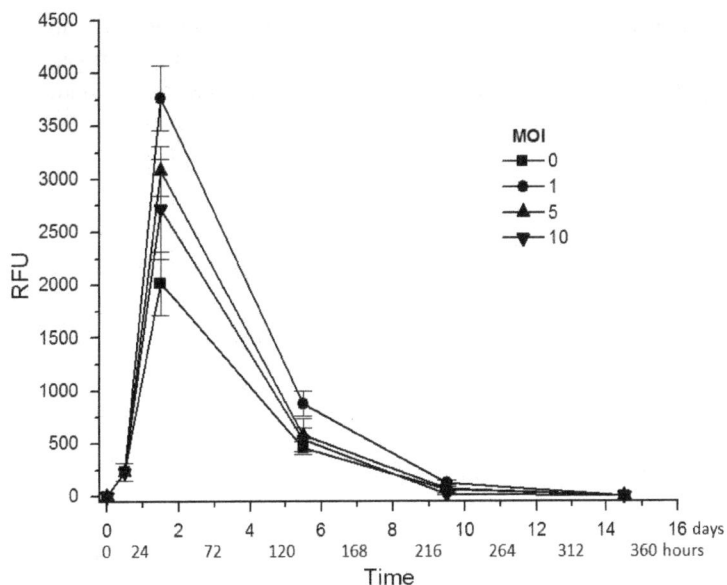

Figure 4. GFP fluorescence intensity over 14.5 days post-transfection in CHO cells, with infection using different MOIs. Fluorescence intensity was measured by the relative fluorescence units (RFU) of equal cell counts. The MOIs used and their characteristic symbols are shown on the Figure

3.2 Gene Expression from the Different Constructs in Cells Infected at MOI of 1 PFU/cell

Expression from the different plasmid constructs was investigated with infection at MOI 1 PFU/cell. The lower MOI allows the side effects encountered at MOIs 5 and 10 PFU/cell to be overcome, which is the massive cellular detachment from the monolayer. All of the plasmids were transfected into CHO cells using Lipofectamine 2000 in two triplicate groups for each plasmid, the first group to be used in RNA and DNA isolation while the second to be used in GFP fluorescence intensity quantification. The medium was changed 6 hours post-transfection to allow the cells to recover from the transfection mix. Twelve hours after transfection, adenovirus was infected into one group of each plasmid. In addition to the zero and 12 hours (considered as the infection's 0 time) collection times, samples were collected at 36, 84 hours, 7.5, 11.5 and 15.5 days post-transfection. Two negative controls were included; the first is CHO cells transfected with pUC19 and used as a control for transfected cells, while the second is pUC19 transfected and infected with the virus.

The isolated DNA was quantified spectrophotometrically and equal amounts were used in a qPCR reaction to determine the copy number per cell of each plasmid in the transfected and infected conditions (Figure 5). As revealed by the graphs and the statistical analysis performed on the data obtained by the two way ANOVA using Tukey's test at a significance level of less than 0.05, all of the plasmids have an insignificant change in their stability in CHO cells with the transfection and the infection conditions. The maximum plasmid copy number was obtained at 12 hours post-transfection with all of the plasmids used and the two conditions. The plasmids' copy numbers remain significantly elevated on 36 hours post-transfection and reach the baseline on day 7.5 post-transfection.

The DNase treated and cleaned RNA was used in an RT reaction followed by qPCR and the copy numbers of GFP mRNA transcripts from each plasmid in the transfection and the infection conditions were calculated per cell (Figure 6). The trend for mRNA transcripts produced from the different plasmids over time in both the transfection and the infection conditions was similar, with a significant increase after 36 to 84 hours. The transcripts' levels almost reached the baseline in day 7.5 and thereafter. Despite having the same trend, the increment with the infected conditions was significantly higher than that of the transfected conditions in the five plasmids, especially over the peak period (36 to 84 hours).

The second collected group was used for GFP quantification by its fluorescence intensity and measured as the RFU in the equal cell counts from the different samples. CHO cells were used as a blank and the two negative controls (pUC19-transfected and pUC19-transfected/infected CHO cell) were included. The RFUs of GFP expressed in each plasmid with both conditions are shown in Figure 7. Three biological replicates were carried out and statistical analysis of the data obtained was performed as mentioned earlier. All of the plasmids showed significant (at $P<0.05$) increase in GFP fluorescence intensity after 12 hours post-transfection, in both the transfection and the infection conditions, and last until day 7.5 post-transfection. Only pMGA showed almost identical expression levels of GFP in both the transfected and the infected conditions. The rest of the plasmids showed significantly higher GFP intensity with the infection over the transfection conditions.

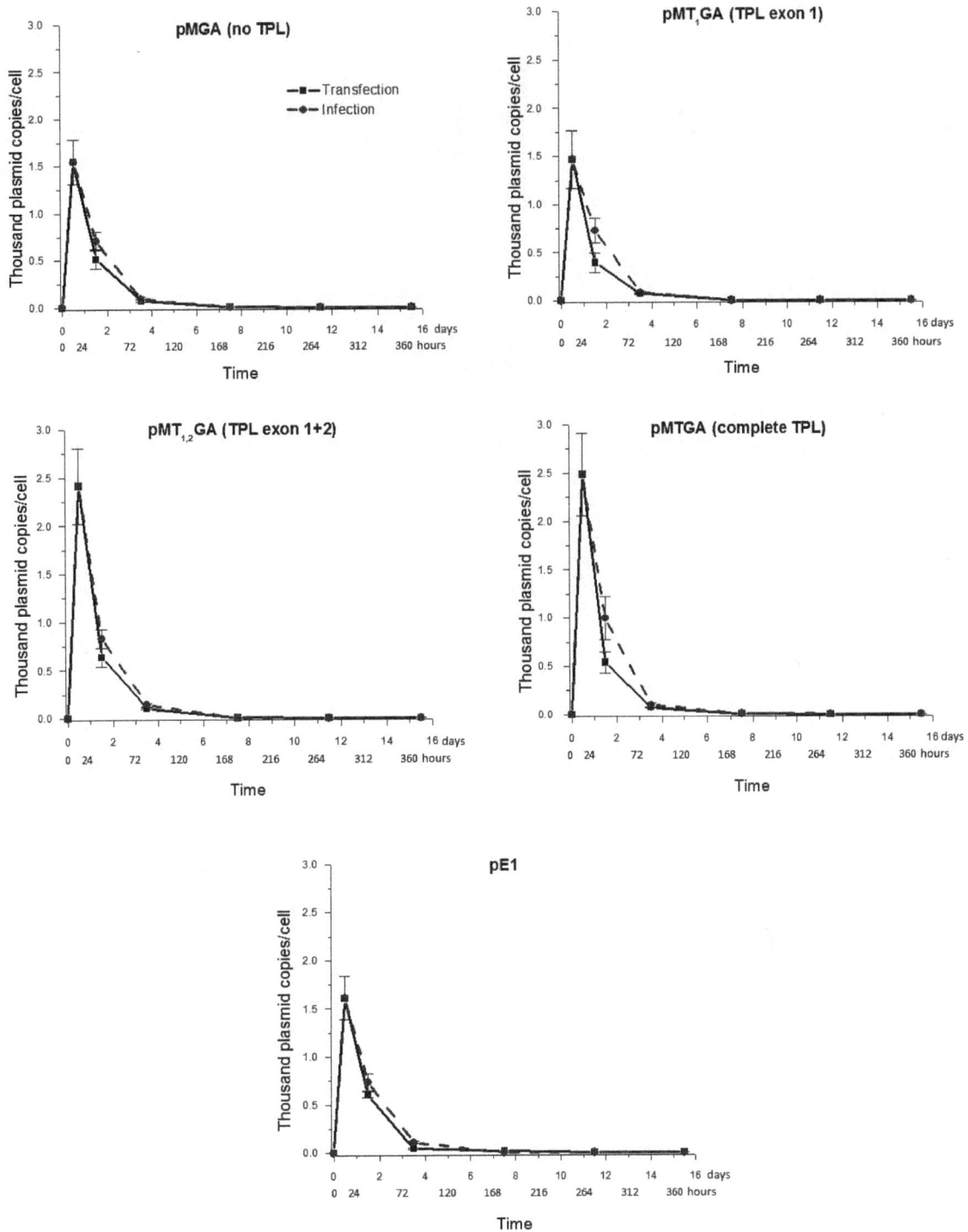

Figure 5. Copy numbers of the different plasmids over 15.5 days post-transfection in CHO cells, with transfection and infection (at MOI of 1 PFU/cell) conditions. Copy numbers were obtained by qPCR using a standard curve of known plasmid DNA concentration. qPCR was performed on equal amounts of DNA isolated from collected samples. Plasmids names are shown on the Figure with transfection (—■—) and infection (- - ● - -) conditions

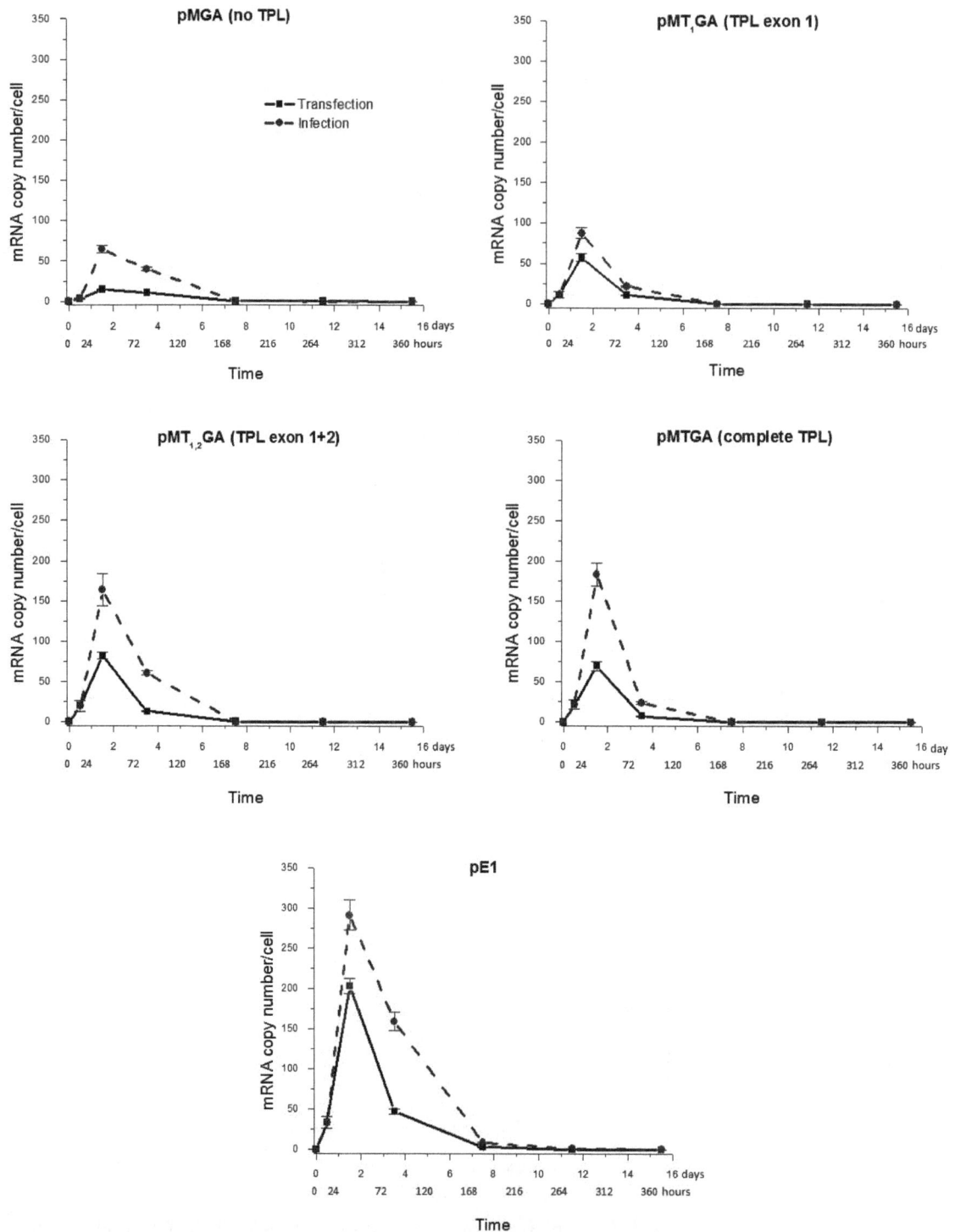

Figure 6. GFP mRNA transcripts from the different plasmids over 15.5 days post-transfection in CHO cells, with transfection and infection (at MOI of 1 PFU/cell) conditions. Copy numbers were obtained by qPCR using a standard curve of known plasmid DNA concentration. qPCR was performed on equal volumes of RT product from equal amounts of RNA isolated from collected samples. Plasmids names are shown on the Figure with transfection (—■—) and infection (- - ● - -) conditions

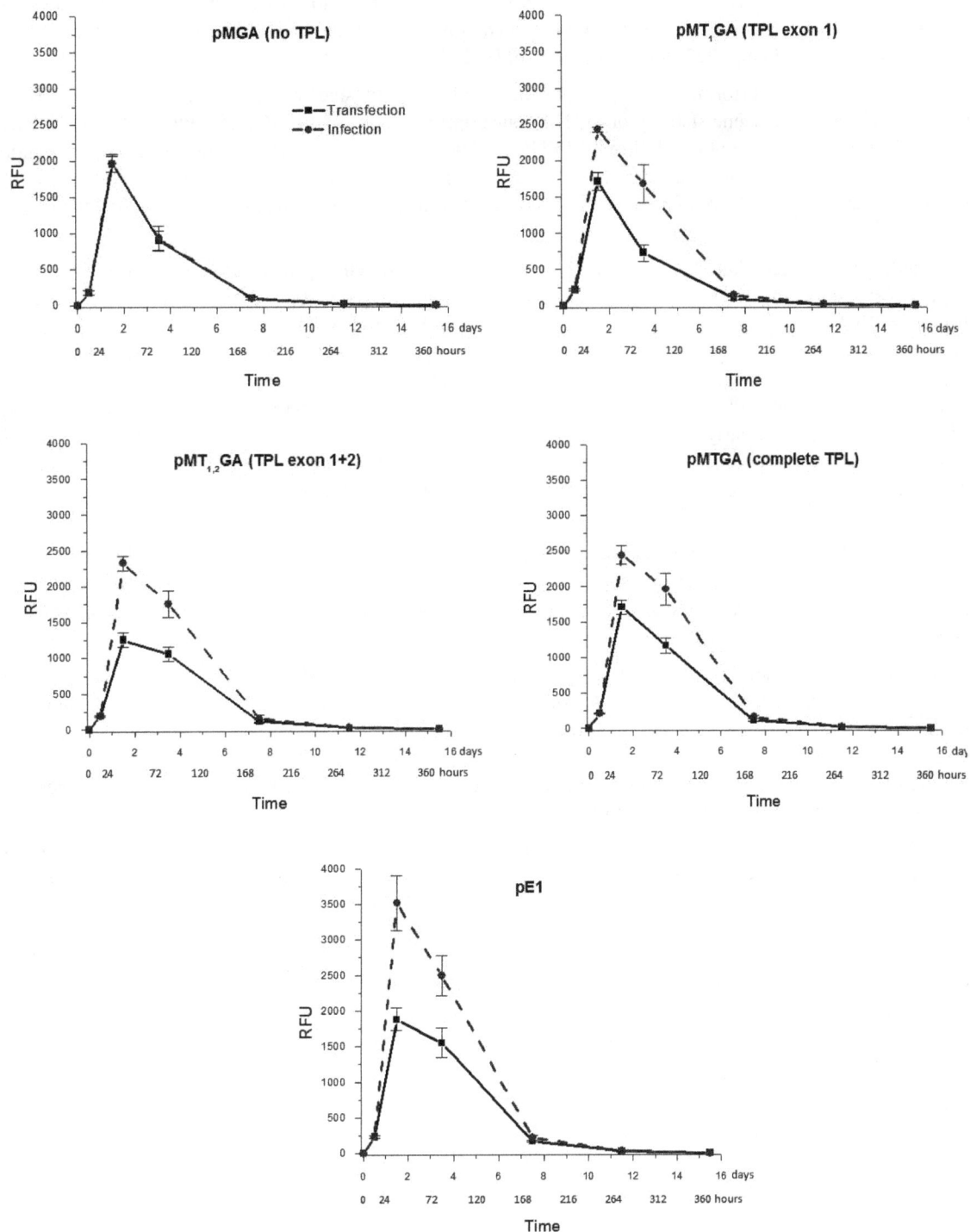

Figure 7. GFP fluorescence intensity over 15.5 days post-transfection in CHO cells, with transfection and infection (at MOI of 1 PFU/cell) conditions. Fluorescence intensity was measured by the relative fluorescence units (RFU) of equal cell counts. Plasmids names are shown on the Figure with transfection (—■—) and infection (- - ● - -) conditions

4. Discussion

Active expression driven by the adenoviral major late promoter (MLP) and tripartite leader sequence (TPL) influenced their in vivo utilization to drive the expression of vaccines (Hammond et al., 2000). The effect of TPL exons as well as E1 genes on transgene expression have not been investigated. In this study, we evaluated mRNA transcription levels and GFP translation from plasmids containing expression cassettes, with MLP and different TPL exons with E1 genes, in the absence and presence of adenovirus infection. We used adenovirus dl309, which has the same properties as the wild type adenovirus serotype 5 with the exception of the E3 region (Bett, Krougliak, & Graham, 1995; Jones & Shenk, 1978, 1979).

MOI optimization was performed for infections of CHO cells transfected with pE1. MOI did not have a measurable effect on transgene stability or mRNA transcription levels. At an MOI of 1 PFU/cell, GFP intensity was significantly higher than at an MOI of 10 PFU/cell. These results concurred with our previous observations of cellular vitality and monolayer attachment. The cells remained attached at an MOI of 1 PFU/cell, similarly to the non-infected control cells, but further increases of MOI resulted in proportional detachment of cells from the monolayer.

These findings indicated that the strength of viral infection and viral protein expression directly affected transgene expression inside the cells. At higher MOIs, it appears that the competition between viral and transgene mRNA translation lowered transgene expression, while high viral loads severely affected host cells. This was in accordance with previously published data obtained from adenovirus infected HEK 293 cells. The data showed that further increases of the MOI over 5 PFU/cell did not elevate viral production (Ferreira, Alves, Gonçalves, & Carrondo, 2005); as cellular machinery had reached its saturation point and could no longer produce more viruses. Similarly, at an MOI of 10 PFU/cell in our experiment, transgene expression was lowered by competing viral proteins, whose expression overtaxed the cellular expression machinery and affected other cellular processes. Therefore, it was concluded that use of lower MOI is more suitable for both transgene expression and cellular vitality.

The use of an MOI of 1 PFU/cell did not severely affect cellular attachment. Therefore, we were able to extend DNA, RNA and protein collection times which provided us with more detailed results. As expected from the previous optimization experiment, adenovirus infection did not affect plasmid stability over time. However, at MOI of 1 PFU/cell, the viral yield reduced by 104 compared to an MOI of 5 PFU/cell (Ferreira et al., 2005).

Viral infection had a positive effect on transgene transcription from the MLP, as GFP mRNA copy numbers increased after the infection. The direct effect of the viral infection on MLP activity was clearly demonstrated with the use of the constructs lacking the TPL sequence. However, the gradual increase in transgene transcript copies with the addition of TPL exons in the expression constructs indicated an additive effect. Although the inclusion of TPL exons in the expression cassette enhanced transgene mRNA expression in non-infected cells, this effect was more pronounced in adenovirus infected cells.

In the absence or presence of the adenoviral infection, about two fold increase in mRNA transcripts was observed when E1 genes were incorporated in the transgene. Most notably, the E1 proteins play a major role in the transactivation of transcription from the MLP as well as mRNA stability. Although adenovirus dl309 also encoded the E1 region, enhanced transgene expression from pMTGA did not reach same levels as pE1 after infection. This indicated that an excess amount of E1 proteins enhances transgene transcription and translation. Also, the physical presence of E1 genes in the same construct as the expression cassette can guarantee the expression of the excess E1 in the same cell at the same time. The pE1 plasmid increased the abundance of these proteins by increasing their gene copy number and hence, GFP expression from this construct was significantly higher than from pMTGA.

The translation of GFP mRNA also increased after viral infection, most likely as a result of the viral encoded E1B 55k and E4 orf6 proteins (Babiss, Ginsberg, & Darnell, 1985; Blackford & Grand, 2009; Bridge & Ketner, 1990; Leppard & Shenk, 1989; Pilder, Moore, Logan, & Shenk, 1986). The only exception was GFP mRNA translation from the pMGA plasmid, which did not increase after infection. This result was expected, since the TPL sequence was absent from this transgene expression construct. On the other hand, the constructs that contained the complete TPL structure (pMTGA and pE1) showed a significant increase in GFP intensity after infection. This demonstrates the importance of the E1B 55k and E4 orf6 viral proteins. In addition, the intensity of GFP expressed from pE1 was significantly higher than that from pMTGA, mainly because of the higher transgene transcription efficiency from pE1.

In conclusion, these results indicate that E1 genes and the complete TPL have positive effect on the transcription levels driven by the MLP. The increased gene expression levels, particularly on the translational levels, in presence of adenoviral infection indicates the role of other viral proteins or elements which interact with the host cell and the transgene to up-regulate its expression. More effort is required to identify such proteins and/or elements that can restore the activity of the MLP in transgene constructs, which will in turn help to utilize these components to drive active transgene expression.

References

Babiss, L. E., Ginsberg, H. S., & Darnell, J. E. (1985). Adenovirus E1B proteins are required for accumulation of late viral mRNA and for effects on cellular mRNA translation and transport. *Mol. Cell Biol, 5*(10), 2552-2558. http://dx.doi.org/10.1128/MCB.5.10.2552

Bett, A. J., Krougliak, V., & Graham, F. L. (1995). DNA sequence of the deletion/insertion in early region 3 of Ad5 dl309. *Virus Res, 39*(1), 75-82. http://dx.doi.org/10.1016/0168-1702(95)00071-W

Blackford, A. N., & Grand, R. J. (2009). Adenovirus E1B 55-kilodalton protein: multiple roles in viral infection and cell transformation. *J. Virol, 83*(9), 4000-4012. http://dx.doi.org/10.1128/JVI.02417-08

Bridge, E., & Ketner G. (1990). Interaction of adenoviral E4 and E1b products in late gene expression. *Virology, 174*(2), 345-353. http://dx.doi.org/10.1016/0042-6822(90)90088-9

Cromeans, T. L., Lu, X., Erdman, D. D., Humphrey, C. D., & Hill, V. R. (2008). Development of plaque assays for adenoviruses 40 and 41. *J. Virol. Methods, 151*(1), 140-145. http://dx.doi.org/10.1016/j.jviromet.2008.03.007

Dobbelstein, M., Roth, J., Kimberly, W. T., Levine, A. J., & Shenk, T. (1997). Nuclear export of the E1B 55-kDa and E4 34-kDa adenoviral oncoproteins mediated by a rev-like signal sequence. *EMBO J, 16*, 4276-4284. http://dx.doi.org/10.1093/emboj/16.14.4276

Dolph, P. J., Huang, J., & Schneider, R. J. (1990). Translation by the adenovirus tripartite leader: elements which determine independence from cap-binding protein complex. *J. Virol, 64*(6), 2669-2677.

Dolph, P. J., Racaniello, V., Villamarin, A., Palladino, F., & Schneider, R. J. (1988). The adenovirus tripartite leader may eliminate the requirement for cap-binding protein complex during translation initiation. *J. Virol, 62*(6), 2059-2066.

Ferreira, T. B., Alves, P. M., Gonçalves, D., & Carrondo, M. J. T. (2005). Effect of MOI and Medium Composition on Adenovirus Infection Kinetics. In F. Gòdia, & M. Fussenegger (Eds.). *Animal cell technology meets genomics* (pp. 329-332). Netherlands: Springer. http://dx.doi.org/10.1007/1-4020-3103-3_65

Flint, S. J., & Gonzalez, R. A. (2003). Regulation of mRNA production by the adenoviral E1B 55-kDa and E4 Orf6 proteins. *Curr. Top. Microbiol. Immunol, 272*, 287-330.

Gonzalez, R. A., & Flint, S. J. (2002). Effects of mutations in the adenoviral E1B 55-kilodalton protein coding sequence on viral late mRNA metabolism. *J. Virol, 76*(9), 4507-4519. http://dx.doi.org/10.1128/JVI.76.9.4507-4519.2002

Gregory, T. R. (2012). *Animal genome size database.* Retrieved January 24 2012, from http://www.genomesize.com

Hammond, J. M., McCoy, R. J., Jansen, E. S., Morrissy, C. J., Hodgson, A. L., & Johnson, M. A. (2000). Vaccination with a single dose of a recombinant porcine adenovirus expressing the classical swine fever virus gp55 (E2) gene protects pigs against classical swine fever. *Vaccine, 18*(11-12), 1040-1050. http://dx.doi.org/10.1016/S0264-410X(99)00347-3

Horridge, J. J., & Leppard, K. N. (1998). RNA-binding activity of the E1B 55-kilodalton protein from human adenovirus type 5. *J. Virol, 72*(11), 9374-9379.

Jones, N., & Shenk, T. (1978). Isolation of deletion and substitution mutants of adenovirus type 5. *Cell, 13*(1), 181-188. http://dx.doi.org/10.1016/0092-8674(78)90148-4

Jones, N., & Shenk, T. (1979). Isolation of adenovirus type 5 host range deletion mutants defective for transformation of rat embryo cells. *Cell, 17*(3), 683-689. http://dx.doi.org/10.1016/0092-8674(79)90275-7

Leppard, K. N., & Shenk, T. (1989). The adenovirus E1B 55 kD protein influences mRNA transport via an intranuclear effect on RNA metabolism. *EMBO J., 8*(8), 2329-2336.

Markrides, S. C. (1999). Components of vectors for gene transfer and expression in mammalian cells. *Protein Expression and Purification, 17*(2), 183-202. http://dx.doi.org/10.1006/prep.1999.1137

Papadakis, E. D., Nicklin, S. A., Baker, A. H., & White, S. J. (2004). Promoters and control elements: designing expression cassettes for gene therapy. *Curr. Gene Ther, 4*(1), 89-113. http://dx.doi.org/10.2174/1566523044578077

Parks, C. L., & Shenk, T. (1997). Activation of the adenovirus major late promoter by transcription factors MAZ and Sp1. *J. Virol, 71*(12), 9600-9607.

Pilder, S., Moore, M., Logan, J., & Shenk, T. (1986). The adenovirus E1B- 55K transforming polypeptide modulates transport or cytoplasmic stabilization of viral and host cell mRNAs. *Mol. Cell Biol, 6*(2), 470-476. http://dx.doi.org/10.1128/MCB.6.2.470

Puck, T. T., Cieciura, S. J., & Robinson, A. (1958). Genetics of somatic mammalian cells: iii. Long-term cultivation of euploid cells from human and animal subjects. *J. Exp. Med., 108*(6), 945-956. http://dx.doi.org/10.1084/jem.108.6.945

Spurgeon, M. E., & Ornelles, D. A. (2009). The adenovirus E1B 55-kilodalton and E4 open reading frame 6 proteins limit phosphorylation of eIF2alpha during the late phase of infection. *J. Virol, 83*(19), 9970-9982. http://dx.doi.org/10.1128/JVI.01113-09

Zhang, Y., Feigenblum, D., & Schneider, R. J. (1994). A late adenovirus factor induces eIF-4E dephosphorylation and inhibition of cell protein synthesis. *J. Virol, 68*(11), 7040-7050.

Ziff, E. B., & Evans, R. M. (1978). Coincidence of the promoter and capped 5' terminus of RNA from the adenovirus 2 major late transcription unit. *Cell, 15*(4), 1463-1475. http://dx.doi.org/10.1016/0092-8674(78)90070-3

2

The Correlation between Resistance to Antimicrobial Agents and Harboring Virulence Factors among Enterococcus Strains Isolated from Clinical Samples

Leila Arbabi[1], Mina Boustanshenas[1], Mohammad Rahbar[1,2], Ali Majidpour[1,3], Nasrin Shayanfar[4], Mastaneh Afshar[1], Maryam Adabi[1], Parwiz Owlia[1], Mahshid Talebi-Taher[1,3]

[1] Antimicrobial Resistance Research Center, Rasoul-e-Akram Hospital, Iran University of Medical Sciences, Tehran, Iran

[2] Department of Microbiology, Reference Health Laboratories Research Center, Deputy of Health, Ministry of Health and Medical Education, Tehran, Iran

[3] Department of Infectious Diseases, Rasoul-e-Akram General Teaching Hospital, Iran University of Medical Sciences, Tehran, Iran

[4] Rasoul-e-Akram Hospital, Iran University of Medical Sciences, Tehran, Iran

Correspondence: Mahshid Talebi-Taher, Antimicrobial Resistance Research Center, Rasoul-e-Akram Hospital, Iran University of Medical Sciences, Tehran, Iran. E-mail: papco2002@yahoo.com

Abstract

Objectives: In Iran as well as throughout the world Enterococci have been rated as the important cause of urinary tract and nosocomial infections. The aim of this study was to evaluate the relationship between high antimicrobial resistance activity and harboring the virulence factors among clinical Enterococcus isolates.

Materials and Methods: Clinical strains were isolated from hospitalized patients. Prevalence of different virulence genes was evaluated by PCR method and the relation between resistance to antibiotics and harboring virulence genes was evaluated by statistical analysis.

Results: The results showed that *E. faecalis* (60%) is more prevalent than *E. faecium* (26%) and harboring more virulence factors. The highest resistance was related to gentamicin in both *E. faecalis* and *E. faecium* isolates with the rate of 88.7% and 93.5% respectively. Harboring *esp*, *ace* and *cyl*A are significantly related to resistance to different antibiotics.

Conclusion: The antimicrobial resistance and virulence pattern of Enterococcus must be constantly monitored in order to choose the best antimicrobial treat and prevent nosocomial infections.

Keywords: antimicrobial resistance, virulence factors, enterococcus, Iran

Highlights:

- The *E. faecalis* is more prevalent than *E. faecium* among hospitalized patients.

- The *E. faecalis* harboring more virulence factors rather than *E. faecium*.

- Gentamicin was detected as the most resistance antibiotics among all isolates.

- The significant correlation between harboring *esp*, *ace* and *cyl*A and resistance to antibiotics has been detected.

1. Introduction

Enterococci are dominant commensal bacteria in the intestinal flora of animals and humans. These bacteria can cause various infections such as endocarditis, septicemia and Urinary Tract Infection (Murray, 1990). Researchers showed that enterococci infections are the second most common cause of bacteremia in United States hospitals, but it is less comprehensive in European countries and in Italian hospitals enterococci infections are the third most common cause of infection and mostly cause UTIs (Bonten, Willems, & Weinstein, 2001; Moro et al., 2001). In Iran as well as throughout the world enterococci have been rated as the second cause of urinary tract infections (Fatholahzadeh et al., 2006). Resistance to antimicrobial agents is an important threat for entercoccal infections control; additionally the emergence of Vancomycin Resistant Enterococci (VRE) raise the threat of nosocomial infection control and also presented the serious challenges for clinicians treating patients with enterococcal infections (Feizabadi et al., 2004). VRE infections among hospital patients especially immuno-compromised are associated with high morbidity and mortality rate in Iran and other countries all over the world (Christidou et al., 2004; Song et al., 2005). Regardless of sporadic reports of enterococcal infections in Iranian hospitals the rate of its mortality and morbidity is on the rise.

Enterococcus has two common species which are responsible for nosocomial infections. The most entercoccal infections are endogenous but cross infection generally occur in hospitalized patients (Cookson et al., 2006).

The prevalence of *Enterococcus faecalis* and *Enterococcus faecium* in human enterococcal infections are about 90% and 10%, respectively (Kayaoglu & Ørstavik, 2004); Nevertheless, other enterococcal species have been reported to cause human infections (Semedo et al., 2003). This high prevalence could be explained by the inherent resistance to various antibiotics and also the presence of different virulence factors which create the greater adaptability in hospital environments for this organism (Soheili et al., 2014; Weng, Ramli, Shamsudin, Cheah, & Hamat, 2013). Epidemiological studies showed that enetrococci are one of the most important reservoirs for transmission of antibiotic resistance genes among different bacteria species (Cetinkaya, Falk, & Mayhall, 2000). Data from previous studies clarify the mechanisms of acquisition of antibiotic resistance; but enterococcal resistance mechanisms and the spread of virulence factors are still ambiguous (M. S. Gilmore, 2002). Enterococci can develop resistance against a wide verity of antibiotics specially glycopeptides such as vancomycin (Lopes et al., 2005). The vancomycin resistance among VRE strains is due to a *van*A gene cluster carried in mobile gene element Tn1546 which can be transferred by conjugated plasmid (Salem-Bekhit, Moussa, Muharram, Alanazy, & Hefni, 2012). Since the enterococci have the ability to transfer the resistance factors to other vancomycin- susceptible species horizontally can cause a serious problem in the treatment of other gram positive bacterial infections (Salem-Bekhit et al., 2012). Enterococci as opportunistic bacteria possess various putative virulence factors, including enterococcal surface protein (Esp), gelatinase (GelE), activator protein A of cytolysin (CylA), collagen-binding protein (Ace), etc. these proteins play an important role in the virulence activity of enterococcal strains. Prevalence of the first three elements in *E. faecalis* is higher than the *E. faecium* even some studies reported that these are specific virulence factors for *E. faecalis* (Vankerckhoven et al., 2004). Gelatinase protein encoded by chromosomal gene *gel*E which has been proven to exacerbate endocarditis in an animal model (Gutschik, MØller, & Christensen, 1979). The *cyl*A is carried by plasmid or integrated into the bacterial chromosome. The cytolysin is composed of two components, lysine part (L) and activator part (A). The cytolysin operan consists of five different genes; cylA is not responsible for the lysine activity but, it is necessary for the expression of whole operon (M. Gilmore, Segarra, & Booth, 1990; Ike, Clewell, Segarra, & Gilmore, 1990). The chromosomal *esp* gen encode the enterococcal surface protein which includes the central core with distinguished tandem repeat units (Vankerckhoven et al., 2004). This protein is associated with colonization, high pathogenic potential and persists in biofilms and urinary tract (Shankar et al., 2001; Toledo-Arana et al., 2001). Collagen-binding protein is encoded by ace gene which is important for adherence the bacterial strains into target cells (Elsner et al., 2000). Multiplex PCR is a rapid and useful assay for simultaneous amplification of target genes which can be more cost effective in both research and clinical laboratories (Henegariu, Heerema, Dlouhy, Vance, & Vogt, 1997). In the present study multiplex PCR was used for detection three virulence factors including *gyl*E, *cyl*A, and *esp* genes.

The purpose of this study was to determine the antimicrobial resistance pattern of the Enterococcus strains were isolated from inpatients of Rasoul-e-Akram Hospital, Tehran, Iran and investigate the prevalence important virulence genes *gyl*E, *cyl*A and *esp* to evaluate the relationship between high antimicrobial resistance activity and harboring the virulence factors.

2. Materials and Methods

A total of 120 Enterococcus strains was collected from inpatients of Rasoul-e-Akram hospital, with an increased risk of infections during March-November 2013 and sent to Antimicrobial Resistance Research Center, Tehran, Iran. The clinical samples consisted of urine, blood, stool and vaginal swabs (n=97 urine, n=11 stool, n=10 blood, n=2 vaginal swabs). All samples were inoculated on MacCankey and Blood agar and incubated at 37 °C for 24-48 hours. For primary screening of VRE strains all isolates first cultured on M-Enterococcus selective agar containing 6 μg. ml^{-1} vancomycin and grown isolates were subjected to biochemical test such as catalase, gram staining, gas production from glucose, bile esculin test and growing in 6.5% NaCl to identify the Enterococcus genus. Other biochemical tests were performed for differentiation of *faecalis* and *faecium* species. Seventy four out of 120 strains isolated from females and 47 isolated from males.

2.1 Molecular Confirmation of E. faecalis and E. faecium Isolates

The confirmation of biochemical identification of isolates was done using polymerase chain reaction (PCR) method. The genome of putative *E. faecalis* and *E. faecium* isolates was extracted by boiling method and used as a DNA template in the PCR assay (Kariyama, Mitsuhata, Chow, Clewell, & Kumon, 2000). The PCR reaction was performed using specific primers, sodA-fcm and sodA-fcl for each species, separately (Table 1). The *E. faecalis* and *E. Faecium* specific primers targeted the *soda* gene, respectively (Kariyama et al., 2000). The PCR reaction was performed in a reaction mixture with total volume of 25 µl, containing 12.5 µl commercial PCR master mix containing Taq polymerase, enzyme buffer, $MgCl_2$ and dNTPs, 0.5 µl from each primers containing 400 nM, 6.5 µl sterile water and 5 µl DNA template. PCR was performed as follows: initial denaturation step at 93°C for 5 min followed by 30 cycles consisting of denaturation (94°C for 1 min), annealing (49°C for 1 min), and extension (72°C for 1 min), followed by a final extension step at 72°C for 5 min. The *E. faecalis* ATCC29212, *E. faecium* ATCC19434 were used as positive controls and *Escherichia coli* ATCC25922 was used as negative control.

Table 1. Primers used in this study

Primer name	Target gene	Oligonucleotide sequences	Amplicon size	Reference
SodA-fcm	*sodA*	F: TTGAGGCAGACCAGATTGACG R: TATGACAGCGACTCCGATTCC	658 bp	(Kariyama et al., 2000)
SodA-fcl	*sodA*	F:ATCAAGTACAGTTAGTCT R:ACGATTCAAAGCTAACTG	941 bp	(Kariyama et al., 2000)
GelE	*gelE*	F:TATGACAATGCTTTTTGGGAT R:ATGACAATGCTTTTTGGGAT	213bp	(Vankerckhoven et al., 2004)
Esp	*esp*	F:AGATTTCATCTTTGATTCTTGG R:AATTGATTCTTTAGCATCTGG	511bp	(Vankerckhoven et al., 2004)
CylA	*cylA*	F:ACTCGGGGATTGATAGGC R:GCTGCTAAAGCTGCGCTT	670bp	(Vankerckhoven et al., 2004)
Ace	*ace*	F:GGAATGACCGAGAACGATGGC R:GCTTGATGTTGGCCTGCTTCCG	616 bp	(Creti et al., 2004)

2.2 Antimicrobial Susceptibility Testing

The antimicrobial agents were selected among those commonly carried by bacterial chromosome and plasmid both and antimicrobial susceptibility testing was performed using Kirby-Bauer disc diffusion method on Mueller- Hinton agar (Bayer, Kirby, Sherris, & Turck, 1966) and for ampicillin (10 µg), penicillin (10 units), gentamicin (10 µg), ciprofloxacin (5 µg), erythromycin(15 µg), vancomycin (30 µg) tetracycline (30 µg) and chloramphenicol (30 µg)results further interpreted according to Clinical and Laboratory Standards Institute (CLSI) guidelines ("Clinical and Laboratory Standards Institute (CLSI). Methods for dilution antimicrobial susceptibility tests for bacteria that grow aerobically approved standard; vol. 29, 18th ed. M07-A8. Wayne, Pa, USA: CLSI; 2014,").

According to CLSI 2014 the MIC value of vancomycin is necessary to recognize the VRE strains. The MIC of vancomycin was determined using agar dilution method ("Clinical and Laboratory Standards Institute (CLSI). Methods for dilution antimicrobial susceptibility tests for bacteria that grow aerobically approved standard; vol. 29, 18th ed. M07-A8. Wayne, Pa, USA: CLSI; 2014,"). Muller-Hinton agar was supplemented with different concentrations of vancomycin and ampicillin from 0 to 256µgml⁻¹. One loop-full bacteria suspension with the turbidity of 0.5 McFarland standard was inoculated in each media with different antibiotic concentration. The plates were further incubated at 37˚C for 18 hours and examined for growth. According to CLSI 2014 guideline the enterococci with vancomycinMIC≥32 µgml⁻¹ were considered as resistant strains for both antibiotics. In all antimicrobial susceptibility tests *E. coli* ATCC 25922 was used as a control strain.

2.3 Genetic Determination of Virulence Factors

All strains were cultured on brain heart infusion agar (BHI) incubated at 37°C overnight. The DNA was extracted using the boiling method. The template DNA was prepared by suspending one loop-full of bacterial cells in 1 ml of sterile DNA/RNase free water. The bacterial suspensions were heated for 10 min at 95°C and centrifuged 10 min at 10000 rpm to remove the debris. The multiplex PCR was performed for three different genes (*gyl*E, *cyl*A and *esp*) using specific primers listed in Table 1. The multiplex PCR mixture was optimized with total volume of 50 µl composed of 25 µl PCR master mix, 400nM of each primer (*gyl*E, *cyl*A and *esp*), 5 µl from extracted DNA and adding sterile DNA/RNase free water up to 50 µl. The PCR process was initiated at 93°C for 5 min and followed by 30 cycles, including; denaturation (94°C for 1 min), annealing (56°C for 1 min), and extension (72°C for 1 min), followed by one cycle consisting of 10 min at 72°C as final extension. For genetic detection of *ace* gene a separate conventional PCR was performed using specific primer and the same condition as described above for Enterococcal species-specific PCR but the annealing temperature was set 59°C.

2.4 Statistical Analysis

The correlation between resistance to different antibiotics and the presence of virulence factors was determined using SPSS software version 22 with Chi-square and the Fisher exact test. A P value of <0.05 was regarded as statistically significant.

3. Results

3.1 Bacterial Strains Analysis

Among a total 120 Enterococci strains *E. faecalis* 72 (60%) was the commonest species isolated followed by *E. faecium* 31 (26%) (Table 2). Forty out of 72 (33%) *E. faecalis* strains isolated from females and 32 (27%) from male and among *E. faecium* 23 out of 31 (19%) and 8 (7%) isolated from females and males respectively. According to the antibiogram results the resistance to all tested antibiotics was higher in *E. faecium* than *E. Faecalis* (Table 3). The highest resistance rate is related to gentamicin in both strains and the lowest one is related to chloramphenicol and vancomycin in *E. faecium* and *E. faecalis,* respectively. According to CLSI 2014 guideline for confirmation of vancomycin resistance and identification of VRE strains the MIC of vancomycin from 0 µgml^{-1} to 256µgml^{-1} measured using the agar dilution method. The vancomycin MIC of 45 (37.5%) out of all *Enterococcus* strains was 32≥ µgml^{-1} which were identified as VRE strains. The agar dilution method indicated that 73 (61.3%) of *E. faecium* and 35 (29.2%) of *E. faecalis* were resistant to vancomycin while according to disc diffusion method these percent were 67.7% and 19.4% respectively.

Table 2. Distribution of *Enterococcus* spp. in different clinical samples

Sample (n=120)	*E.faecalis* N	*E. faecium* N	*Enterococcus* spp. N	Total N (%)
Urine	62	27	8	97 (80.8)
Stool	3	2	6	11 (9.2)
Blood	7	2	1	10 (8.4)
Vagina	0	0	2	2 (1.6)
Total	72	31	17	120 (100)

Table 3. Percent of resistance to different antibiotics among *E. faecium and E. faecalis* isolates

Antibiotics	*E .faecalis* N	*E. faecium* N
Penicillin	52.8	80.6
Ampicillin	33.3	67.7
Gentamicin	88.7	93.5
Ciprofloxacin	70.8	87.1
Erythromycin	83.3	87.1
Vancomycin	29.2	61.3
chloramphenicol	31.9	48.4
Tetracycline	75	77.4

3.2 Prevalence of Virulence Factors in E. faecium and E. faecalis

The presence of genes encoding for potential virulence factors was evaluated by multiplex PCR (Figure 1). Sixty three (52.5%) out of 120strains showed to be positive for harboring one or several tested virulence factors. The results showed that *esp* gene (40%) which code enterococcal surface protein is the most frequently detected genes followed by ace (38%), *cyl*A (23.3%) and *gel* E (3.3%). The main virulence profile which observed among isolates was related to virulence profile type D (*ace*+, *esp*+, *cyl*A-) with 12.5% frequency. The virulence profile of all *E. faecium* and *E. faecalis* has been shown in Table 4.

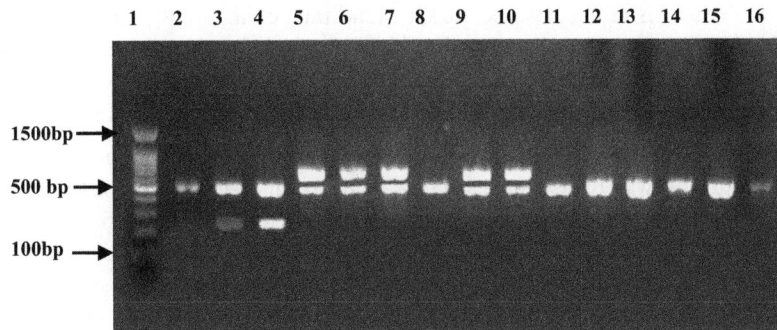

Figure 1. Agarose gel electrophoresis of amplified *cyl* A, *esp* and *gel*E by multiplex PCR and *ace* by simple PCR. Lane 1: 1 kb DNA ladder; lanes 2, 3 positive controls for *esp* (510 b and *gel*E (210bp); Lane 4: isolate positive for *gel*E (213 bp) and *esp* (510bp); lane 5: positive controls for *esp* (510bp) and *cyl* A (670bp); lanes 6, 7: isolates positive for *esp* (510bp) and *cyl* A (670bp); Lane 11-15: isolate positive for *ace* (616 bp); lane 16: positive control for *ace* gene (616 bp)

Table 4. Distribution of different virulence profiles among *Enterococcus* isolates

Virulence profiles type	Virulence genes	strains			
		E. faecium N	*E. faecalis* N	Entercoccus spp. N	Total N (%)
A	ace, esp, cyl A	2	12	0	14 (11.6)
B	ace,esp	1	14	0	15 (12.5)
C	esp, cylA	3	3	0	6 (5)
D	cylA, ace	0	3	0	3 (2.5)
E	gelE, ace	1	1	0	2 (1.7)
F	gelE, cylA	1	0	0	1 (0.8)
G	ace	3	5	1	9 (7.5)
H	esp	4	5	0	9 (7.5)
I	gylA	1	0	1	2 (1.7)
J	gelE	1	1	0	2 (1.7)
K	None	14	28	15	57 (47.5)
Total		31	72	17	120 (100)

3.3 Relation between Antimicrobial Resistance and Virulence Factors

According to statistical analysis a significant relation between some antibiotics and existence the virulence factors were observed. Among all tested isolates significant association was observed between the resistant to ampicillin, vancomycin and penicillin with harboring the *ace* and *esp* genes (P< 0.05). Furthermore, other correlations were observed as follows: significant correlation between resistant to tetracycline and gentamicin with the *esp* gene (P< 0.05), significant correlation between resistant to chloramphenicol and ciprofloxacin and harboring *ace* gene (P< 0.05), significant correlation between resistant to erythromycin and harboring *cyl* and *esp* genes (P< 0.05).

4. Discussion

Enterococcus spp. is one of the most important bacterial species which is caused hospital acquired infection, especially in intensive care units all over the world. Infections specially Bacteremia caused by Enterococcus species are becoming more serious because of developing and increased prevalence of MDR strains and VRE and a growing immune-suppressed population (Fisher & Phillips, 2009 ; Ma, Xu, & Ma, 2005   ). VRE are often concomitantly resistant to multiple antimicrobial classes.

According to the results of the present study the prevalence of *E. faecalis* among infectious clinical samples was about 2 fold higher than *E. faecium*, similar results have been reported from other studies in Iran (Jabalameli et al., 2009; Sharifi et al., 2013), Malaysia (Sharifi et al., 2013), central and south India (Bhat, Paul, & Ananthakrishna, 1998; Fernandes & Dhanashree, 2013; Menditatta et al., 2008); while in some other studies the prevalence of *E. faecium* has been reported higher than *E. faecalis* (Kapoor, Randhawa, & Deb, 2005). Fernandes and Dhanashree (2013) suggested that possessing the hemolysin and gelatinase might be the reason of higher prevalence of *E. faecalis* in infectious samples (Fernandes & Dhanashree, 2013).

Antimicrobial susceptibility test showed that all isolates were resistance at least to one tested antibiotics and the most common resistant profile was related to resistance to all eight tested antibiotics which it means the high resistance rate among Enterococcus isolates. This result indicate the much higher resistance among Enterococcus isolates in compare with other studies in European countries (Hällgren et al., 2001), The prevalence of antimicrobial resistance especially resistance to vancomycin among Enterococcus spp. have been increased in Iran recently. According to previous studies the rate of VRE strains was reported about 7% (Fatholahzadeh et al., 2006), 8% (Jabalameli et al., 2009), 18.6% (Sharifi et al., 2013) and 29.3% (Javadi et al., 2008) in different periods of time which is revealed the increasing of VRE prevalence. In our study the observed resistance was about 37.5% among Enterococcus spp. isolated from different clinical samples. This high increase in prevalence of VRE strains presented a serious challenge for the Iranian medical community.

In our study the highest resistant is related to gentamicin which is in concurrence with other studies carried out in Iran or other countries (Agarwal, Kalyan, & Singh, 2009; Fernandes & Dhanashree, 2013; Jabalameli et al., 2009; Sharifi et al., 2013). In the case of resistance different results have been reported; some studies reported the highest resistance to rifampicin, amikacin or erythromycin (Agarwal et al., 2009; Fernandes & Dhanashree, 2013). This disparity in resistance rate might be because of differences in strain properties, sample size, sex, age or even methods of studies. Overall resistance to antibiotics was higher among *E. faecium* rather than *E. faecalis*, which is in agreement with other reports (Bhat et al., 1998; Fernandes & Dhanashree, 2013; Menditatta et al., 2008).

A few studies have been performed on correlation between resistance and virulence factors in enterococci in Iran but studies in all over the world suggested some kinds of correlation between resistance to antibiotics and harboring different virulence determinants (Padilla & Lobos, 2013; Terkuran et al., 2014). In the present study the prevalence of different important virulence factors has been recognized. The correlation between antimicrobial resistance and the presence of virulence determinants have been evaluated using statistical analysis. The results showed *E. faecalis* carried more virulence genes than *E. faecium* while the vice versa results have been reported in Turkey (Terkuran et al., 2014) and also Vankerckhoven and colleagues (2008) have reported that *E. faecium* strains were generally free of virulence factors (Vankerckhoven et al., 2008). The result showed that the most prevalence factor among Enterococcus isolate was *esp* gene with the rate of 40% among isolate while in the study by Padilla and Lobos (2013) performed in Chile it was reported as 70.5% (Padilla & Lobos, 2013) and in another study by Sharifi and colleagues which was performed in Iran (2013) this prevalence was reported as 52.1% (Sharifi et al., 2013). Vankerckhoven and colleagues (2004) investigated the prevalence of some virulence genes in eight different European countries. The results showed that *gel*E and *cyl*A were not detected among enterococcus isolates and the prevalence of *esp* was about 65% (Vankerckhoven et al., 2004) while we detected *gel*E and *cyl*A genes among clinical isolates with the rate of 3.3% and 23.3% respectively .This hypothesis can be challengeable that geographic properties might be the reason of disparity among different rates of virulence factors and antimicrobial resistant profile. In accordance to statistical analysis the significant correlation was observed between resistance to vancomycin and harboring the *esp* gene (p <0.05) and this finding is in agreement with the study by Vankerckoven and colleagues which covering the large geographic area (Vankerckhoven et al., 2008). In our study we recognized that harboring the virulence gene *esp* significantly related to resistance to ampicillin, vancomycin, penicillin, tetracycline, gentamicin and erythromycin (p< 0.05). The possessing of cyl gene significantly related to resistance to erythromycin (p< 0.05) and harboring ace gene significantly related to resistance to ampicillin, vancomycin, penicillin, chloramphenicol and ciprofloxacin (p< 0.05). The *gel*E virulence gene did not exihibit significant relation with any tested antibiotic in our study while in the study by Padila and Lobos (2013) possessing of *gel*E was reported to be significantly related to resistance to chloramphenicol, gentamicin, tetracycline ampicillin and gentamicin (Padilla & Lobos, 2013).

5. Conclusion

In conclusion our study demonstrated that *E. faecali*s is more prevalent and harboring more virulence factors than *E. faecium* and harboring the virulence factors might play an important and effective role of resistance to different antibiotics. The antimicrobial resistance and virulence pattern of local Enterococcus is a subject of great importance and that must be constantly monitored to have new information to allow knowing specific characters of these bacteria in order to choose the best antimicrobial treat and prevent nosocomial infections.

Acknowledgements

This study has been supported by Iran University of Medical Sciences with grant number of 23139.

References

Agarwal, J., Kalyan, R., & Singh, M. (2009). High-level aminoglycoside resistance and Beta-lactamase production in enterococci at a tertiary care hospital in India. *Jpn J Infect Dis, 62*, 158-159.

Bayer, A., Kirby, W., Sherris, J., & Turck, M. (1966). Antibiotic susceptibility testing by a standardized single disc method. *Am J clin pathol, 45*(4), 493-496.

Bhat, K. G., Paul, C., & Ananthakrishna, N. (1998). Drug resistant enterococci in a south Indian hospital. *Tropical doctor, 28*(2), 106-107.

Bonten, M. J., Willems, R., & Weinstein, R. A. (2001). Vancomycin-resistant enterococci: why are they here, and where do they come from? *The Lancet infectious diseases, 1*(5), 314-325.

Cetinkaya, Y., Falk, P., & Mayhall, C. G. (2000). Vancomycin-resistant enterococci. *Clinical microbiology reviews, 13*(4), 686-707.

Christidou, A., Gikas, A., Scoulica, E., Pediaditis, J., Roumbelaki, M., Georgiladakis, A., & Tselentis, Y. (2004). Emergence of vancomycin ‐ resistant enterococci in a tertiary hospital in Crete, Greece: a cluster of cases and prevalence study on intestinal colonisation. *Clinical microbiology and infection, 10*(11), 999-1005.

Clinical and Laboratory Standards Institute (CLSI). Methods for dilution antimicrobial susceptibility tests for bacteria that grow aerobically approved standard; vol. 29, 18th ed. M07-A8. Wayne, Pa, USA: CLSI; 2014.

Cookson, B., Macrae, M., Barrett, S., Brown, D., Chadwick, C., French, G., . . . Wade, J. (2006). Guidelines for the control of glycopeptide-resistant enterococci in hospitals. *Journal of Hospital Infection, 62*(1), 6-21.

Creti, R., Imperi, M., Bertuccini, L., Fabretti, F., Orefici, G., Di Rosa, R., & Baldassarri, L. (2004). Survey for virulence determinants among Enterococcus faecalis isolated from different sources. *Journal of medical microbiology, 53*(1), 13-20.

Elsner, H.-A., Sobottka, I., Mack, D., Laufs, R., Claussen, M., & Wirth, R. (2000). Virulence factors of Enterococcus faecalis and Enterococcus faecium blood culture isolates. *European Journal of Clinical Microbiology and Infectious Diseases, 19*(1), 39-42.

Fatholahzadeh, B., HASHEMI, F. B., EMANEINI, M., ALIGHOLI, M., NAKHJAVANI, F. A., & KAZEMI, B. (2006). Detection of vancomycin resistant enterococci (VRE) isolated from urinary tract infections (UTI) in Tehran, Iran. *DARU Journal of Pharmaceutical Sciences, 14*(3), 141-145.

Feizabadi, M. M., Asadi, S., Aliahmadi, A., Parvin, M., Parastan, R., Shayegh, M., & Etemadi, G. (2004). Drug resistant patterns of enterococci recovered from patients in Tehran during 2000–2003. *International journal of antimicrobial agents, 24*(5), 521-522.

Fernandes, S. C., & Dhanashree, B. (2013). Drug resistance & virulence determinants in clinical isolatesof Enterococcus species. *The Indian journal of medical research, 137*(5), 981.

Fisher, K., & Phillips, C. (2009). The ecology, epidemiology and virulence of Enterococcus. *Microbiology, 155*, 1749-1757.

Gilmore, M., Segarra, R., & Booth, M. (1990). An HlyB-type function is required for expression of the Enterococcus faecalis hemolysin/bacteriocin. *Infection and immunity, 58*(12), 3914-3923.

Gilmore, M. S. (2002). *The enterococci: pathogenesis, molecular biology, and antibiotic resistance*: Zondervan.

Gutschik, E., MØller, S., & Christensen, N. (1979). Experimental endocarditis in rabbits. *Acta Pathologica Microbiologica Scandinavica Section B Microbiology, 87*(1 - 6), 353-362.

Hällgren, A., Abednazari, H., Ekdahl, C., Hanberger, H., Nilsson, M., Samuelsson, A., . . . Group, S. I. S. (2001). Antimicrobial susceptibility patterns of enterococci in intensive care units in Sweden evaluated by different MIC breakpoint systems. *Journal of Antimicrobial Chemotherapy, 48*(1), 53-62.

Henegariu, O., Heerema, N., Dlouhy, S., Vance, G., & Vogt, P. (1997). Multiplex PCR: critical parameters and step-by-step protocol. *Biotechniques, 23*(3), 504-511.

Ike, Y., Clewell, D., Segarra, R., & Gilmore, M. (1990). Genetic analysis of the pAD1 hemolysin/bacteriocin determinant in Enterococcus faecalis: Tn917 insertional mutagenesis and cloning. *Journal of bacteriology, 172*(1), 155-163.

Jabalameli, F., Emaneini, M., Shahsavan, S., Sedaghat, H., Abdolmaliki, Z., & Aligholi, M. (2009). Evaluation of antimicrobial susceptibility patterns of Enterococci isolated from patients in Tehran University of Medical Sciences Teaching Hospitals. *Acta Medica Iranica, 47*(4), 325-328.

Javadi, A., Ataei, B., Khorvash, F., Toghyani, S., Mobasherzadeh, S., & Soghrati, M. (2008). Prevalence of vancomycin resistant Enterococci colonization in gastrointestinal tract of hospitalized patients. *Archives of Clinical Infectious Diseases, 3*(3).

Kapoor, L., Randhawa, V., & Deb, M. (2005). Antimicrobial resistance of enterococcal blood isolates at a pediatric care hospital in India. *Jpn J Infect Dis, 58*(2), 101-103.

Kariyama, R., Mitsuhata, R., Chow, J. W., Clewell, D. B., & Kumon, H. (2000). Simple and reliable multiplex PCR assay for surveillance isolates of vancomycin-resistant enterococci. *Journal of Clinical Microbiology, 38*(8), 3092-3095.

Kayaoglu, G., & Ørstavik, D. (2004). Virulence factors of Enterococcus faecalis: relationship to endodontic disease. *Critical Reviews in Oral Biology & Medicine, 15*(5), 308-320.

Lopes, M. d. F. S., Ribeiro, T., Abrantes, M., Marques, J. J. F., Tenreiro, R., & Crespo, M. T. B. (2005). Antimicrobial resistance profiles of dairy and clinical isolates and type strains of enterococci. *International journal of food microbiology, 103*(2), 191-198.

Ma, L., Xu, S., & Ma, J. (2005). Detection on part of the pathogenic genes and phenotyp of Enterococcus. . *Chinese J. Clin. Lab. Sci., 28*, 529-533.

Mendiratta, D., Kaur, H., Deotale, V., Thamke, D., Narang, R., & Narang, P. (2008). Status of high level aminoglycoside resistant Enterococcus faecium and Enterococcus faecalis in a rural hospital of central India. *Indian journal of medical microbiology, 26*(4), 369.

Moro, M. L., Gandin, C., Bella, A., Siepi, G., Petrosillo, N., & Istituto Superiore di Sanita', R. (2001). A national survey on the surveillance and control of nosoconial infection in public hospital in Italy.

Murray, B. E. (1990). The life and times of the Enterococcus. *Clinical microbiology reviews, 3*(1), 46-65.

Padilla, C., & Lobos, O. (2013). Virulence, bacterocin genes and antibacterial susceptibility in Enterococcus faecalis strains isolated from water wells for human consumption. *SpringerPlus, 2*(1), 43.

Salem-Bekhit, M., Moussa, I., Muharram, M., Alanazy, F., & Hefni, H. (2012). Prevalence and antimicrobial resistance pattern of multidrug-resistant enterococci isolated from clinical specimens. *Indian journal of medical microbiology, 30*(1), 44.

Semedo, T., Santos, M. A., Martins, P., Lopes, M. F. S., Marques, J. J. F., Tenreiro, R., & Crespo, M. T. B. (2003). Comparative study using type strains and clinical and food isolates to examine hemolytic activity and occurrence of the cyl operon in enterococci. *Journal of Clinical Microbiology, 41*(6), 2569-2576.

Shankar, N., Lockatell, C. V., Baghdayan, A. S., Drachenberg, C., Gilmore, M. S., & Johnson, D. E. (2001). Role of Enterococcus faecalissurface protein ESP in the pathogenesis of ascending urinary tract infection. *Infection and immunity, 69*(7), 4366-4372.

Sharifi, Y., Hasani, A., Ghotaslou, R., Naghili, B., Aghazadeh, M., Milani, M., & Bazmany, A. (2013). Virulence and antimicrobial resistance in enterococci isolated from urinary tract infections. *Advanced pharmaceutical bulletin, 3*(1), 197.

Soheili, S., Ghafourian, S., Sekawi, Z., Neela, V., Sadeghifard, N., Ramli, R., & Hamat, R. A. (2014). Wide distribution of virulence genes among Enterococcus faecium and Enterococcus faecalis clinical isolates. *The Scientific World Journal, 2014.*

Song, J. Y., Hwang, I. S., Eom, J. S., Cheong, H. J., Bae, W. K., Park, Y. H., & Kim, W. J. (2005). Prevalence and molecular epidemiology of vancomycin-resistant enterococci (VRE) strains isolated from animals and humans in Korea. *The Korean journal of internal medicine, 20*(1), 55-62.

Terkuran, M., Erginkaya, Z., ÜNAL, E., GÜRAN, M., Kızılyıldırım, S., Ugur, G., & Köksal, F. (2014). The relationship between virulence factors and vancomycin resistance among Enterococci collected from food and human samples in Southern Turkey. *Ankara Üniversitesi Veteriner Fakültesi Dergisi, 61*(2), 133-140.

Toledo-Arana, A., Valle, J., Solano, C., Arrizubieta, M. a. J., Cucarella, C., Lamata, M., . . . Lasa, I. (2001). The enterococcal surface protein, Esp, is involved in Enterococcus faecalis biofilm formation. *Applied and environmental microbiology, 67*(10), 4538-4545.

Vankerckhoven, V., Huys, G., Vancanneyt, M., Snauwaert, C., Swings, J., Klare, I., . . . Lammens, C. (2008). Genotypic diversity, antimicrobial resistance, and virulence factors of human isolates and probiotic cultures constituting two intraspecific groups of Enterococcus faecium isolates. *Applied and environmental microbiology, 74*(14), 4247-4255.

Vankerckhoven, V., Van Autgaerden, T., Vael, C., Lammens, C., Chapelle, S., Rossi, R., . . . Goossens, H. (2004). Development of a multiplex PCR for the detection of asa1, gelE, cylA, esp, and hyl genes in enterococci and survey for virulence determinants among European hospital isolates of Enterococcus faecium. *Journal of Clinical Microbiology, 42*(10), 4473-4479.

Weng, P. L., Ramli, R., Shamsudin, M. N., Cheah, Y.-K., & Hamat, R. A. (2013). High genetic diversity of Enterococcus faecium and Enterococcus faecalis clinical isolates by pulsed-field gel electrophoresis and multilocus sequence typing from a hospital in Malaysia. *BioMed research international, 2013.*

3

New Advances Reconstructing the Y Chromosome Haplotype of Napoléon the First Based on Three of his Living Descendants

Gérard Lucotte[1] & Peter Hrechdakian[2]

[1] Institute of Molecular Anthropology, 44 Monge Street, Paris 75 005, France

[2] Unifert Group S.A., 54 Louise Avenue, Immeuble Stéphanie, Bruxelles 1050, Belgium

Correspondence: Gérard Lucotte, Institute of Molecular Anthropology, 44 Monge Street, Paris 75 005, France.
E-mail: lucotte@hotmail.com

Abstract

This paper describes the findings of the complete reconstruction of the lineage Y chromosome haplotype of the French Emperor Napoléon I. In a previous study (Lucotte et al., 2013) we reconstructed, for more than one hundred Y-STRs (Y–short tandem repeats), the Y-chromosome haplotype of Napoléon I based on data comparing STR allelic values obtained from the DNA of two of his living descendants: Charles Napoléon (C.N.) and Alexandre Colonna Walewski (A.C.W.); in the present study we compare STR allelic values of C.N. and A.C.W. to those of Mike Clovis (M.C.), a living fifth generation descendant of Lucien (one of Napoléon's brothers). When compared between M.C., C.N. and A.C.W., STR allelic values are identical for a total of 93 STRs; that permits us to propose those values, for which the three living descendants are identical, as expected allelic values of Napoléon I's Y-chromosome haplotype. For seven STRs, allele values are variable between M.C., C.N. and A.C.W.; we propose for three of them (DYS442, DYS454 and DYS712) expected allelic values, based on data concerning the allele distributions of these STRs in the population.

Keywords: Napoléon the First, lineage reconstruction, Y-chromosome haplotype

1. Introduction

The French Emperor Napoléon the First (1769-1821), was the son of Charles Bonaparte (1746-1785) and Letizia Ramolino (1750-1836). He had four brothers: Joseph (1768-1844), Lucien (1775-1840), Louis (1778-1846) and Jérôme (1784-1860).

In a first study (Lucotte et al., 2011) we determined the Y-chromosome non-recombinant part (NRY) haplogroup of Napoléon, based on genomic DNA extracted from two islands of follicular sheats associated with his beard hairs conserved in the Vivant-Denon reliquary (Lucotte, 2010). This haplogroup, established by the study of 10 NRY-SNPs (single nucleotide polymorphisms), is E1b1b1c1*; an "oriental" haplogroup of origin, as shown by the frequency map of M34 in contemporary European populations (Lucotte and Diéterlen, 2014), the antepenultimate SNP of the E1b1b1c1* differentiation.

In this same first study (Lucotte et al., 2011) we studied the buccal smear DNA of Charles Napoléon (C.N.), the living fourth generation of male descent from Jérôme, for the first 37 NRY-STRs (short tandem repeats) of the Family Tree DNA (FT DNA) kit; that permits us to establish a first Y-STR profile of C.N. This profile is highly indicative of the E1b1b1 haplogroup, because of STR allelic values at the discriminant (from Athey, 2006) Y-markers DYS19 (allele *13*) and at DYS464.a, .b,.c and .d (alleles *13*, *14*, *15* and *16* respectively); moreover allele values (of *13*) at DYS19 and at YCaII.a and .b (*19* and *22*) are the same for Napoléon (N) and for C.N.

In a second study (Lucotte et al., 2013) we established a more complete (because based on the FTDNA-111STRs kit) Y-STR profile of C.N., and the 111-Y-STRs profile of Alexandre Colonna Walewski (A.C.W.), the fifth generation descendant of Alexandre Walewski (1810-1868) who was the son born of the union between Napoléon I and Countess Maria Walewska (1786-1817). Comparisons at the time between the two STRs profiles were realized for a total number of 130 STRs, six of them (DYS454; DYS481; DYS635 = Y-GATA-C4; DYS712; DYS724 = CDY.a and DYF397.2) having different allelic values between C.N. and A.C.W. At that time we only had three direct determinations available on real allele values of Napoléon (for DYS19 = *13*, and for the palindromic YCAII.a= *19* and .b = *22*).

We then proposed (Lucotte et al., 2013) a first reconstruction of the Y-chromosome haplotype of Napoléon, based on the expected STR allelic values obtained from the 124 identical STRs between C.N. and A.C.W.

We have obtained now (Lucotte & Bouin Wilkinson, 2014) sixteen supplementary allelic direct determinations (on a lock of hair dandruff dating from 1811) for Napoléon I STRs, in order; DYS 19 = *13*; palindromic DYS385.a and b. = *16*; DYS389.i = *14*, .ii = *31*; DYS390 = *24*; DYS391 = *10*; DYS392 = *11*; DYS393 = *14*; DYS438 = *10*; DYS439 = Y-GATA-A4 = *12*; DYS448 = *20*; DYS456 = *15*; DYS458 = *16*; Y-GATA-C4 = *23* and Y-GATA-H4 = *11*. These results confirm our previous ones for allele *13* at DYS19; moreover all these other direct determinations (except for Y-GATA-C4) are in accordance with our previous direct predictions (Lucotte et al., 2013) concerning the expected values for the corresponding STRs.

Mike Clovis (M.C.) is the fifth generation descendant (Figure 1) of Lucien; to visualize the generations of the two male descent from the Walewski and the Jérôme lines, see the first figure of the Lucotte et al., 2013 article. In order to realize a triangular comparison between three living males related to Napoléon I (a direct descendant: A.C.W.; an indirect descendant from his brother Jérôme: C.N.; and M.C., an indirect descendant from his brother Lucien), we study now in the present article the Y-STR profile of M.C. by means of the FT-DNA – 111 STRs kit; and we compare this STR profile to those of C.N. and A.C.W.

2. Methods

Mike Clovis (M.C.) is the *propositus* (Figure 1) for this study. Buccal swab samples for this DNA donor were collected with informed consent. DNA extraction and STRs typing ("upgrade" for 111 genetic markers) were done according to FTDNA recommendations.

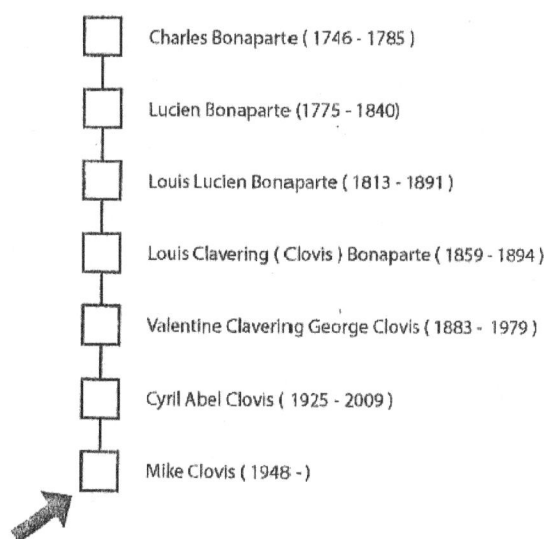

Figure 1. Chain of transmission (seven successive generations of paternal ancestry) from the ancestor Charles Bonaparte (Napoléon's father) to the *propositus* (arrow) Mike Clovis

3. Results

3.1 Comparisons of STR Allelic Values Between M.C. and C.N. – A.C.W.

Table 1 compares, for a total number of 106 STRs, allelic values obtained for M.C. to those of Charles Napoléon (C.N.) and Alexandre Colonna Walewsky (A.C.W.).

Seven STRs show different alleles between these three individuals, in order: DYS442 with an allele value = *11* for M.C. compared to *12* for both C.N. and A.C.W.; DYS447 with an allele value = *22* for M.C. compared to *21* for both C.N. and A.C.W.; for DYS454 the allele value = *11* for M.C. is identical to that of A.C.W., C.N. having an allele value = *7*; for DYS481 the allele value = *27* for M.C. is identical to that of C.N., A.C.W. having an allele value = *28*; for Y-GATA-C4 the allele value = *22* for M.C. is identical to that of C.N., A.C.W. having an allele value = *21*; for DYS712 the allele value = *23* for M.C. is identical to that of C.N., A.C.W. having an allele value = *25*; and for the palindromic CDY.a the allele value of M.C. = *34* is identical to that of A.C.W., C.N. having an allele value = *35*.

Table 1. Allelic values at 106 Y-STRs (numbers refer to the Y-markers of the FTDNA 111 "upgrade" for Mike Clovis (M.C.), Charles Napoléon (C.N.) and Alexandre Colonna Walewski (A.C.W.) NRY-DNAs. Asterisks indicate the seven differential markers (in italics) between M.C., and C.N. and A.C.W.

Numbers	Y-STRs	Allelic values					
		Napoléon I	M.C.	C.N.	A.C.W.	Napoléon I	expected
3	DYS19 = DYS394	13	13	13	13	13	(direct determination)
5	DYS385.a (palindromic)	16	16	16	16	16	(direct determination)
6	.b	16	16	16	16	16	(direct determination)
8	DYS388		12	12	12	12	
10	DYS389.i	14	14	14	14	14	(direct determination)
12	.ii	31	31	31	31	31	(direct determination)
2	DYS390 = DYS708	24	24	24	24	24	(direct determination)
4	DYS391	10	10	10	10	10	(direct determination)
11	DYS392 (located in the untranslated region of the transcription unit TTTY10)	11	11	11	11	11	
1	DYS393= DYS395	14	14	14	14	14	(direct determination)
49	DYS413.a (palindromic)		22	22	22	22	
50	.b		22	22	22	22	
48	DYS425 (one copy of DYF371)		0	0	0	0	
7	DYS426 = DYS483		11	11	11	11	
105	DYS434		9	9	9	9	
53	DYS436		12	12	12	12	
19	DYS437= DYS457		14	14	14	14	
37	DYS438 (located in the untranslated region of the USP9 Y gene)	10	10	10	10	10	(direct determination)
9	DYS439 = Y-GATA-A4		12	12	12	12	(direct determination)
89	DYS441		14	14	14	14	
36	DYS442*	*11*	12	12	12		
57	DYS444= DYS542		11	11	11	11	
86	DYS445		11	11	11	11	
60	DYS446		12	12	12	12	
18	DYS447*	*22*	21	21	21?		
20	DYS448 (located in the P3 loop)	20	20	20	20	20	(direct determination)
21	DYS449		28	28	28	28	
56	DYS450		7	7	7	7	
85	DYS452		30	30	30	30	
17	DYS454*= DYS639		11	7	11	11	
16	DYS455 (located in the intron 2 of the TBL1 Y gene)		11	11	11	11	
30	DYS456	15	15	15	15	15	(direct determination)
13	DYS458	16	16	16	16	16	(direct determination)

14	DYS459.a (palindromic)	9	9	9	9
15	.b	9	9	9	9
26	DYS460= Y-GATA-A7.1	10	10	10	10
106	DYS461= Y-GATA-A7.2	11	11	11	11
84	DYS462	12	12	12	12
88	DYS463	18	18	18	18
22	DYS464.a (palindromic)	14	14	14	14
23	.b	15	15	15	15
24	.c	16	16	16	16
25	.d	17	17	17	17
45	DYS472	8	8	8	8
58	DYS481*	*27*	*27*	*28*	*28 or 27?*
69	DYS485	15	15	15	15
63	DYS487= DYS698	14	14	14	14
54	DYS490	12	12	12	12
66	DYS492= DYS604	10	10	10	10
80	DYS494	9	9	9	9
71	DYS495	15	15	15	15
103	DYS497	14	14	14	14
96	DYS504= DYS660	16	16	16	16
75	DYS505	13	13	13	13
104	DYS510	17	17	17	17
47	DYS511	10	10	10	10
97	DYS513= DYS605	12	12	12	12
59	DYS520= DYS654	18	18	18	18
79	DYS522	12	12	12	12
38	DYS531= DYS600	10	10	10	10
94	DYS532	11	11	11	11
81	DYS533	11	11	11	11
55	DYS534	15	15	15	15
43	DYS537	12	12	12	12
72	DYS540	11	11	11	11
77	DYS549	12	12	12	12
76	DYS556	12	12	12	12
51	DYS557	21	21	21	21
98	DYS561	15	15	15	15
67	DYS565	11	11	11	11
62	DYS568	12	12	12	12
33	DYS570 (located in the untranslated region of the TBL1 Y gene)	19	19	19	19
64	DYS572	11	11	11	11

83	DYS575		8	8	8	8	
32	DYS576		18	18	18	18	
39	DYS578		8	8	8	8	
101	DYS587		22	22	22	22	
78	DYS589		11	11	11	11	
42	DYS589		7	7	7	7	
92	DYS593		16	16	16	16	
52	DYS594		11	11	11	11	
31	DYS607		12	12	12	12	
61	DYS617		13	13	13	13	
70	DYS632		8	8	8	8	
100	DYS635*= Y-GATA-C4	23	22	22	21	23	(direct determination)
82	DYS636		11	11	11	11	
65	DYS640= DYS606		13	13	13	13	
44	DYS641		11	11	11	11	
102	DYS643		12	12	12	12	
93	DYS650		18	18	18	18	
68	DYS710		31	31	31	31	
91	DYS712*		23	23	*25*	23	
73	DYS714		24	24	24	24	
95	DYS715		23	23	23	23	
74	DYS716		28	28	28	28	
34	DYS724 = CDY.a* (palindromic) = gene		34	35	34	35 ?	
35	.b		36	36	36	36	
99	DYSS726		15	15	15	15	
87	Y-GATA-A10		12	12	12	12	
27	Y-GATA-H4	11	11	11	11	11	(direct determination)
90	Y-GGATT-1B07		13	13	13	13	
28	YCAII.a (palindromic)	19	19	19	19	19	(direct determination)
29	.b		22	22	22	22	(direct determination)
40	DYF395S1.a (palindromic)	22	15	15	15	15	
41	.b		15	15	15	15	
46	DYF406S1		10	10	10	10	

Because you know exactly how many generations ago the ancestor lived (Figure 1), it is interesting to see how statistics / probabilities compare with reality. In this particular case: Mike > Cyril > Valentine-Louis Clavering > Louis > Lucien > Charles Bonaparte (= Carlo Buonaparte), the probabilities based on the calculations of the time of the most recent common ancestor (the TMRCA) calculations (from Walsh, 2001) are incorrect: comparing the STRs showing only 6 mismatches, it only estimates the probability that Mike Clovis (328303) and Alexandre Colonna Walewski (218983) shared a common ancestor within the last 1 generation = 0.07%, 2 generations = 1.13%, 3 generations = 4.93%, 4 generations = 12.48%, 5 generations = 23.38%, 6 generations = 36.19%, 7 generations = 49.27%,…; so, about 36 to 50% probability six generations back.

3.2. STRs with Identical Values

A total number of 82 STRs have identical allelic values between M.C., C.N. and A.C.W.: in order, the 71 non-palindromic STRs DYS388 = *12*; DYS425 = *0*; DYS426 = *11*; DYS434 = *9*; DYS436 = *12*; DYS437 = *14*; DYS441 = *14*; DYS444 = *11*; DYS445 = *11*; DYS446 = *12*; DYS449 = *28*; DYS450 = *7*; DYS452 = *30*; DYS455 = *11*; DYS460 = *1O*; DYS461 = *11*; DYS462 = *12*; DYS463 = *18*; DYS472=*8*; DYS485 = *15*; DYS487 =*14*; DYS490 = *12*; DYS492 = *10*;DYS494 =*9*;DYS495 =*15*;DYS497 = *14*; DYS504 = *16*; DYS505 = *13*; DYS510 = *17*;DYS511 = *10*; DYS513 = *12*; DYS520 = *18*; DYS522 = *12*; DYS531 = *10*; DYS532 = *11*; DYS533 = *11*; DYS534 = *15*; DYS537 = *12*; DYS540 = *11*; DYS549 = *12*; DYS556 = *12*; DYS557 = *21*; DYS561 = *15*; DYS565 = *11*; DYS568 = *12*; DYS570 = *19*; DYS572 = *11*; DYS575 = *8*;DYS576 = *18*; DYS578 = *8*; DYS587 = *22*; DYS589 = *11*; DYS590 = *7*; DYS593 = *16*; DYS594 = *11*; DYS607 = *12*; DYS617 = *13*; DYS632 = *8*; DYS636 = *11*; DYS640 = *13*; DYS641 = *11*; DYS643 = *12*; DYS650 = *18*; DYS710 = *31*; DYS714 = *24*; DYS715 = *23*; DYS716 = *28*; DYS726 = *15*; Y-GATA-A10 = *12*; Y-GGAAT-1B07 = *13* and DYF40651=*10*.

Likewise for the 11 palindromic STRs: DYS413.a = *22*,.b = *22*; DYS459.a = *9*,.b=*9*; DYS464.a=*14*,.b=*15*,. c=*16*,. d=*17*, CDY.b=*36* and DYF395S1.a=*15*,.b =*15*,which have identical values between M.C., C.N. and A.C.W.

Because of this identity, we can reasonably infer that the 93 allelic values of the above genetic markers correspond to those expected for Napoléon I (because they have remained unchanged for 5/6 generations of remote ancestry).

3.3 Differential STRs

Table 2 lists and characterizes the seven STRs that differentiate between M.C., C.N., and A.C.W. Only one of them (CDY.a) is palindromic. The mutation rates, when known (Burgarella & Navascués, 2011), of these differential alleles are in the 10^{-3} range (except for DYS447). These rates are impossible to evaluate for the palindromic STR CDY. a, and unknown for the moment for DYS712.

Table 2. Allele values for the seven differencial Y-STRs between Mike Clovis (M.C.), Charles Napoléon (C.N.) and Alexandre Colonna Walewski (A.C.W.). Expected allele values for **N** (Napoléon I) are established for the three Y-STRs DYS454 = *11*, DYS712 = *23* and DYS442 = *12*

Numbers	Y-STRs	Palindromic	Mutation rates	N (direct determination)	Allele values M.C.	Allele values C.N.	Allele values A.C.W.	Racial background	N (deduced / expected values)
1	**Y-GATA-C4**		2.832x 10-3	23	22	22	21		
2	**DYS454**		2.182x10⁻³		11	7	11		**11**
3	**DYS712**		?		23	23	25		**23**
4	**DYS442**		1.926x10⁻³		11	12	12		**12**
5	**DYS447**		7.414x10⁻⁴		22	21	21	+?	**21?**
6	**DYS481**		6.937x10⁻³		27	27	28	+	**27-28?**
7	**CDY.a**	+			34	35	34		**35?**

We already know (Lucotte & Bouin-Wilkinson, 2014) the real allele value = *23* of Y-GATA-C4 for Napoléon I. Figure 2 shows the bimodal distribution of Y-GATA-C4 alleles in the population; Napoléon I (N) value corresponds to that of the second modal class. Allele values (=*22*) of Charles Napoléon (C.N.) and Mike Clovis (M.C.) can be explained admitting one-step (*minus* 1) mutations, and that (=*21*) of Alexandre Colonna Walewsky (A.C.W.) admitting a two-step (*minus* 2) mutation.

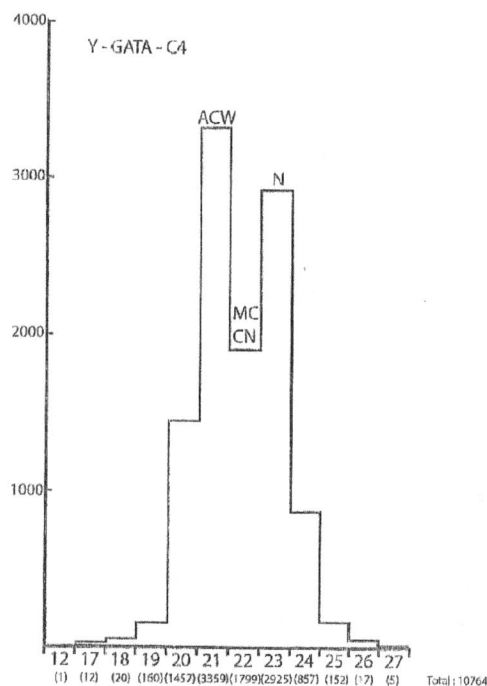

Figure 2. Y-GATA-C4 allelic distribution
(www.genebase.com/in/dnaMarkerDetail.php?t=y&d
=DYS635), based on values from 10 764 subjects

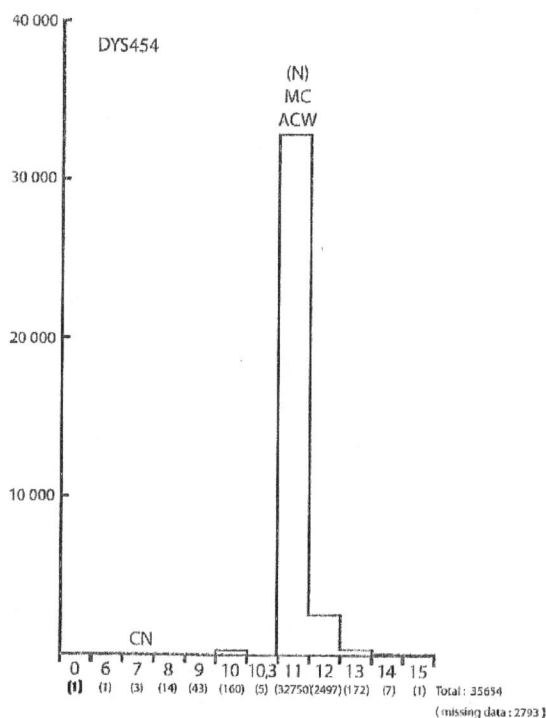

Figure 3. DYS454 allelic distribution (CEPH database),
based on values from 35 654 subjects

The distribution of DYS454 alleles in the population is shown in Figure 3. According to Redd *et al.* (2002) DYS454 is one of the most stable (with a pre-eminent modal class = *11*) of the marker set. Because both A.C.W. and M.C. alleles belong to this modal class, the most parsimonious interpretation is that allele *11* at this marker is the ancestral form - that of Napoléon I (N) - and that the allele value = *7* for C.N. represents a derived one, which happened during one of the five generations separating C.N. from the common ancestor Carlo Buonaparte. It is probable that this variant *7* (characteristic of the Jérôme line) is due to a multistep deletion, a rare event which often results in a most stable allele (Lucotte et al., 2013).

Although based on a relatively low number of subjects studied, figure 4 shows a representative allele distribution for DYS712. The interest of this recently described marker is that it certainly represents one of the most variable STR of the panel (its modal class corresponds to allele *22*). Alleles of C.N. and M.C. = *23* and the A.C.W. allele = *25*. The expected N value is probably *23*, because this class corresponds to the second one in importance; and the A.C.W. value = *25* must be due to a two-step (*plus* 2) mutation.

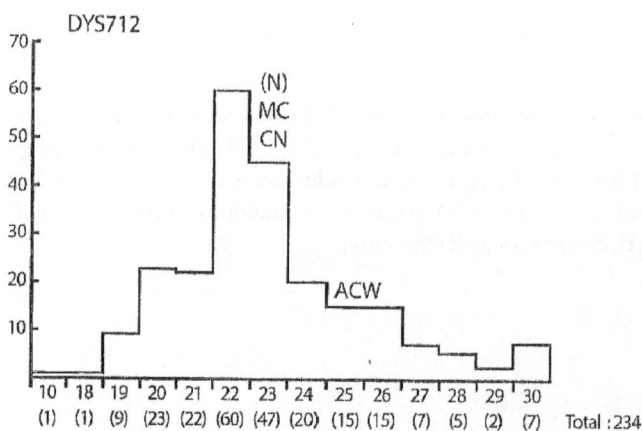

Figure 4. DYS712 allelic distribution (K. Norved, personal communication) of 234 American Caucasians

Figure 5 shows the modal distributions of alleles for DYS442 and DYS447, based on a sample of 1000 European subjects (English: 56, Germans: 59, French Parisians: 191, French Basques: 97, Corsicans: 328, North Italians: 46, Central-Italians: 112, Sardinians: 111; from Diéterlen and Lucotte, 2005). For DYS442 the allele value = *12* of C.N. and A.C.W. corresponds to the modal class; so it is probable that the expected N value is *12*. The value = *11* for M.C. must correspond to a one-step (*minus* 1) mutation.

The pattern of variations is more complicated to interpret for DYS447, because none of the allele values is located at the modal (=*25*) nor at the adjacents (*26* and *24*) classes: allele values for C.N. and A.C.W. = *21*, and *22* for M.C. We presume that the DYS447 distribution presents a small peak in frequencies (possibly due to racial background) at the left tail of the modal class. In this hypothesis, but it is highly speculative, the expected N value could be = *21* (because of the two *21* allelic values of C.N. and A.C.W.).

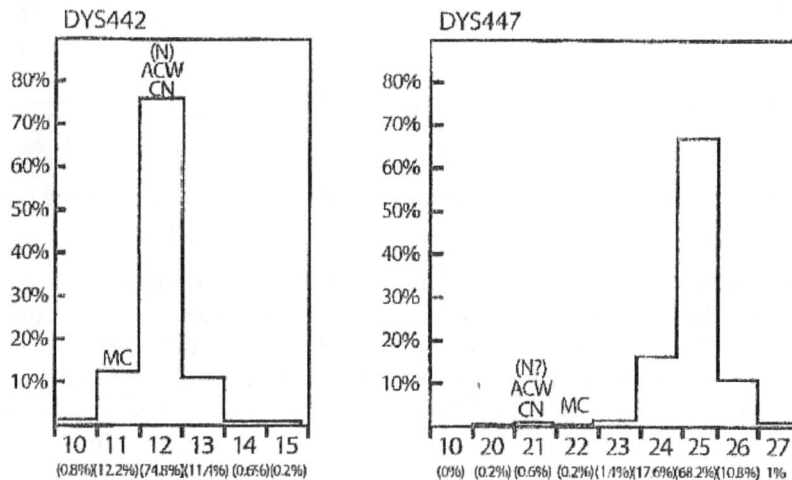

Figure 5. DYS442 and DYS447 allelic distributions, based on our sample of 1 000 unrelated European Caucasians

The existence of this racial background is evident for DYS481 (English: 102, Indians: 83, Africa: 94; from D'Amato et al., 2010), where the three European, Asiatic and African distributions are superposed on the graph (Figure 6): the *27* (for C.N. and M.C.) and *28* (for A.C.W.) classes are relatively well represented at the right tail of the Asiatic distribution, but none (for *27*) or few (for *28*) for the European distribution; but in any case we cannot decide if the expected N value is *27* or *28*.

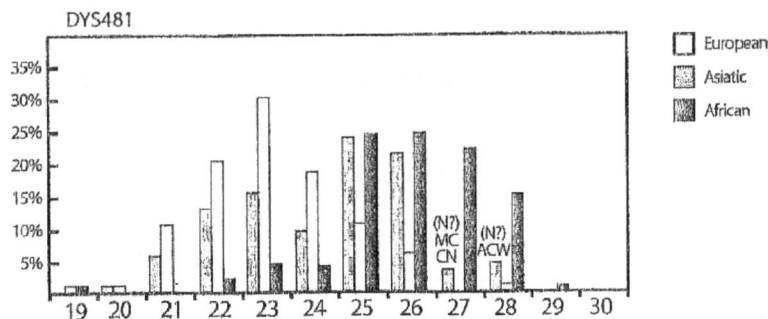

Figure 6. DYS481 allelic distribution (from d' Amato et al., based on samples of 102 European subjects (English), 83 Indians and 94 Xhosas

This sort of racial context intervening for some STR alleles determining the Y-haplotype is interesting to consider, because of the oriental origin (Lucotte & Diéterlen, 2014) of the E1b1b1 haplogroup of Napoléon I, more precisely known now (Lucotte et al., 2013) as E1b1b1b2a1 L792[+] haplogroup.

Figure 7 shows the modal distributions – based on our sample of 1000 European subjects – of allelic classes for the palindromic markers CDY.a and CDY.b. It is because of the identity of allele values = *36* between C.N., A.C.W.

and M.C. that we proposed that the expected N value for CDY.b is *36.* (though it corresponds to the fourth frequency class in importance, located at the left tail of the modal class).

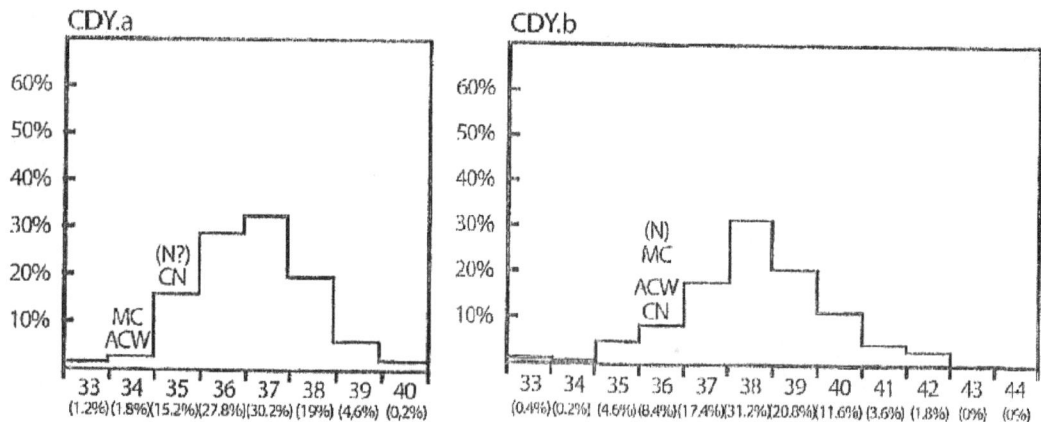

Figure 7. CDY.a and .b allelic distributions, based on the sample of 1 000 unrelated European Caucasians

Predictions about the variation of palindromic Y-STRs, even in the more simple situation of a two-copy marker like CDY, is a very hazardous matter (Lucotte et al., 2013) because we do not know exactly the precise mechanisms involved. For CDY.a the C.N. allelic value = *35,* and the A.C.W. and M.C. values = *34.* We retain here the possibility, but it is also highly speculative, that the expected N value could be = *37* (because, as for CDY.b, it corresponds to the third frequence class, located at the left tail of the modal one). Certainly the *35* alleles for A.C.W. and M.C. cannot be explained by such a simple mechanism as that of one-step (*minus* 1) mutation.

4. Discussion

In the goal to establish the Y-chromosome haplotype of Napoléon the First we determined initially, in his genomic DNA extracted from two islands of follicular sheats associated with his beard hairs conserved in the Vivant-Denon reliquary (Lucotte et al., 2011), allelic values for the three Y-STRs DYS19 and for two palindromic STRs YcaII.a and .b. Subsequently (Lucotte and Bouin-Wilkinson, 2014), based on genomic DNA of his hair dandruff dating from 1811, we determined allelic values for 16 STRs: DYS19 (for which we confirmed the first allelic value previously obtained), the palindromic STRs DYS385.a and .b, DYS389.i and ii, DYS390, DYS391, DYS392, DYS393, DYS438, the variable STR-Y-GATA-A4, DYS448, DYS456, DYS458, Y-GATA-C4 and Y-GATA-H4. The corresponding allele values for these 18 STRs correspond to the real allelic values of the Napoléon I Y-haplotype.

Because of the identity of allelic values of STRs between Charles Napoléon (the living fourth generation of male descent from Jérôme (Napoléon I's youngest brother) and Alexandre Colonna Walewski (a direct living sixth generation descendant from Napoléon I), we proposed (Lucotte et al., 2013) expected allelic values of Napoléon I for a total number of 109 STRs (33 of them being palindromics). For some of the six variables (between Charles and Alexandre) STRs: DYS454, DYS481, Y-GATA-C4, DYS712, CDY.a (palindromic) and DYF397.2 (palindromic), we proposed as expected allelic values for Napoléon I the most probable allelic forms according to STR distributions; the allele value of DYS454 = *7* for Charles Napoléon appeared then as a highly discordant one.

Mike Clovis is a living, previously unknown, fifth descendant of Lucien (another brother of Napoléon I). The objective of the present study is to compare, for a total number of 106 STRs, allelic values between him, Charles Napoléon and Alexandre Colonna Walewski. Identity of allelic values between the three was confirmed for 82 non-palindromic STRs and for 11 palindromic STRs; that confirms, in a triangular form, that these 93 STR allelic values are definitely those previously proposed as expected allele values of the Napoléon I Y-haplotype.

These comparisons between Mike Clovis, Charles Napoléon and Alexandre Colonna Walewsky permit us to clarify some of the questions asked by the variable values between them: for DYS454, the allele value = *11* for Mike Clovis is the expected allelic value of Napoléon I, as previously proposed. For DYS712, the allele value = *23* for Mike Clovis corresponds also to the expected allelic value of Napoléon I already proposed; however in this case it is not the modal class of distribution of DYS71 values that is concerned, but the nearest one of the right edge of this distribution.

Compared to Charles Napoléon and Alexandre Colonna Walewski, Mike Clovis had different allele values for DYS442 = *11* and DYS447 = *22*. For DYS442, as proposed previously, allele value = *12* is probably the expected allelic value of Napoléon I because it corresponds to the modal class of the distribution; and allele value = *11* for Mike Clovis results from a single mutational event (one-step, *minus* 1).

It is impossible to predict some expected allelic value of Napoléon I for DYS447, because the three obtained allele values (that of Mike Clovis = *22* could be the result of a one-step *plus* 1 mutational event) are all located at the left tail of the distribution. It is impossible also to predict some expected allelic value of Napoléon I for DYS481, even when interpreted in the context of the oriental origin of the E1b1b1c1 haplogroup (Lucotte and Diéterlen, 2014), because all the three obtained allele values are now located at the right tail of the distribution.

We ignore, for the moment, what is the Y-chromosome haplotype of Carlo Buonaparte; but it seems highly probable, because of the similarities between the Y-STR values presently obtained, that he was the biological father of Lucien, Napoléon and Jérôme (all these three having the same Y-haplogroup). As a by-product of such studies, we established that the allele value = *7* for DYS545 is highly characteristic of the Jérôme line; possibly, as shown here, the allele value = *11* for DYS442 could be characteristic of the Lucien line. It remains a possibility that allele values of *25* for DYS712 and of *28* for DYS481 could be characteristics of the direct Napoléon I line, at least for the Walewski descent.

References

Athey, W. T. (2006). Haplogroup prediction from Y-STR values using a Bayesian allele frequency approach. *J. Genet. Geneal., 2*, 34-39.

Burgarella, C., & Navascués, M. (2011). Mutation rate estimates for 110 Y-chromosome STRs combining population and father-son pair data. *Eur. J. Hum. Genet., 19*, 70-75. http://dx.doi.org/10.1038/ejhg.2010.154

D'Amato, M. E., Ehrenreich, L., Cloete, K.,Benjeddou, M., & Davison, S. (2010). Characterization of the highly discriminatory loci DYS449, DYS481, DYS518, DYS612, DYS626, DYS644 and DYS710. *Forens. Sci. Int. Genet., 4*, 104-110. http://dx.doi.org/10.1016/j.fsigen.2009.06.011

Diéterlen, F., & Lucotte, G. (2005). Haplotype XV of the Y-chromosome is the main haplotype in West-Europe. *Biomed. Pharmacother, 59*, 269-272. http://dx.doi.org/10.1016/j.biopha.2004.08.023

Lucotte G., & Bouin-Wilkinson, A. (2014). An autosomal STR profile of Napoléon the First. *Op. J. Genet., 4*, 292-299. http://dx.doi.org/10.4236/ojgen.2014.44027

Lucotte, G. (2010). A rare variant of mtDNA HSV1 sequence in the hairs of Napoléon's family. *Invest. Genet., 1*, 1-4. http://dx.doi.org/10.1186/2041-2223-1-7

Lucotte, G., & Diéterlen, F. (2014). Frequencies of M34, the ultimate genetic marker of the terminal differenciation of Napoléon the First's Y-chromosome haplogroup E1b1b1c1 in Europe, Northern Africa and the Near East. *Int. J. Anthropol., 29*(1-2), 27-41

Lucotte, G., Macé, J., & Hrechdakian, P. (2013). Reconstruction of the lineage Y chromosome haplotype of Napoléon the First. *Int. J. Sciences., 9*, 127-139.

Lucotte, G., Thomasset, T. & Hrechdakian, P. (2011). Haplogroup of the Y chromosome of Napoléon the First. *J. Mol. Biol. Res., 1*, 12-19. http://dx.doi.org/10.5539/jmbr.v1n1p12

Redd, A. J., Agellon, Al B., Kearney, V. A., Contreras, V. A., Karafet, T., Park, H., ... Hammer, M. F. (2002). Forensic value of 14 novel STRs on the human Y chromosome. *Forens. Sci. Int., 130*, 97-111. http://dx.doi.org/10.1016/S0379-0738(02)00347-X

Walsh, B. (2001). Estimating the time of the most recent ancestor for the Y chromosome or mitochondrial DNA for a pair of individuals. *Genetics, 158*, 897-912.

4

Methodical Approaches to the Study of Human Chromosomal Q-Heterochromatin Variability

Ibraimov A. I.[1]

[1] Institute of Balneology and Physiotherapy, Bishkek and Laboratory of Human Genetics, National Center of Cardiology and Internal Medicine, Bishkek, Kyrgyzstan

Correspondence: Ibraimov A. I., Institute of Balneology and Physiotherapy, Bishkek and Laboratory of Human Genetics, National Center of Cardiology and Internal Medicine, Bishkek, Kyrgyzstan. E-mail: ibraimov_abyt@mail.ru

Abstract

In spite of the fact that chromosomal Q-heterochromatin regions (Q-HRs) in the genome have been opened almost half a century ago, we still know extremely few of their possible roles in the human life activity. One of the reasons of such state is the lack of methodical approaches mostly suitable to the nature and features for chromosomal Q-HRs. In the present work the existing methodical approaches of the human chromosomal Q-HRs has been analyzed, beginning from empirical observations up to the analytical approaches, aimed to detect regularities of Q-HRs distribution and possible effects at population level, in norm and pathology. It is appeared, that all depends on how we consider the nature of chromosomal Q-HRs, namely, whether they are structurally uniform formations in genome or their possible effects depend on features of Q-HRs localization on this or that chromosome in the human karyotype.

Keywords: Q-heterochromatin, C-heterochromatin, genetic system, human body heat conductivity, human adaptation

1. Introduction

The fact that chromosomes of the higher eukaryotes consists of two important components - euchromatin and heterochromatin - has been determined in the thirties years of the 20th century (Heitz, 1928, 1934, 1935). In the seventies of the last century it has been found out, that there are, at least, two kinds of constitutive heterochromatin: C-heterochromatin, existing in genome in all of the higher eukaryotes and Q-heterochromatin, which can be found out in genome only in three higher primacies (*Homo s. sapiens, Pan troglodytes and Gorilla gorilla*) (Caspersson et al., 1970; Paris Conference, 1971, 1975; Pearson, 1973, 1977; ISCN, 1978).

There is the huge literature devoted to study of microscopic structure, molecular composition, methods of identification and a quantitative estimation of chromosomal C-heterochromatin areas (C-HRs) in karyotype, detailed in a number of reviews (Schmid, 1967; Prokofyeva-Belgovskaya, 1986; Verma & Dosik, 1980a; Stahl & Hartung, 1981; Ibraimov & Mirrakhimov, 1985; John, 1988; Verma, 1988; Bhasin, 2007). The present work considers the same questions regarding human chromosomal Q- heterochromatin regions (Q-HRs). Therefore we will start with results of the first empirical observations which have been based into all subsequent analytical works, aimed at searches of a possible biological role of chromosomal Q-HRs in the human genome.

2. Empirical researches

The early 1970's witnessed the development of several new cytochemical methods for studying microscopically certain types of constitutive heterochromatin directly in chromosome preparations (Paris Conference, 1971, 1975; ISCN, 1978). One of the important results of such studies was the discovery of genetic polymorphism at the chromosomal level due to the high variability of the heterochromatic regions. We owe the discovery of Q-heterochromatin polymorphisms to a group of scientists at the Karolinska Institute in Stockholm headed by the noted Swedish cytologist T. Caspersson (Caspersson et al., 1970), who were able to show that by using appropriate methods to stain chromosome preparations with quinacrine mustard a remarkable picture could be seen under a fluorescent microscope: each homologous pair of metaphase chromosomes has an individual pattern of differential fluorescence by which it can be identified. These authors showed that individual areas of certain chromosomes vary considerably in their intensity of fluorescence; specifically, that there are bands with particularly brilliant fluorescence in the centromeric region of chromosome 3, the short arms of acrocentric chromosomes 13-15, and the distal portion of the long arm of the Y chromosome. Evans and co-workers (1971) found similar brilliantly fluorescent areas on autosome pairs 21 and 22. It is now firmly established that brilliantly fluorescent polymorphic areas can be found only on seven autosomes – 3, 4, 13-15, 21, 22, and the Y chromosome. Later, at the Paris Conference on Standardization in Human Cytogenetics it was recommended that this method be termed Q-staining (from quinacrine mustard), and that fluorescence chromosomal bands be called Q-bands (Paris Conference, 1971, 1975; ISCN, 1978).

Unlike C-heterochromatin, Q-heterochromatin is not always apparent. C-heterochromatin is known to be present in absolutely all the 46 chromosomes of the human karyotype, varying only in size and location. In contrast, Q-heterochromatin may be completely absent in any of these chromosomes without any appreciable pathologic or other phenotypic consequences to the carrier, whereas complete absence of C-heterochromatin even in one chromosome is an extremely uncommon occurrence (Paris Conference, 1971, 1975).

Appropriate studies have shown that Q- and C-heterochromatin variants sometimes differ in both their location and their size in those chromosomes where they are located in the same regions (the Y chromosome, chromosome 3, and the short arms of acrocentric chromosomes (Verma & Dosik, 1980b). C-heterochromatin is known to account for 15-29% of the human genome. According to some authors, heterochromatic Q-band variants of a certain class may be completely absent in a considerable portion of a human population (Yamada & Hasegawa, 1978; Al-Nassar et al., 1981; Ibraimov & Mirrakhimov, 1982a,b,c, 1985; Ibraimov et al., 1982, 1986). Thus, there is ample evidence of significant qualitative and quantitative differences between regions of Q- and C-heterochromatin.

The chief morphologic expression of Q-heterochromatin polymorphism in differences among individuals in a given population in the presence, fluorescence intensity, size, and location of Q-heterochromatin in 12 polymorphic loci of seven autosomes (3p11q11, 4p11q11, 13p11, 13p13, 14p11, 14p13, 15p11, 15p13, 21p11, 21p13, 22p11, 22p13) and in the q12 band of the chromosome Y. It should be emphasized that there is no individual in human population who has Q-heterochromatin in all the 13 potentially polymorphic loci. The number of Q-heterochromatin variants usually range from 0 to 10 (Ibraimov, 2010, 2015; Ibraimov & Mirrakhimov, 1982a,b,c, 1985; Ibraimov et al., 1982; 1986; 2013; 2014). After Q-staining, brilliant fluorescence (Q-heterochromatin) is usually seen on one-half to two-thirds of the long arm of the Y. It is the variable size of the brilliantly fluorescent band (q12) of the long arm of the Y that mainly accounts for the polymorphism in this chromosome (Paris Conference, 1971; 1975).

At the stage of empirical researches basically the questions of identification, the account and registration of variants of chromosomal Q-HRs have been studied. The matter is that the discovery of chromosomal polymorphism in man after Q-staining necessitated the development of a rational and unified system of recording Q-heterochromatin variants, primarily quantitative, for subsequent statistical analysis. Comparison of the frequency of various types of Q-variants in different populations (normal, pathologic, age group, ethnic groups, etc.) and assessment of their possible adaptive value are the most important problems in any study on human chromosomal polymorphism.

Currently the following quantitative measures of chromosomal Q-heterochromatin polymorphism in human populations:

- The frequency of Q-HRs in 12 potentially polymorphic loci of seven autosomes. It is usually expressed in percentages of the number of chromosomes analyzed, separately for each polymorphic locus;

- The distribution of Q-HRs in a population, i.e., distribution of individuals having different numbers of Q-HRs in the karyotype regardless of the location (distribution of Q-HRs), which also reflected the range of Q-HRs variability in the population genome;

- The mean number of chromosomal Q-HRs per individual, as determined by dividing the total number of Q-HRs detected in a given sample by the number of individuals studied;

- The size of the Y chromosome, being (a) large (Y = F), (b) medium (F > Y > G), and (c) small (Y = G).

Various aspects of the heritability of human chromosomal Q-HRs variability detected at this stage of researches. The heritability of Q-HRs variants has been the object of several studies. Phillips (1977) studied the pattern of inheritance of Q- and C-HRs variants in 36 subjects in three unrelated families and found that 50% of the offspring inherited chromosomes having polymorphic variants. Robinson et al., (1976) studied the segregation pattern of polymorphic chromosomes in 32 families and found that, on the whole, there was complete agreement of actual data with those predicted by Mendelian law. Similar conclusions were also drawn by other investigators (McKenzie & Lubs, 1975; Tupitsina & Stobetsky, 1980).

Thus, at the stage of empirical observations it was possible to find out: 1) localization of chromosomal Q-HRs in the human karyotype and methods of their identification; 2) existence of morphological variants of chromosomal Q-HRs in a human population; 3) mode chromosomal Q-HRs inheritance in the raw of generations; 4) to standardize the variants of chromosomal Q-HRs (Paris Conference, 1971; 1975; ISCN, 1978), required for comparative population researches. Thus, thanks to empirical observations it was possible to find out existence of wide variability of chromosomal Q-HRs in genome of human populations. Though these works have not opened the role of chromosomal Q-HRs in the human life activity, they have allowed beginning the systematic researches aimed to find out biological role of Q-HRs in evolution and development of the higher primacies.

3. Analytical researches

There is no agreement as to the nature of chromosomal Q-HRs variability, although all arguments are based on the 'selectionist' hypothesis. One approach, implying that Q-heterochromatin with different locations is basically similar in structural and functional features, as defined by us (Ibraimov et el., 1986; 1990) in the following manner: 'of primary importance to an individual is the dose and not the location of Q-variants'. The term 'dose' is defined as the amount of Q-heterochromatin material in the genome regardless of its location in any chromosome. In other words, this approach is based on the assumption that chromosomal Q-HRs lack locus-specificity. Those favoring the alternative approach believe that derivations from expected Q-HRs frequencies, observed in any loci, reflect some structural and functional features of these loci and are due either to selection or to non-fortuitous segregation of chromosomes bearing the given Q-HRs (Geraedts & Pearson, 1974; Mikelsaar et al., 1975; Nazarenko, 1987).

Let's begin with the analysis of the second as we conditionally name, a locus specific approach. The supporters of such approach in the obvious or latent form mean, that Q-HRs, localized on different chromosomes in the human karyotype, represent different by its properties and biological effects hereditary structures. This is evidenced by the researches, where the authors pay paramount attention to frequencies of homo (+ / + or -/-) and heterozygotes (-/+) of chromosomal Q-HRs in population; namely the agreement of observed homo- and heterozygote frequencies in a population with those predicted by the law of Hardy-Weinberg. Though there are no direct instructions in the texts of these researches, that authors at this draw an analogy with the well-known HbS (as at sickle-cell anemia), nevertheless, judging by the method of the selected statistical analysis it is difficult to exclude such possibility.

It should be noted that frequencies of homo- and heterozygous chromosomal Q-HRs variants were not studied by all investigators in terms of the Hardy-Weinberg law. Nevertheless, there are many published studies in which the authors have performed a detailed analysis of the agreement of observed frequencies and those predicted by this fundamental law of population genetics (Schnedl, 1971; Inuma et al., 1973; Mikelsaar et al., 1974; Geraedts & Pearson, 1974; Muller et al., 1975; Buckton et al., 1976; Van Dyke et al., 1977; Al-Nassar et al., 1981; Ibraimov & Mirrakhimov, 1982a,b,c; Ibraimov et al., 1982; Stanyon et al., 1987; Kalz et al., 2005; Decsey et al., 2006). Agreement was found in most cases. However, several authors found some discrepancy between observed and predicted frequencies of certain Q-HRs variants in the seven autosomes. We decided not to examine these observations in detail, since the reasons for these discrepancies between observed homo- and heterozygote frequencies and those predicted by the Hardy-Weinberg law are unknown. Therefore, we shall only list the hypothetical explanations for these discrepancies suggested by the authors of these studies: 1. Methodological difficulties involved in the calculation of Q-HRs variants; 2. Sample sizes, since the Hardy-Weinberg law is known to be valid only for a large, randomly mated population; 3. Natural selection in the case of excess heterozygotes in a population in certain polymorphic loci (Geraedts & Pearson, 1974; Mikelsaar et al., 1974; Müller et al., 1975; Bobinson et al., 1976; Tupitsina & Stobetsky, 1980).

We found no statistically significant deviations between observed homo- and heterozygote frequencies and those predicted by the Hardy-Weinberg law even in those cases where: 1) the sample consisted of no more than 40 subjects; 2) the population had a long term exposure to extreme climate conditions (the extreme North of the Eastern Siberia or high altitudes of Pamir and Tien-Shan); or 3) a relatively high level of inbreeding had occurred (Ibraimov & Mirrakhimov, 1982a,b,c; 1985; Ibraimov et al., 1982, 1986). Jacobs (1977) is of the opinion that it is premature to test the agreement of observed homo- and heterozygote frequencies with those predicted by the Hardy-Weinberg law, since methods of calculation of the Q-band variants are not sufficiently accurate because of the continuous nature of the size distribution of the Q-HRs in a population.

In order to investigate the possible role of Q-HRs variants in the development of malignant diseases, Kivi and Mikelsaar (1981) studied 37 female patients with ovarian or breast cancer and 150 normal female subjects of the same age. The following parameters were evaluated in both groups: the frequency of Q-HRs variants in seven autosomes, the mean number of brilliantly fluorescent Q-bands, and the frequency of inverted chromosome 3. Their results showed that the presence of Q-polymorphic variants was not associated with the high risk of developing ovarian or breast cancer. Mikelsaar et al. (1981) attempted to establish a correlation between Q-heterochromatin chromosomes and the variability in stature by studying Q-polymorphism in 589 normal Estonian subjects (180 males and 409 females). The subjects were classified homo- or heterozygotes according to each Q-polymorphic band analyzed, and each Q-polymorphic band was investigated in relation to stature. The authors were unable to find any statistically significant correlation between Q-HRs polymorphism and stature in females. However, such a correlation was observed in males for bands 13p11, 21p11, 22p11, and 22p13. Also heterozygous males were consistently taller on the average than homozygous males.

The study history of inversion of Q-heterochromatin segment of the human chromosome 3 is the other bright example of the locus specific approach to find out the chromosomal Q-HRs possible role in the human life activity. Allderdice (1973) was the first to describe pericentric inversion of the Q-heterochromatin band in chromosome 3 (*inv 3*) (p15q12) in two phenotypically normal subjects. Later, Soudek et al. (1974) found such an inverted chromosome 3 in three mentally deficient patients. After further investigations on sibs of these patients with *inv 3* authors proposed to regard this variable chromosome 3 as a polymorphic feature. Subsequently, Soudek and Sroka (1978) studied the frequency of *inv 3* in 370 mentally deficient patients and 222 mentally normal subjects. The frequency of *inv 3* was found to be 4.05% in the former and 4.32% in the latter. These authors investigated the segregation of *inv 3* in six families and found no significant deviation from the mendelian distribution. Mikelsaar et al. (1978) came to the same conclusion after a comparative study of 102 normal newborns and 45 mentally deficient subjects of Estonian nationality.

Fogle and McKenzie (1980) studied a large black family consisting of 83 members and covering four generations and found *inv 3* in 23 of them, this inversion occurring in in a homozygous form in three of those affected. Those authors found no noticeable mental or other pathologic deviations in any of the 23 members of this family who exhibited *inv 3* both in hetero- and the homozygous form. Kivi and Mikelsaar (1981) studied 74 female subjects with malignant breast and ovarian tumors and 80 healthy female subjects and found no differences in this parameter (8.1% and 7.5, respectively), a finding that is in good agreement with data obtained previously in their laboratory during evaluation of the frequency of *inv 3* in an Estonian population.

Verma and Dosik (1980b) performed a comparative analysis of Q- and C-bands in chromosome 3 after QFQ and CBG staining (ISCN, 1978). In several cases the authors found complete agreement in the location and sizes of Q- and C-HRs variants. However, they observed cases where this polymorphic area of chromosome 3 failed to show any intense fluorescence despite the fact that CBG staining revealed the presence of a C-band and *vice versa*. This observation led the authors to conclude that these two methods of staining (QFQ and CBG) of chromosomes yield different information of heterochromatin polymorphisms of chromosome 3.

Soudek and Sroka (1978) were right in pointing out that this interesting cytogenetic marker does not always receive due attention in comparative population studies. Therefore, in all the populations of Eurasia and Africa studied we accurately recorded the frequency of *inv 3*. Among all the populations we studied, the frequency of *inv 3* proved to be highest in Russians (6.0%), this inversion occurring even in homozygous forms (0.5%) in this sample. Among the eight populations of Asia Mongoloids studied, this inversion was observed in only five of them, with a frequency ranging from 0.3% to 3.0%. We explained this fact by the presence of a European "admixture" in their gene stock that was due either to their ethnic composition (the Kazakhs and the Kyrgyz) or to a definite stage of their political history in the past (Mongolians, the Chukchi, the Yakut), i.e., to the forefather effect (Ibraimov, 2010; Ibraimov & Mirrakhimov, 1982a,b,c, 1985; Ibraimov et al., 1982, 1986). There is no other explanation for the fact that among 400 Japanese (Yamada & Hasegawa, 1978), 124 Chinese, 120 Khakass, and two Kyrgyz populations of Pamir and Tien-Shan not a single case of such inversion was found. Furthermore, this inversion was absent in 148 Mozambiquan, 132 Angolan, 34 Zimbabwean and 13 Guinea-Bissau natives whom we studied. However, we found *inv 3* among 52 Ethiopians (2.9%) in whose gene stock the existence of a Europoid component is universally recognized.

Kurmanova (1991) studied 277 mountaineers of the Russian nationality at whom *inv 3* was met at 23 individuals, and its frequency in the sampling was comparable to that in the control (6.8 % and 6.0 %, accordingly). Thus, under this quantitative characteristic of chromosomal Q-HRs variability these two samplings of the Russian did not differ essentially. In other two groups of mountaineers ("metises" and the mixed group) *inv 3* was met with frequency of 4.8 % (3 individuals) and 4.4 % (2 individuals), accordingly. Unfortunately, in none case she did not manage to carry out the family analysis. Nevertheless, the existence of *inv 3* in the metises group it is possible to explain for that they have, at least, Russian as one of the parents. Both individuals with *inv 3* in the mixed group as appeared to be Caucasoids, that once again testifies in favor of the assumption that *inv 3* is the original "Caucasoid" marker.

In connection with the aforementioned, we consider that the inverted chromosome 3 (having in mind its pericentric inversion of Q-heterochromatin segment) is the original "Caucasoid" cytogenetic marker; and under frequency of its occurrence in the sampling, it is possible to judge of existence of European "impurity" in this or that population (Ibraimov & Mirrakhimov, 1982a,b,c, 1985; Ibraimov et al., 1990, 1991). Our conclusions were completely confirmed by Rossi (1985) at research of some the Mexican populations. Thereupon, we think it is interesting to use this cytogenetic marker of ethnic anthropology in study of intensity of process metisation in a modern society where there is a mixture of various racial, national and ethnic groups as it is possible to define precisely the *inv 3* frequencies in initial populations.

Thus, to summarize the results of researches where for finding-out the possible role of chromosomal Q-HRs in the human life activity we attached paramount significance to a place of their localization in karyotype, it is possible to consider, that as a whole they were useful and instructive.

Now, let's proceed to the approach where chromosomal Q-HRs is considered as a single structural-and-functional system in the human genome and regard as of paramount importance the total quantity, instead of localization of Q-heterochromatin segments in the karyotype. At this, if it concerns a concrete individual, then the important criterion to estimate variability of chromosomal Q-HRs their number (from 0 to 10) is considered in karyotype and if it concerns population, then the mean number of chromosomal Q-HRs per individual, as determined by dividing the total number of Q-HRs detected in a given sample by the number of individuals studied.

The assumption of absence of Q-HRs locus specificity, revealed by Q-staining, we stated on the basis of that observation, that at increase or decrease of an mean number of Q-HRs per an individual in populations there is a proportional increase or decrease in frequency of Q-variants on all Q-polymorphic loci simultaneously (Ibraimov, 1993; Ibraimov et al., 1986). In particular, it has appeared, that if to range all samples as the mean number of Q-HRs increase; and to execute the same operation with frequencies of Q-variants on seven Q-polymorphic autosomes, then distribution of populations on frequencies of Q-variants in these autosomes, as a whole, corresponds to their distribution on the mean number of Q-HRs calculated per individual in populations. The same was defined by us at the analysis of data from all other researchers (Buckton et al., 1976; Lubs et al., 1977; Kalz et al., 2005), revealed in the conditions of the same laboratory the interpopulation distinctions on the mean numbers of Q-HRs (Ibraimov, 1993, 2010).

In favor of the assumption about locus specificity absence in chromosomal Q-HRs variants the following testify: 1) despite the fact that in the human karyotype there are 25 loci where chromosomal Q-HRs could potentially be found, in reality the maximal number of Q-HRs does not exceed 10 (Yamada & Hasegawa, 1978; Al-Nassar et al., 1981; Ibraimov & Mirrakhimov, 1985); 2) in human populations the number of Q-HRs in the karyotype usually ranges from 0 to 10 (Ibraimov & Mirrakhimov, 1985) without visible phenotypic effects; 3) distribution of Q-HRs in a population is near normal (Ibraimov et al., 1986, 1990, 1991); 4) at the population level the distribution of Q-HRs on seven Q-polymorphic autosomes is uneven, the greatest number of Q-HRs is found on chromosomes 3 and 13 (over 50%), the rest are more or less evenly distributed on the other five autosomes (Ibraimov, 1993, 2010); 5) human populations do not differ from each other in the relative content of Q-HRs on seven autosomes (the portions of Q-HRs on autosomes 3, 4, 13, 14, 15, 21 and 22 on average are 25.5%, 3.5%, 30.7%, 8.6%, 12%, 10.6% and 9.1%, respectively) (Ibraimov, 1993; 2010; 2011); 6) the quantitative content of chromosomal Q-HRs in the population genome is best determined by the value of the mean number of Q-HRs per individual (x) (McKenzie & Lubs, 1975; Yamada & Hasegawa, 1978; Al-Nassar et al., 1981; Ibraimov & Mirrakhimov, 1982a,b,c, 1985; Ibraimov et al., 1986, 1990, 1991); 7) decreases and increases in x in a population are due to simultaneous but proportional decreases or increases in the absolute number of Q-HRs on all the seven Q-polymorphic autosomes (Ibraimov, 1993; Ibraimov et al., 1986; 1990); 8) there are significant interpopulation differences in the quantitative content of chromosomal Q-HRs in the population genome (Buckton et al., 1976; Lubs et al., 1977; Ibraimov & Mirrakhimov, 1982a,b,c; Ibraimov et al., 1982, 1990, 1991; Kalz et al., 2005); 9) these differences proved to be related to features of the ecological environment of the place of permanent residence and not to the racial and ethnic composition of the populations (Ibraimov and Mirrakhimov, 1982 a, b, c; 1985; Ibraimov et al., 1982, 1986, 1990, 1991, 2013); 10) changes in the amount of Q-HRs in the population genome have a tendency towards a decrease from southern geographical latitudes to northern ones, and from low-altitude latitudes to high-altitude ones (Ibraimov & Mirrakhimov, 1982a,b,c, 1985; Ibraimov et al., 1982, 1990, 1991, 1997, 2013); 11) both decreases and increases of the x value are as a rule accompanied by narrowing or widening of the range of variability in the number of Q-HRs in a population (Ibraimov, 1993, 2010); 12) segregation of individuals in a human population with different number of Q-HRs in the karyotype (from 0 to 10) is due to the fact that Q-HRs are unevenly distributed on seven potentially Q-polymorphic autosomes (Ibraimov, 1993, 2010); 13) males in a population differ from each other in the size of the Q-heterochromatin segment of the Y chromosome (Paris Conference, 1971, 1975; Yamada & Hasegawa, 1978; Ibraimov & Mirrakhimov, 1985); 14) in different age groups x values differ, the greatest number of Q-HRs is characteristic of newborns, while the least number - in elderly subjects (Buckton et al., 1976; Ibraimov et al., 2014); 15) in the first days, weeks, months and years of life, ceteris paribus, among healthy children the infants often die with the greatest number of Q-HR in genome (Ibraimov & Karagulova, 2006); 16) individuals that are capable to adapt to the extreme climate of high altitudes (e.g. mountaineers) and to that of the Far North (e.g. borers - oil industry workers of the Jamal peninsula, Eastern Siberia) have extremely low numbers of Q-HRs in their genome (Ibraimov et al., 1986, 1990, 1991); 17) individuals with a lesser number of Q-HRs in their genome proved to be prone to alcoholism and obesity, while those with great number of Q-HRs - to drug addiction (Ibraimov, 2010; 2015); 18) Q-HR on the Y chromosome is the largest in the human karyotype, and its size, on average, is twice larger than all the Q-HRs on autosomes, taken together (ISCN, 1978; Ibraimov & Mirrakhimov, 1985); 19) Q-heterochromatin on the Y chromosome, being the largest in the human genome, −mehow "restricts" the total content of Q- HR on the autosomes in males. On the population level the value of x is influenced by the amount of Q-HR on the Y chromosome - for example in samples of males with great blocks of Q-heterochromatin on the Y chromosome the mean number of Q-HRs on their autosomes is lower and *vice versa* (Ibraimov et al., 2000); 20) there is some mechanism that compensates for the deficiency of Q-heterochromatin material in the female genome due to the lack of Y chromosome in their karyotype by increasing the amount of Q-HRs on autosomes. This pattern persists regardless of age and racial-ethnic characteristics of human populations (Ibraimov et al., 2014a); and at last, 21) individuals in population truly differ from each other in body heat conductivity and its level depends on the amount of chromosomal Q-HRs in human genome (Ibraimov et al., 2014b).

As it is seen from above stated, the non-locus specific approach has allowed clarifying some regularities in the distribution of chromosomal Q-HRs at level of populations, and some effects of this type constitutive heterochromatin upon the human life activity.

Of course we are far from the idea that the above listed regularities and effects are limited by the role of chromosomal Q-HRs in the human genome. It will not be surprising if it turns out that Q-HRs has many other important effects on life activity of three higher primates.

4. Concluding remarks

Despite the fact that chromosomal C- and Q-heterochromatin are defined by a single term, "constitutive heterochromatin", they are undoubtedly significantly different intrachromosomal structures. There are several significant differences between them: C-heterochromatin is found in the chromosomes of all the higher eukaryotes, while Q-heterochromatin – only in man (*Homo s. sapiens*), the chimpanzee (*Pan troglodytes*) and gorilla (*Gorilla gorilla*) (Pearson, 1973, 1977). C-HRs is known to be invariably present in all the chromosomes of man, varying mainly in size and location (inversion). Q-HRs variability can be found in man only on seven autosomes (3, 4, 13, 14, 15, 21 and 22), as well as on chromosome Y. Chimpanzees have Q-HRs on five autosomes (14, 15, 17, 22 and 23), while in gorillas they are present on eight (3, 12, 13, 14, 15, 16, 22 and 23) and on chromosome Y (Paris Conference, 1971; 1975; Chiarelli & Lin, 1972; Pearson, 1973, 1977; Dutrillaux et al., 1981). Chromosomal Q-HRs is subject to considerably greater variability in any population as compared to C-HRs. Erdtmann (1982) emphasized that "recent analyses... show a great population and evolutionary stability of C-band homeomorphisms... From interpopulation comparisons, C-band means show a tendency to maintain a constant amount of constitutive heterochromatin".

The most remarkable in all this, to our opinion, is the fact of Q-heterochromatin detected in genome of only three higher primacies. This circumstance, in its time, made some researchers to search for relation between Q-HRs variants and the human intelligence. Thus, Lubs et al. (1977) were unable to find any correlation between IQ and the distribution pattern of Q-HRs variants in a group of children aged 7–8 years. Schwinger and Wehner (1976) were also unable to find any correlation with the frequency of Q-HRs in seven autosomes of 89 randomly selected normal subjects and 244 patients with various types of mental disorders. Tharapel and Summit (1978) also found no statistically significant differences in Q-HRs polymorphism between 200 mentally deficient subjects and 200 controls. The results of Matsuura et al. (1979), obtained in patients with different forms of mental deficiency, are not at variance with these observations.

In due time we have supposed to consider chromosomal Q-HRs as a single structural-and-functional system. Usually the system is meant as an aggregate singled out real or imagined elements in any way from other world (Bertalanffy, 1968). It is possible to consider this aggregate as a biological system, if: 1) each of elements is indivisible further (not in the physical sense, but conditionally); 2) it co-operates with world around as whole; 3) the relations existing between elements of the singled out aggregate are set, and at evolution between them an unequivocal conformity remains, that is there is some orderliness.

Indivisibility of Q-HRs, located in 12 polymorphic loci of seven pairs of autosomes and on the distal part of Y chromosome results from specificity of cytochemical methods used now for their revealing (Caspersson et al., 1970; Paris Conference, 1971; 1975; ISCN, 1978). As for the second requirement it basically is reduced to the following: as closer to a place of human dwelling to the northern pole, as well as with increase in height of the location above the sea level, the amount of Q-heterochromatin decreases in the human population genome, irrespective of race-ethnic features of populations, and reduction of the content of Q-HRs occurs proportionally on all Q-polymorphic autosomes. Thus, if to require, that elements of the assumed system would co-operate with the world around as a whole unit, the facts of proportional decrease or increase in quantity of Q-HRs on Q-polymorphic autosomes in a human population gene pool, depending on features of the ecological environment round their dwelling, apparently, convincingly enough testify in favor of the aforementioned representation (Ibraimov, 1993, 2010, 2011, 2015; Ibraimov et al., 1986).

Concerning the third requirement, probably, it is worth to notice once again, that:

1) portion of Q-HRs variants in 12 Q-polymorphic loci of seven pairs of autosomes in the populations, expressed in percentage of number of the Q-variants revealed in this or that sampling, remains comparable, irrespective of the values of mean numbers of Q-HRs calculated per individual in populations (Ibraimov, 1983, 1993, 2010, 2011);

2) even in those populations where the mean number of Q-HRs per individual are very low, there is not the disappearance of Q-heterochromatin on the autosomes with low frequency of Q-variants and, moreover, the contribution of Q-high polymorphic, and the Q-low polymorphic chromosomes in the total pool of Q-heterochromatin material remains quite comparable in different samplings, regardless their racial or ethnic origin or characteristics of their permanent residence place (Ibraimov et al., 1986, 1990, 1991, 2013).

3) the distribution of the numbers of Q-HRs in the population always has the form close to a normal distribution (Ibraimov, 1993, 2010). All this is possible if the studied genetic material in the human genome will behave at the level of populations as a single self-sustaining structural and functional system (Ibraimov, 1993; Ibraimov et al., 1990).

The bulk of the data obtained on over of half a century empirical observations and analytical studies, described above, have allowed us to hypothesize that 'of primary importance to an individual is the dose and not the location of Q-variants' and to assume that the possible effect of chromosomal Q-HRs in the human body depends on its total amount in the genome (Ibraimov et al., 1986). And really it turned out that, for example, the level of human body heat conductivity directly depends on amount of chromosomal Q-HRs in his genome (Ibraimov et al., 2014b). Thus, we have concluded that all of the existing methods to analyze chromosomal Q-HRs is the most fruitful approach when this type of constitutive heterochromatin is considered as a single self-sustaining structural and functional genetic system in human genome.

Acknowledgements

I apologize to those authors whose work is not cited or cited only through reviews. The reason for this is only the space limitations.

References

Allderdice, P. W. (1973). Identification of the location of a crossover in a pericentric inversion heterozygote which resulted in a duplication deficient chromosome 3. *Am J Hum Genet, 25*, 11.

Al-Nassar, K. E, Palmer, C. G., Connealy, P. M., & Pao-Lo, Y. (1981). The genetic structure of the Kuwaiti population. II. The distribution of Q-band chromosomal heteromorphisms. *Hum Genet, 57*, 423-427.

Bertalanfy, V. L. (1968). *General System Theory*. Foundations, Development, Applications. New York.

Bhasin, M. K. (2007). Human population cytogenetics. A review. In V, Bhasin & M. K. Bhasin (Eds), *Anthropology today: trends, scope and applications*. Delhi, India: Kamla-Raj Enterprises.

Buckton, K. E., O'riordan, M. L., Jacobs, P. A., Robinson, J. A., Hill, R., & Evans, H. J. (1976). C - and Q - band polymorphisms in the chromosomes of three human populations. *Annals of human genetics, 40*(1), 99-112.

Caspersson, T., Zech, L., & Johansson, C. (1970). Differential binding of alkilating fluorochromes in human chromosomes. *Exp Cell Res, 60*, 315-319.

Chiarelli, B., & Lin, C. C. (1972). Comparison of fluorescence patterns in human and chimpanzee chromosomes. *Genet Phaenen, 15*, 103-106.

Décsey, K., Bellovits, O., & Bujdoso, G. M. (2006). Human chromosomal polymorphism in Hungarian sample. *Int J Hum Genet, 6*(3), 177-183.

Dutrillaux, B., Counturier, J., & Viegas-Pèquignot, E. (1981). Chromosomal evolution in primates. In M. D. Bennet, M. Boboraw, & G. M. Herwitt (Eds.), *Chromosomes today* (vol 7, pp. 176-191). New-York.

Erdtmann, B. (1982). Aspects of evaluation, significance, and evolution of human C-band heteromorphism. *Hum Genet, 61*, 281-294.

Evans, H. J., Buckton, K., & Sumner, A. T. (1971). Cytological mapping of human chromosomes: results obtained with quinacrine fluorescence and the acetic-saline-giemsa techniques. *Chromosoma, 35*, 310.

Fogle, T. A., & McKenzie, W. H. (1980). Cytogenetic study of a large black kindred: inversions, heteromorphisms and segregation analysis. *Hum. Genet., 55*, 345.

Geraedts, J. P. M., & Pearson, P. L. (1974). Fluorescent chromosome polymorphism: frequencies and segregation in a Dutch population. *Clin Genet, 6*, 247-257.

Heitz, E. (1934). Die Somatische heteropyknose bei Drosophila melanogaster und ihre genetische Bedeutung. *Z. Zellfosch, 20*, 237-287.

Heitz, E. (1935). Chromosomen structur und Gene. *Ztschr induct Abstammungs und vererbungslehre, 70*, 402-447.

Heitz, E. (1928). Das Heterochromatin der Moose. *J Jahrb Wisenschs Bot, 69*, 762-818.

Ibraimov, A. I, Akanov, A. A., Meymanaliev, T. S., Smailova, R. D., & Baygazieva, G. D. (2014a). Chromosomal Q-heterochromatin and age in human population. *J Mol Biol Res, 4*(1), 1-9.

Ibraimov, A. I. (1993). The origin of modern humans: a cytogenetic model. *Hum Evol, 8*, 81-91.

Ibraimov, A. I. (2010). Chromosomal Q-heterochromatin regions in populations and human adaptation. In: MK Bhasin, C Susanne (Eds.): Anthropology Today: Trends and Scope of Human Biology. Delhi: Kamla- Raj Enterprises, pp. 225-250.

Ibraimov, A. I. (2011). Origin of modern humans: a cytogenetic model. *Hum Evol, 26*(1-2), 33-47.

Ibraimov, A. I. (2014). Chromosomal Q-heterochromatin and sex in human population. *J Mol Biol Res, 4*(1), 10-19.

Ibraimov, A. I. (2015). Heterochromatin: The visible with many invisible effects. *Global Journal of Medical Research (C), 15*(3), 7-32 (Version 1.0).

Ibraimov, A. I., & Kagagulova, G. O. (2006). Chromosomal Q-heterochromatin variability in neonates deceased during first year of age. *Int J Hum Genet, 6*(4), 281-285.

Ibraimov, A. I., & Mirrakhimov, M. M. (1982a). Human chromosomal polymorphism. III. Chromosomal Q-polymorphism in Mongoloids of northern Asia. *Hum Genet, 62*, 252-257.

Ibraimov, A. I., & Mirrakhimov, M. M. (1982b). Human chromosomal polymorphism. IV. Chromosomal Q-polymorphism in Russians living in Kyrghyzia. *Hum Genet, 62*, 258-260.

Ibraimov, A. I., & Mirrakhimov, M. M. (1982c). Human chromosomal polymorphism. V. Chromosomal Q-polymorphism in African populations. *Hum Genet, 62*, 261-265.

Ibraimov, A. I., & Mirrakhimov, M. M. (1985). Q-band polymorphism in the autosomes and the Y chromosome in human populations. In: Progress and Topics in Cytogenetics. The Y chromosome. Part A. Basic Characteristics of the Y chromosome. Ed. by A. A. Sandberg. Alan R. Liss Inc., New York, pp.213-287.

Ibraimov, A. I., Akanov, A. A., Meimanaliev, T. S., Sharipov, K. O., Smailova, R. D., & Dosymbekova, R. (2014b). Human Chromosomal Q-heterochromatin Polymorphism and Its Relation to Body Heat Conductivity. *Int J Genet, 6*(1), 142-148.

Ibraimov, A. I., Akanov, A. A., Meymanaliev, T. S., Karakushukova, A. S., Kudrina, N. O., Sharipov K. O., & Smailova, R. D. (2013). Chromosomal Q-heterochromatin polymorphisms in 3 ethnic groups (Kazakhs, Russians and Uygurs) of Kazakhstan. *Int J Genet, 5*(1), 121-124.

Ibraimov, A. I., Axenrod, E. I., Kurmanova, G. U., & Turapov, D. A. (1991). Chromosomal Q-heterochromatin regions in the indigenous population of the northern part of West Siberia and new migrants. *Cytobios, 67*, 95-100.

Ibraimov, A. I., Karagulova, G. O., & Kim, E. Y. (1997). Chromosomal Q-heterochromatin regions in indigenous populations of the Northern India. *Ind J Hum Genet, 3*, 7-81.

Ibraimov, A. I., Karagulova, G. O., & Kim, E. Y. (2000). The relationship between the Y chromosome size and the amount of autosomal Q-heterochromatin in human populations. *Cytobios, 102*, 35-53.

Ibraimov, A. I., Kurmanova, G. U., Ginsburg, E. K. h., Aksenovich, T. I., & Axenrod, E. I. (1990). Chromosomal Q-heterochromatin regions in native highlanders of Pamir and Tien-Shan and in newcomers. *Cytobios, 63*, 71-82.

Ibraimov, A. I., Mirrakhimov, M. M., Axenrod, E. I., & Kurmanova, G. U. (1986). Human chromosomal polymorphism. IX. Further data on the possible selective value of chromosomal Q-heterochromatin material. *Hum Genet, 73*, 151-156.

Ibraimov, A. I., Mirrakhimov, M. M., Nazarenko, S. A., Axenrod, E. I., & Akbanova, G. A. (1982). Human chromosomal polymorphism. I. Chromosomal Q-polymorphism in Mongoloid populations of Central Asia. *Hum Genet, 60*, 1-7.

Inuma, K., Matsunaga, E., & Nakagome, J. (1973). Polymorphism of C and Q bands in human chromosomes. *Annu Rep Natl Genet (Japan), 23*, 112.

ISCN. (1978). An international system for human cytogenetic nomenclature. Report of the standing committee on human cytogenetic nomenclature. *Cytogenet Cell Genet, 21*, 313(1)-404(92).

Jacobs, P. A. (1977). Human chromosome heteromorphisms (variants). In: Progress in Medical Genetics, Vol. II. Steinberg A.G., Beam A.G. and Motulsky A.G. (eds). Philadelphia. Saunders, pp. 251-272.

John, B. (1988). The biology of heterochromatin. In: Heterochromatin: Molecular and Structural Aspects. Edited by R. S. Verma. Cambridge University Press Cambridge, New York, New Rochelle, Melbourne, Sydney.

Kalz, L., Kalz-Fuller, B., Hedge, S., & Schwanitz, G. (2005). Polymorphism of Q-band heterochromatin; qualitative and quantitative analyses of features in 3 ethnic groups (Europeans, Indians, and Turks). *Int J Hum Genet, 5*(2), 153-163.

Kivi, S. Y., & Mikelsaar, A. V. N. (1981). Variants of chromosomal Q- and C-bands in patients with breast or ovarian cancer. In Prokofyeva-Belgovskaya AA, Zakharov AF 9eds): Human Chromosomal Polymorphism. Moscow, Meditsina, p.196.

Lubs, H. A., Patil, S. R., Kimberling, W. J., Brown, J., Hecht, F., Gerald, P., & Summitt, R. L. (1977). Racial differences in the frequency of Q- and C-chromosomal heteromorphism. *Nature, 268*, 631-632.

Matsuura, J. S., Mayer, M., & Jacobs, P. A. (1979). A cytogenetic survey of an institution for the mentally retarded. III. Q-band chromosome heteromorphisms. *Hum Genet, 52*, 203-210.

McKenzie, W. H., & Lubs, H. A. (1975). Human Q and C chromosomal variations: distribution and incidence. *Cytogenet Cell Genet, 14*, 97-115.

Mikelsaar, A. V. N., Ilus, T., & Kivi, S. (1978). Variant chromosome 3 (inv. 3) in normal newborns and parents and in children with mental retardation. *Hum Genet, 41*, 109-113.

Mikelsaar, A. V. N., Kivi, S. Y., & Ilus, T. A. (1981). Chromosomal variants and human growth. In A. A. Prokofyeva-Belgovskaya & A. F. Zakharov (Eds.), *Human Chromosomal Polymorphism*. Moscow, Meditsina, p.107.

Mikelsaar, A. V. N., Kääosaar, M. E., & Tüür, S. J. (1975). Human karyotype polymorphism. III. Routine and fluorescence microscopic investigation of chromosomes in normal adults and mentally retarded children. *Humangenetik, 26*, 1-23.

Mikelsaar, A. V. N., Viikmao, M. N., Tuur, S. J., & Kaosaar, M. E. (1974). Human karyotype polymorphism. II. The distribution of individuals according to the presence of brilliant bands in chromosomes 3, 4 and 13 in a normal adult population. *Humangenetik, 23*, 59-63.

Müller, H. J., Klinger, H. P., & Glasser, M. (1975). Chromosome polymorphism in a human newborn population. II. Potentials of polymorphic chromosome variants for characterizing the idiograms of an individual. *Cytogenet Cell Genet, 15*, 239-255.

Nazarenko, S. A. (1987). Age dynamics of fluorescent polymorphism in human chromosomes. *Cytol Genet (Russian), 21*, 183-186.

Paris Conference. (1971) and Supplement (1975). Standartization in human cytogenetics. Birth Defects: Original Article Series, XI, 1-84. The National Foundation, New York.

Pearson, P. L. (1973). Banding patterns chromosome polymorphism and primate evolution. *Progress in Medical Genetics, 2*, 174-197.

Pearson, P. L. (1977). The uniqueness of the human karyotype. In T. Caspersson & L. Zech (Eds.), *Chromosome identification: technique and applications in biology and medicine.* New York, London: Academic Press.

Phillips, R. B. (1977). Inheritance of Q and C-band polymorphisms. *Can J Genet Cytol, 19*, 405- 415.

Prokofyeva-Belgovskaya, A. A. (1986). Heterochromatin Regions of Chromosomes (Russian). Nauka, Moscow.

Robinson, J. A., Buckton, K. E., & Spowart G. (1976). The segregation of human chromosome polymorphisms. *Ann Hum Genet, 40*, 113-121.

Rossi, S. L. (1985). Polymorphismos chromosomicos: metodologia y aplicacion en diagnostico citogenetico prenatal: Tesis Profecional. Mexico.

Schmid, W. (1967). Heterochromatin in mammals. *Arch. Julius Klaus-Stiftung Vererb, 42*, 1-60.

Schnedl, W. (1971). Banding pattern of human chromosomes. *Nature, 233*, 93.

Schwinger, E., & Wehner, H. (1976). Frequency of chromosomal fluorescence polymorphism in normal persons and in clinical patients with diagnosed chromosome aberrations. *Hum Genet, 32*, 115-119.

Soudek, D., & Sroka, H. (1978). Inversion of "fluorescent" segment in chromosome 3: A polymorphic trait. *Hum Genet, 44*, 109-115.

Soudek, D., O.'Shaunghnesky, L. P., & McCreary, B. D. (1974). Pericentric inversion of "fluorescent" segment in chromosome No 3. *Humangenetik, 22*, 343-346.

Stahl, A., & Hartung, M. I. (1981). L'heterochromatine. *Ann. Genet., 24*, 69-77.

Stanyon, R., Studer, M., Dragone, A., De Benedicts, G., & Brancati, C. (1988). Population cytogenetics of Albanians in the province of Cosenza (Italy): frequency of Q and C band variants. *Int J Anthropol, 3*(1), 14-29.

Tharapel, A. T., & Summit, R. L. (1978). Minor chromosome variations and selected homeomorphisms in 200 unclassifiable mentally retarded patients and 200 normal controls. *Hum Genet, 41*, 121-130.

Tupitsina, L. P., & Stobetsky, V. I. (1980). Segregation of chromosomal Q-polymorphic variants. *Genetics (USSR), 16*, 727.

van Dyke, D. L., Palmer, C. G., Nance, W. E., & Pao-Lo, Y. (1977). Chromosome polymorphism and twin zygosity. *Am J Hum Genet, 29*, 431-447.

Verma, R. S. (1988). Heteromorphism of heterochromatin. In: Heterochromatin: Molecular and Structural Aspects. Edited by R. S. Verma. Cambridge University Press, Cambridge, New York, New Rochelle, Melbourne, Sydney.

Verma, R. S., & Dosik, H. (1980a). Human chromosomai heteromorphisms: nature and clinical significance. *Int Rev Cytol, 62*, 361-383.

Verma, R. S., & Dosik, H. (1980b). Human chromosomal heteromorphisms in American blacks. I. Structural variability of chromosome 3. *J Hered, 71*, 441.

Yamada, K., & Hasegawa, T. (1978). Types and frequencies of Q-variant chromosomes in a Japanese population. *Hum Genet, 44*, 89-98.

Kurmanova, G. U. (1991). Izmenchivosti Q-heterochrominovykh rayonov khromosom v svyazy s adaptatsiei cheloveka k vysokogoryu. Thesis. SB AS USSR. Novosibirsk.

5

Divalent Cations Affect the Stability and Structure of Dad2p, a Subunit of the *Candida Albicans* Kinetochore Dam1 Complex

Jennifer Turner Waldo[1], Tsering Dolma[1] & Emily Rouse[1]

[1] State University of New York at New Paltz, USA

Correspondence: Jennifer Turner Waldo, Biology Department, State Uniersity of New York at New Paltz, New Paltz, NY, USA. E-mail: waldoj@newpaltz.edu

Abstract

The heterodecameric Dam1 complex is involved in establishing and maintaining the connection between the kinetochore and the mitotic spindle during mitosis. Biochemical studies of the reconstituted complex have shed light upon how it interacts with microtubules. However, little information about the biochemical properties of the isolated subunits has been available. This report examines the stability and structure of Dad2p, one of the Dam1 complex subunits isolated from *Candida albicans*. By employing differential scanning fluorimetry, protease protection and hydrodynamic analyses, we show that Dad2p is specifically responsive to the presence of divalent cations. This observation may be important for understanding the dynamic structure and regulation of the Dam1 complex in fungal cells.

Keywords: Mitosis, kinetochore, Dam1 complex, *Candida albicans*

1. Introduction

The process of mitosis has been studied extensively. While many fundamental mysteries remain to be unraveled, traditional biochemical and cell biological studies, as well as more recent genomic and proteomic approaches have resulted in a rather lengthy list of protein players necessary for correct chromosome segregation (Gascoigne & Cheeseman, 2011). Despite a tremendous amount of effort, the ability to reconstitute mitosis *in vitro*—the "gold standard" for a detailed biochemical understanding of the process—has been technically unfeasible (Akiyoshi & Siggins, 2012). Therefore, unlike other fundamental cellular processes like DNA replication or transcription, our understanding of the detailed dynamics of the mitotic machinery remains comparatively limited.

Mitosis requires the formation of a highly regulated connection between the microtubules of the mitotic spindle and the DNA of each sister chromatid. This contact is mediated by the kinetochore, a dynamic protein complex that is assembled at epigenetically marked regions of the chromosome known as centromeres (Allshire & Karpen, 2008). Scores of proteins have been identified as present in and important for kinetochore function (Gascoigne & Cheeseman, 2011; Roy, Varshney, Yadav, & Sanyal, 2013). An emerging portrait of the structure of the kinetochore includes an increasingly detailed understanding of the biochemical properties of several multi-protein complexes, including those that contact the DNA directly, and those that appear to solely function in binding to microtubules (Alushin & Nogales, 2011).

Many groups, including our own, have focused on one of these kinetochore constituents, the Dam1 complex (also called the DASH complex) (Buttrick & Millar, 2011). The Dam1 complex is involved in establishing and maintaining the connection between the chromosome and the mitotic spindle. It is comprised of ten different proteins, and was first identified in *S. cerevisiae*. Elegant biochemical and biophysical analyses have shown that multiple copies of the complex are capable of forming rings around microtubules *in vitro* (Miranda, De Wulf, Sorger, & Harrison, 2005; Miranda, King, & Harrison, 2007; Ramey et al., 2011; Wang et al., 2007; Westermann et al., 2005; Westermann et al., 2006). Although this structure provides a satisfying model which may explain how kinetochores remain tethered to the depolymerizing microtubule, the necessity for Dam1 encircling the microtubule *in vivo* remains somewhat controversial (Asbury, Gestaut, Powers, Franck, & Davis, 2006; Grishchuk et al., 2008; Nogales & Ramey, 2009; Westermann et al., 2006).

Interestingly, the ten proteins that make up this complex are absolutely required for cell survival in fungal species that employ a mechanism of attachment between the mitotic spindle and the kinetochore that involves a ratio of 1 microtubule: 1 kinetochore. This is the case in *S. cerevisiae* (Cheeseman et al., 2001; Janke, Ortiz, Tanaka, Lechner, & Schiebel, 2002; Li, Li, & Elledge, 2005) and *C. albicans* (Burrack, Applen, & Berman, 2011; Thakur & Sanyal, 2011), but not so for *S. pombe* (Sanchez-Perez et al., 2005). This difference has been postulated to reflect the more stringent requirement for maintaining connection with the depolymerizing microtubule when a single point of attachment is present; thus, under these circumstances, the ring may provide necessary stability (Burrack et al., 2011; Thakur & Sanyal, 2011). Outside of the yeasts, sequence-based homologues of the Dam1 complex proteins are not identifiable, though a variety of evidence suggests that the functional homologue in metazoans may be the Ska complex (Gaitanos et al., 2009; Guimaraes & Deluca, 2009; Hanisch, Sillje, & Nigg, 2006; Jeyaprakash et al., 2012; Welburn et al., 2009).

The internal structure of the Dam1 complex has been studied by electron microscopy (Miranda et al., 2005; Wang et al., 2007), sub-complex formation (Legal, Zou, Sochaj, Rappsilber, & Welburn, 2016; Miranda et al., 2007) and yeast-two hybrid analysis (Ikeuchi, Nakano, Kamiya, Yamane, & Kawarasaki, 2010; Shang et al., 2003). Together, these studies provide a low-resolution model of the Dam1 complex, which includes putative sites of interaction with the microtubule. In the commonly studied *S. cerevisiae* system, the ten individual proteins are not amenable to soluble expression in bacterial systems (J. Waldo & Scherrer, 2008; Westermann et al., 2005), so the opportunity to study the complex using biochemical reconstitution and structural analysis of the subunits has not been available. We have been examining the individual subunits of the Dam1 complex from *Candida albicans*, as some of these proteins are capable of being expressed in bacteria. For example, the *C. albicans* Dad1p has been shown to behave as an intrinsically disordered protein when isolated (J. T. Waldo, Greagor, Iqbal, Gittens, & Grant, 2010). Based on these studies, a working model is that Dad1p undergoes a structural transition upon binding to the other components of the complex.

This study examines another isolated subunit of the *C. albicans* Dam1 complex, Dad2p. Unlike Dad1p, the evidence presented here suggests that Dad2p is not a disordered protein, but that its stability and structure are altered by the addition of divalent cations. This insight may have implications for future work in developing *in vitro* Dam1 complex reconstitution, for considering how the activity of the complex may be regulated *in vivo*, and for guiding the development of novel anti-fungal compounds.

2. Materials and Methods

2.1 Construction of Bacterial Expression Vectors

The *C. albicans* DAD2 gene was cloned by PCR amplification of genomic DNA. PCR products were treated with the restriction enzymes BamHI and NgoMIV (New England Biolabs, Massachusettes, USA) and ligated into pST44-2 treated with the same enzymes (Tan, Kern, & Selleck, 2005). Plasmids were verified by DNA sequencing (MWG Biotech, Texas, USA).

2.2 Expression and Purification of Dad2p

Gene expression was performed in BL21(DE3) *E. coli* cells grown in auto-inducing media (Studier, 2005). Cells were harvested by centrifugation at 4°C, resuspended in Buffer A (20mMTris pH 7.5, 500mM NaCl) and frozen at -80°C. Thawed cells were lysed by sonication and the clarified lysate was applied to a 10ml chelating-sepharose (GE Healthcare Lifesciences, New Jersey, USA) column charged with nickel sulfate. The column with developed with Buffer A supplemented with 300 mM imidazole. Fractions containing the protein of interest were identified through SDS-PAGE, pooled and dialyzed overnight vs. 20 mM Tris pH 7.5, 2.5 mM EDTA. The protein solution was centrifuged, applied to a 1ml MonoQ column and eluted with a linear gradient of NaCl (0 to .5M). Fractions containing protein were identified by SDS-PAGE, pooled and stored at -80°C. All chromatography steps were performed on an AKTA-FPLC (GE Healthcare Lifesciences, New Jersey, USA) at 4°C.

2.3 Differential Scanning Fluorescence

Purified samples of Dad2p were incubated in a final volume of 20µl, with various additives and Sypro Orange (Sigma Aldrich, Missouri, USA) at a final concentration of 5X. Samples were mixed and incubated at room temperature for 15 minutes prior to the addition of Sypro Orange. A CFX-96 RT-PCR system (BioRad, California, USA) was used to increase the incubation temperature of the samples from 20°C to 75°C at a rate of 1°C per minute, taking a fluorescence reading every 0.2°C using a LED/photodiode set matched to the excitation and emission wavelengths of Sypro Orange. As the proteins unfold, Sypro Orange binds to the revealed hydrophobic amino acids and increases fluorescence emissions. Data was recorded using the included system software; the first derivative of the melting curves results in a negative peak, which is recorded as the protein's melting point.

For assays in which the goal was to screen with a commercially available matrix, 5μl of each sample from Wizard Crystal Screen I (Rigaku, Texas, USA) was added to make the final volume 20 μl. For assays that explored the effect of different salts, 200mM of the indicated salt was used and the analysis was repeated 8 times.

2.4 Protease Assays

Samples contained 10μg Dad2p, 0.2μg thermolysin, 20 mM Tris pH 7.5, 50mM NaCl and 500mM of additional salt solutions as indicated. Proteolysis took place during a 15 minute incubation at 30°C and was stopped with the addition of 0.2% SDS. Samples were visualized on a 15% SDS-PAGE stained with BioSafe Coomassie Stain (BioRad, California, USA).

2.5 Size Exclusion Chromatography

Samples of Dad2p containing 150μg protein were diluted 1:1 with column buffer containing either 20mM Tris pH 7.5, 2.5mM EDTA, 500mM NaCl; or 20mM Tris pH 7.5, 25mM NaCl, 500mM MgCl$_2$. Samples were applied to a Superdex 200 column (GE Healthcare Lifesciences, New Jersey, USA) equilibrated with the appropriate buffer at 4°C. Protein standards (BioRad, California, USA) of 670, 158, 44, 17 and 1.35kDa were run under conditions identical to the experimental samples.

2.6 Protein Concentration Determination

Dad2p was incubated with or without MgCl$_2$ for 15min at the indicated temperatures in a final volume of 100μl. Samples were removed and added to 200μl Bradford Reagent (BioRad, California, USA) and absorbance was measured at 595nm. To calculate protein concentration, a standard curve generated with BSA was utilized. Alternatively, following incubation, protein concentration was determined by applying 2μl of sample to a Nanodrop (Thermo Fisher, Massachusettes, USA) spectrophotometer and monitoring absorbance at 280nm.

3. Results and Discussion

The *Candida albicans DAD2* gene was placed in a pET-based expression plasmid and the protein was purified to homogeneity (Figure 1A). As a first step towards characterizing Dad2p, we utilized a high-throughput differential scanning fluorimetry (DSF) assay to rapidly screen for conditions and/or chemical additives that had a stabilizing effect on the protein. This assay uses the environmentally sensitive dye Syrpo Orange, which has been documented to fluoresce upon binding to hydrophobic residues on proteins (DeSantis, 2012; Niesen, Berglund, & Vedadi, 2007; Vedadi et al., 2006). Following incubation, the protein samples were placed in an RT-PCR system that allowed the temperature to be slowly increased as fluorescence is monitored. As proteins denature, the signal from Syrpo Orange binding can be detected (Figure 1B). Analysis of these melting curves by calculating the first derivative of the fluorescence intensity plots allows a melting temperature (T_m) to be determined (Figure 1C).

Hundreds of conditions can rapidly be tested in this manner in order to screen for molecules that affect a protein's structure. Results from a typical scan are presented in Figure 1D. The 48 conditions reported here are from the Rigaku Wizard I crystal screen kit. The red bars show how the T_m of Dad2p changes as a result of incubation of the protein in each condition. An increase in T_m is generally understood to be reflective of stabilization of the protein structure. Of the 48 experimental conditions presented here, four showed an increase in the T_m greater than 4°C. These were: #23 (15% ethanol, 200mM MgCl$_2$ 100mM imidazole pH8.0), #25 (30% PEG-400, 200mM MgCl$_2$, 100mM Tris pH8.5), #44 (30% PEG-400, 200mM c, 100mM Sodium Acetate pH4.5) and #40 (10% isopropanol, 200mM Calcium Acetate, 100mM MES pH6.0). The identity of the other conditions tested can be found at www.rigakureagents.com. To illustrate the specificity of this assay, results from an identical run with lysozyme, a completely unrelated protein of similar size, are shown (blue bars in Figure 1D). As the profiles for Dad2p and lysozyme are markedly different from each other, there is no evidence that any of the Dad2p stabilizing conditions picked up in this screen work as non-specific protein stabilizers.

A unifying feature of the four conditions that were observed to specifically stabilize Dad2p is the presence of MgCl$_2$ and Calcium Acetate, salts of similar ionic composition. Comparable results were seen in other screens as well (data not shown). In order to follow up on this observation, the individual components present in the four solutions were titrated into Dad2p and T_m was determined. Addition of MgCl$_2$ and Calcium Acetate clearly increased T_m (Figures 2A,B), while the other additives found in the initial conditions, PEG-400, isopropanol and ethanol did not appreciably affect the T_m (Figure 2C,D). Buffer composition and pH level also had no effect (data not shown).

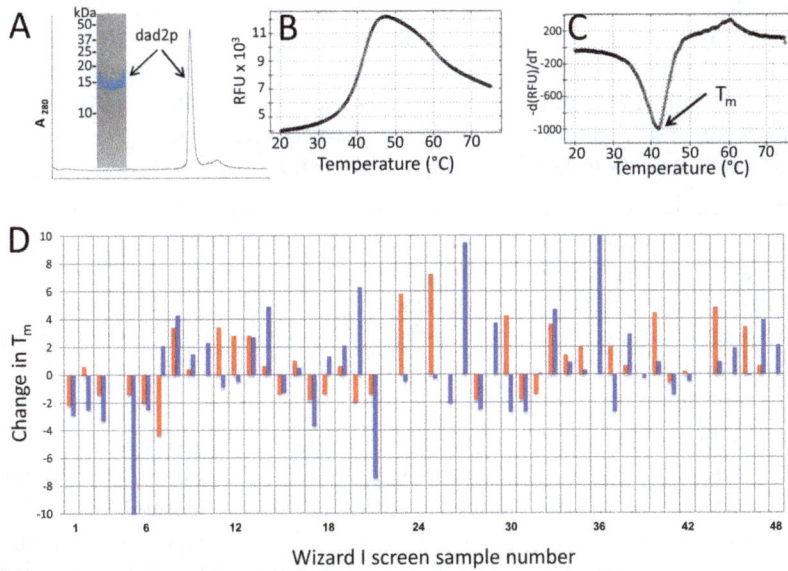

Figure 1. Screen for conditions that stabilize Dad2p. A) Following purification, Dad2p elutes as a single peak from a Superdex 200 gel filtration column, and runs as a single band on SDS-PAGE (insert). B) Fluorimetry assay with Dad2p. Samples were incubated with Sypro Orange and monitored for fluorescence upon increasing temperature. As the protein unfolds, fluorescence increases. C) The data in Panel B are replotted to include the first derivative of the fluorescence measurements. The minima reflects the melting point (T_m). D) Samples of Dad2p (red bars) and lysozyme (blue bars) were incubated with the 48 solutions comprising a commercially available crystallization screen and T_m of the protein was determined and compared to the T_m of the protein in a solution containing only Tris buffer (pH 7.5) and 100mM NaCl

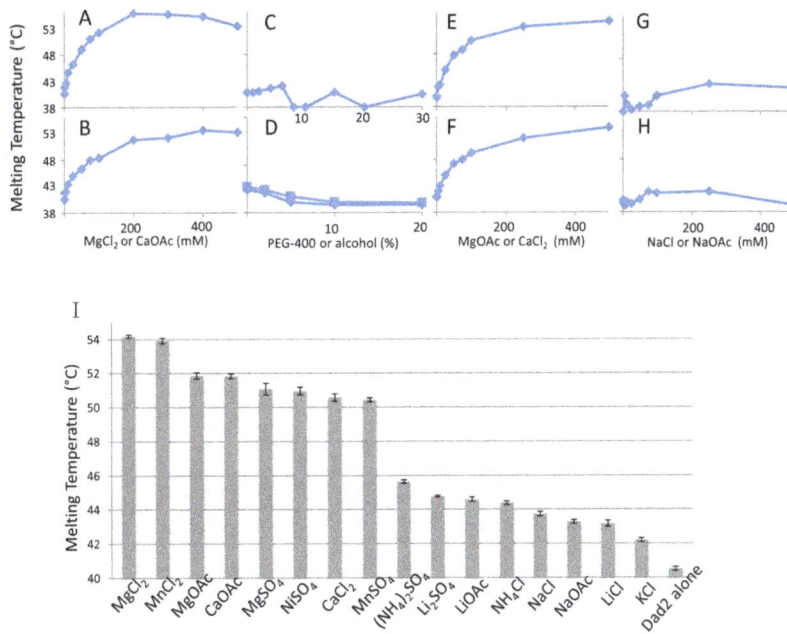

Figure 2. Melting temperature of Dad2p as a function of additive titration. Dad2p, Sypro Orange and the potentially stabilizing additives identified in Figure 1 were incubated prior to T_m determination. Increasing amounts of A) $MgCl_2$ and B) Calcium Acetate resulted in a higher T_m, while C) PEG-400 and D) ethyl (triangles) or isopropyl (squares) alcohol did not. Addition of E) Magnesium Acetate and F) $CaCl_2$ increased T_m, but the monovalent salt solutions of G) NaCl and H) Sodium Acetate did not. I) The melting point assay was repeated for a variety of salt solutions, each present at a final concentration of 200mM in a solution containing 20mM Tris pH 7.5. Each assay was repeated eight times, the values reported are mean +/- SEM

We next explored the specificity of the requirement for these salt solutions. Incubating Dad2p with Magnesium Acetate or CaCl$_2$, effectively switching the cation and anion components of the salts described above, resulted in a similar increase in thermal stability (Figure 2E,F), indicating that the effect is not dependent upon the specific combination of anion and cation. However, monovalent cation salts with the same anion, NaCl and Sodium Acetate, did not elevate the T$_m$ (Figure 2G,H). Together these observations suggest that the specific agents responsible for the increased stability of Dad2p are the divalent cations Mg^{2+} and Ca^{2+}.

To determine if other salts could similarly stabilize Dad2p, we performed the melting point assay in the presence of a variety of divalent and monovalent cation solutions (Figure 2I). The divalent cations magnesium, manganese, calcium, and nickel were shown to increase melting temperature more than the monovalent cations sodium, lithium, ammonium and potassium. Addition of iron, zinc, copper and cobalt solutions resulted in an inability to determine any melting point in this assay (data not shown). Therefore, there appears to be a general trend whereby divalent cations other than those initially identified impact the structure of Dad2p in a way that is measurably different than monovalent cations.

We next explored the ability of additional biochemical assays to detect, and thereby confirm, this effect. To visualize how the structure of Dad2p changes in the presence of divalent cations, we first employed a protease protection assay. Incubation of Dad2p with the broad-specificity protease thermolysin results in near complete degradation, with two distinct proteolytic fragments identifiable under these experimental conditions (blunt arrows, Figure 3A). When this assay is repeated in the presence of divalent cations (Mg^{2+}, Ca^{2+}), there is a marked decrease in protease activity, while a monovalent cation solution (Na$^+$) did not afford the same protection. As a control, cleavage of lysozyme was not impacted by the addition of any of these solutions (data not shown).

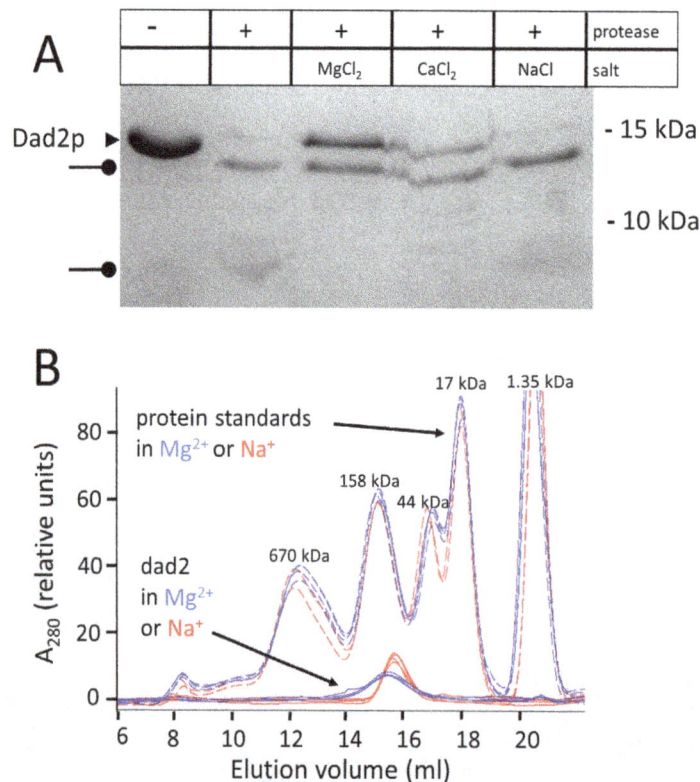

Figure 3. Protease sensitivity and hydrodynamic properties are also affected by divalent cations. A) Dad2p was incubated with or without the protease thermolysin and 500mM salt solutions as indicated. Protein fragments were analyzed in a Coomassie Blue stained SDS-PAGE. The arrow indicates the position of intact Dad2p, the blunt arrows indicate the position of proteolytic fragments. B) Samples of protein standards (dotted lines) and Dad2p (solid lines) were applied to a Superdex200 gel filtration column in the presence of 500mM NaCl (red traces) or MgCl$_2$ (blue traces). Each samples was run in triplicate and all nine chromatograms are layered in this figure

Next, the hydrodynamic properties of Dad2p were explored via a gel filtration assay. Figure 3B presents overlaid chromatograms of three repeats each of a set of protein standards (dotted lines) or Dad2p (solid lines) run with buffer containing NaCl (red traces) or $MgCl_2$ (blue traces). Interestingly, purified Dad2p elutes at a position significantly earlier than would be expected for the monomeric protein (observed molecular weight via gel filtration ~120,000 Da, expected molecular weight = 14,000 Da). This may reflect either protein oligomerization or an extended or unusual protein conformation. In the presence of $MgCl_2$, two things happen: there is a shift of the Dad2p peak to a slightly earlier elution position, and the peak becomes noticeably spread out. Neither of these observations hold true for the protein standards, so this change in protein structure dependent upon $MgCl_2$ addition appears to be specific to Dad2p.

While working with Dad2p we observed that the protein consistently gave lower protein concentration values than expected when using the Bradford reagent system. In particular, we found that boiling the protein resulted in a ~500% increase in the protein concentration calculated from the Bradford readings as compared to the value calculated at room temperature, while a similar impact was not observed when protein concentration was determined by UV spectroscopy (Figure 4A). This suggests that the amino acids reactive with Bradford reagent (primarily Arginine) are concealed by the protein's tertiary or quaternary structure, and that this structure is not disrupted by low pH conditions (~2, in the Bradford reagent), but it is disrupted at high temperatures. In addition, the residues that typically react to result in absorbance at 280nm (primarily Tryptophan and Tyrosine) must not be similarly concealed and revealed in a temperature-dependent manner.

To determine if this property would be affected by the presence of divalent cations, we incubated Dad2p at a variety of temperatures and then subjected the samples to Bradford quantitation over a series of increasing $MgCl_2$ additions (Figure 4B). The Bradford reactivity increased as temperature was raised. The midpoint of these cures is analogous to the T_m observed in the DSF assays. Increasing $MgCl_2$ resulted in a shift of the curves to the right; thus, higher concentrations of $MgCl_2$ are increasing the apparent melting temperature and reflect a stabilization of the protein's structure. Further, the T_m observed in this experiment with Bradford reagent in the absence (~40C) and the presence (~55C) of divalent cations are virtually the same as those seen in the Sypro-Orange monitored DSF experiments.

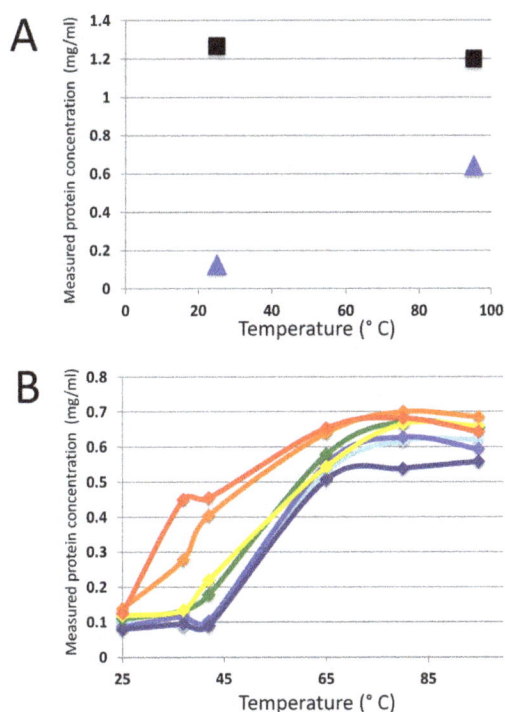

Figure 4. Temperature affects Dad2p reactivity with the Bradford reagent. A) Dad2p was incubated at room temperature or 100°C for 15 minutes. Aliquots were either added to Bradford reagent (blue triangles) or applied directly to a small-volume UV spectrophotometer (black squares) to determine protein concentration. B) Dad2p was incubated with 0 mM (red), 50 mM (orange), 75 mM (yellow), 100 mM (green), 150 mM (royal blue), 250 mM (cyan), 500 mM (indigo) $MgCl_2$ at the indicated temperature for two minutes. Protein concentration was then determined by incubation with the Bradford reagent

Taken together, the DSF experiments, protease protection assay, gel filtration analysis and Bradford reagent reactivity, confirm that the structure of Dad2p is specifically altered by the addition of divalent cations. The mechanism behind this observation remains to be elucidated. Sequence analysis does not provide any obvious metal binding motifs (data not shown), but analysis of the crystal structure, when available, may either reveal a co-crystallizing divalent cation, or provide the structural information necessary to mount a search for possible sites of interaction (Brylinski & Skolnick, 2011). This observation should provide important information for the development of models for how the Dam1 complex may be regulated or assembled, as divalent cations have been shown to play important roles in other multi-protein complexes (Huet, Conway, Letellier, & Boulanger, 2010; Rubin, 2007; Tiwari, Askari, Humphries, & Bulleid, 2011; Weinreb et al., 2012).

In addition, this report provides an example of a relatively novel and practical way of uncovering fundamental biochemical and biophysical properties of a protein. Many labs have access to RT-PCR systems, and the utility of this approach should be of interest to many, as it doesn't require protein modification or antibody production.

Acknowledgments

This work was supported by an NSF MRI grant (1039966) and by a SUNY New Paltz Academic Year Undergraduate Research Experience (AYURE) Award.

References

Akiyoshi, B., & Siggins, S. (2012). Reconstituting the kinetochore-microtubule interface: what, why, and how. *Chromosoma, 121*(3), 235-250. http://dx.doi.org/10.1007/s00412-012-0362-0

Allshire, R. C., & Karpen, G. H. (2008). Epigenetic regulation of centromeric chromatin: old dogs, new tricks? *Nat Rev Genet, 9*(12), 923-937. http://dx.doi.org/10.1038/nrg2466

Alushin, G., & Nogales, E. (2011). Visualizing kinetochore architecture. *Curr Opin Struct Biol, 21*(5), 661-669. http://dx.doi.org/10.1016/j.sbi.2011.07.009

Asbury, C. L., Gestaut, D. R., Powers, A. F., Franck, A. D., & Davis, T. N. (2006). The Dam1 kinetochore complex harnesses microtubule dynamics to produce force and movement. *Proc Natl Acad Sci U S A*. http://dx.doi.org/10.1073/pnas.0602249103

Brylinski, M., & Skolnick, J. (2011). FINDSITE-metal: integrating evolutionary information and machine learning for structure-based metal-binding site prediction at the proteome level. *Proteins, 79*(3), 735-751. http://dx.doi.org/10.1002/prot.22913

Burrack, L. S., Applen, S. E., & Berman, J. (2011). The requirement for the Dam1 complex is dependent upon the number of kinetochore proteins and microtubules. *Curr Biol, 21*(10), 889-896. http://dx.doi.org/10.1016/j.cub.2011.04.002

Buttrick, G. J., & Millar, J. B. (2011). Ringing the changes: emerging roles for DASH at the kinetochore-microtubule Interface. *Chromosome Res, 19*(3), 393-407. http://dx.doi.org/10.1007/s10577-011-9185-8

Cheeseman, I. M., Brew, C., Wolyniak, M., Desai, A., Anderson, S., Muster, N., ... Barnes, G. (2001). Implication of a novel multiprotein Dam1p complex in outer kinetochore function. *J Cell Biol, 155*(7), 1137-1145. http://dx.doi.org/10.1083/jcb.200109063

DeSantis, K., Reed, A, Rahhal, R and Reinking, J. (2012). Use of differential scanning fluorimetry as a high-throughput assay to identify nuclear receptor ligands. *Nuclear Receptor Signaling, 10*(e002), 1-5. http://dx.doi.org/10.1621/nrs.10002

Gaitanos, T. N., Santamaria, A., Jeyaprakash, A. A., Wang, B., Conti, E., & Nigg, E. A. (2009). Stable kinetochore-microtubule interactions depend on the Ska complex and its new component Ska3/C13Orf3. *EMBO J, 28*(10), 1442-1452. http://dx.doi.org/10.1038/emboj.2009.96

Gascoigne, K. E., & Cheeseman, I. M. (2011). Kinetochore assembly: if you build it, they will come. *Curr Opin Cell Biol, 23*(1), 102-108. http://dx.doi.org/10.1016/j.ceb.2010.07.007

Grishchuk, E. L., Spiridonov, I. S., Volkov, V. A., Efremov, A., Westermann, S., Drubin, D., ... McIntosh, J. R. (2008). Different assemblies of the DAM1 complex follow shortening microtubules by distinct mechanisms. *Proc Natl Acad Sci U S A, 105*(19), 6918-6923. http://dx.doi.org/10.1073/pnas.0801811105

Guimaraes, G. J., & Deluca, J. G. (2009). Connecting with Ska, a key complex at the kinetochore-microtubule interface. *EMBO J, 28*(10), 1375-1377. http://dx.doi.org/10.1038/emboj.2009.124

Hanisch, A., Sillje, H. H., & Nigg, E. A. (2006). Timely anaphase onset requires a novel spindle and kinetochore complex comprising Ska1 and Ska2. *EMBO J, 25*(23), 5504-5515. http://dx.doi.org/10.1038/sj.emboj.760 1426

Huet, A., Conway, J. F., Letellier, L., & Boulanger, P. (2010). In vitro assembly of the T=13 procapsid of bacteriophage T5 with its scaffolding domain. *J Virol, 84*(18), 9350-9358. http://dx.doi.org/10.1128/JVI. 00942-10

Ikeuchi, A., Nakano, H., Kamiya, T., Yamane, T., & Kawarasaki, Y. (2010). A method for reverse interactome analysis: High-resolution mapping of interdomain interaction network in Dam1 complex and its specific disorganization based on the interaction domain expression. *Biotechnol Prog, 26*(4), 945-953. http://dx.doi.org/10.1002/btpr.403

Janke, C., Ortiz, J., Tanaka, T. U., Lechner, J., & Schiebel, E. (2002). Four new subunits of the Dam1-Duo1 complex reveal novel functions in sister kinetochore biorientation. *EMBO J, 21*(1-2), 181-193. http://dx.doi.org/10.1093/emboj/21.1.181

Jeyaprakash, A. A., Santamaria, A., Jayachandran, U., Chan, Y. W., Benda, C., Nigg, E. A., & Conti, E. (2012). Structural and functional organization of the Ska complex, a key component of the kinetochore-microtubule interface. *Mol Cell, 46*(3), 274-286. http://dx.doi.org/10.1016/j.molcel.2012.03.005

Legal, T., Zou, J., Sochaj, A., Rappsilber, J., & Welburn, J. P. (2016). Molecular architecture of the Dam1 complex-microtubule interaction. *Open Biol, 6*(3). http://dx.doi.org/10.1098/rsob.150237

Li, J. M., Li, Y., & Elledge, S. J. (2005). Genetic analysis of the kinetochore DASH complex reveals an antagonistic relationship with the ras/protein kinase A pathway and a novel subunit required for Ask1 association. *Mol Cell Biol, 25*(2), 767-778. http://dx.doi.org/10.1128/MCB.25.2.767-778.2005

Miranda, J. J., De Wulf, P., Sorger, P. K., & Harrison, S. C. (2005). The yeast DASH complex forms closed rings on microtubules. *Nat Struct Mol Biol, 12*(2), 138-143. http://dx.doi.org/10.1038/nsmb896

Miranda, J. J., King, D. S., & Harrison, S. C. (2007). Protein arms in the kinetochore-microtubule interface of the yeast DASH complex. *Mol Biol Cell, 18*(7), 2503-2510. http://dx.doi.org/10.1091/mbc.E07-02-0135

Niesen, F. H., Berglund, H., & Vedadi, M. (2007). The use of differential scanning fluorimetry to detect ligand interactions that promote protein stability. *Nat Protoc, 2*(9), 2212-2221. http://dx.doi.org/10.1038/nprot. 2007.321

Nogales, E., & Ramey, V. H. (2009). Structure-function insights into the yeast Dam1 kinetochore complex. *J Cell Sci, 122*(Pt 21), 3831-3836. http://dx.doi.org/10.1242/jcs.004689

Ramey, V. H., Wang, H. W., Nakajima, Y., Wong, A., Liu, J., Drubin, D., ... Nogales, E. (2011). The Dam1 ring binds to the E-hook of tubulin and diffuses along the microtubule. *Mol Biol Cell, 22*(4), 457-466. http://dx.doi.org/10.1091/mbc.E10-10-0841

Roy, B., Varshney, N., Yadav, V., & Sanyal, K. (2013). The process of kinetochore assembly in yeasts. *FEMS Microbiol Lett, 338*(2), 107-117. http://dx.doi.org/10.1111/1574-6968.12019

Rubin, H. (2007). The logic of the Membrane, Magnesium, Mitosis (MMM) model for the regulation of animal cell proliferation. *Arch Biochem Biophys, 458*(1), 16-23. http://dx.doi.org/10.1016/j.abb.2006.03.026

Sanchez-Perez, I., Renwick, S. J., Crawley, K., Karig, I., Buck, V., Meadows, J. C., ... Millar, J. B. (2005). The DASH complex and Klp5/Klp6 kinesin coordinate bipolar chromosome attachment in fission yeast. *EMBO J, 24*(16), 2931-2943. http://dx.doi.org/10.1038/sj.emboj.7600761

Shang, C., Hazbun, T. R., Cheeseman, I. M., Aranda, J., Fields, S., Drubin, D. G., & Barnes, G. (2003). Kinetochore protein interactions and their regulation by the Aurora kinase Ipl1p. *Mol Biol Cell, 14*(8), 3342-3355. http://dx.doi.org/10.1091/mbc.E02-11-0765

Studier, F. W. (2005). Protein production by auto-induction in high density shaking cultures. *Protein Expr Purif, 41*(1), 207-234. http://dx.doi.org/10.1016/j.pep.2005.01.016

Tan, S., Kern, R. C., & Selleck, W. (2005). The pST44 polycistronic expression system for producing protein complexes in Escherichia coli. *Protein Expr Purif, 40*(2), 385-395. http://dx.doi.org/10.1016/j.pep.2004. 12.002

Thakur, J., & Sanyal, K. (2011). The essentiality of the fungus-specific Dam1 complex is correlated with a one-kinetochore-one-microtubule interaction present throughout the cell cycle, independent of the nature of a centromere. *Eukaryot Cell, 10*(10), 1295-1305. http://dx.doi.org/10.1128/EC.05093-11

Vedadi, M., Niesen, F. H., Allali-Hassani, A., Fedorov, O. Y., Finerty, P. J., Jr., Wasney, G. A., ... Edwards, A. M. (2006). Chemical screening methods to identify ligands that promote protein stability, protein crystallization, and structure determination. *Proc Natl Acad Sci U S A, 103*(43), 15835-15840. http://dx.doi.org/10.1073/pnas.0605224103

Waldo, J. T., Greagor, S. A., Iqbal, A. J., Gittens, A. S., & Grant, K. K. (2010). The Dad1 subunit of the yeast kinetochore Dam1 complex is an intrinsically disordered protein. *Biochem Biophys Res Commun, 400*(3), 313-317. http://dx.doi.org/10.1016/j.bbrc.2010.08.050

Waldo, J., & Scherrer, M. (2008). Production and initial characterization of Dad1p, a component of the Dam1-DASH kinetochore complex. *PLoS One, 3*(12), e3888. http://dx.doi.org/10.1371/journal.pone.0003888

Wang, H. W., Ramey, V. H., Westermann, S., Leschziner, A. E., Welburn, J. P., Nakajima, Y., ... Nogales, E. (2007). Architecture of the Dam1 kinetochore ring complex and implications for microtubule-driven assembly and force-coupling mechanisms. *Nat Struct Mol Biol, 14*(8), 721-726. http://dx.doi.org/10.1038/nsmb1274

Weinreb, P. H., Li, S., Gao, S. X., Liu, T., Pepinsky, R. B., Caravella, J. A., ... Woods, V. L., Jr. (2012). Dynamic Structural Changes Are Observed Upon Collagen and Metal Ion Binding to the Integrin alpha1 I Domain. *J Biol Chem.* http://dx.doi.org/10.1074/jbc.M112.354365

Welburn, J. P., Grishchuk, E. L., Backer, C. B., Wilson-Kubalek, E. M., Yates, J. R., 3rd, & Cheeseman, I. M. (2009). The human kinetochore Ska1 complex facilitates microtubule depolymerization-coupled motility. *Dev Cell, 16*(3), 374-385. http://dx.doi.org/10.1016/j.devcel.2009.01.011

Westermann, S., Avila-Sakar, A., Wang, H. W., Niederstrasser, H., Wong, J., Drubin, D. G., ... Barnes, G. (2005). Formation of a dynamic kinetochore- microtubule interface through assembly of the Dam1 ring complex. *Mol Cell, 17*(2), 277-290. http://dx.doi.org/10.1016/j.molcel.2004.12.019

Westermann, S., Wang, H. W., Avila-Sakar, A., Drubin, D. G., Nogales, E., & Barnes, G. (2006). The Dam1 kinetochore ring complex moves processively on depolymerizing microtubule ends. *Nature, 440*(7083), 565-569. http://dx.doi.org/10.1038/nature04409

6

Identification and *In-vivo* Characterization of a Novel OhrR Transcriptional Regulator in *Burkholderia xenovorans* LB400

Tinh T. Nguyen[1], Ricardo Martí-Arbona[1], Richard S. Hall[1], Tuhin Maity[1], Yolanda E. Valdez[1], John M. Dunbar[1], Clifford J. Unkefer[1] & Pat J. Unkefer[1]

[1] Bioscience Division, Los Alamos National Laboratory, Los Alamos, NM, USA

Correspondence: Pat J. Unkefer, Los Alamos National Laboratory, P.O. Box 1663, MS E529, Los Alamos, New Mexico 87545, USA. E-mail: punkefer@lanl.gov

Abstract

Transcriptional regulators (TRs) are an important and versatile group of proteins, yet very little progress has been achieved towards the discovery and annotation of their biological functions. We have characterized a previously unknown organic hydroperoxide resistance regulator from *Burkholderia xenovorans*LB400, Bxe_B2842, which is homologous to *E. coli's* OhrR. Bxe_B2842 regulates the expression of an organic hydroperoxide resistance protein (OsmC). We utilized frontal affinity chromatography coupled with mass spectrometry (FAC-MS) and electrophoretic mobility gel shift assays (EMSA) to identify and characterize the possible effectors of the regulation by Bxe_B2842. Without an effector, Bxe_B2842 binds a DNA operator sequence (DOS) upstream of *osmC*. FAC-MS results suggest that 2-aminophenol binds to the protein and is potentially an effector molecule. EMSA analysis shows that 2-aminophenol attenuates the Bxe_B2842's affinity for its DOS. EMSA analysis also shows that organic peroxides attenuate Bxe_B2842/DOS affinity, suggesting that binding of the TR to its DOS is regulated by the two-cysteine mechanism, common to TRs in this family. Bxe_B2842 is the first OhrR TR to have both oxidative and effector-binding mechanisms of regulation. This paper reveals further mechanistic diversity TR mediated gene regulation and provides insights into methods for function discovery of TRs.

Keywords: transcriptional regulator (TR), transcriptional regulator effector (TRE), fontal affinity chromatography coupled to mass spectrometry (FAC-MS), *Burkholderia xenovorans* LB400, organic hydroperoxide resistance regulator (OhrR), derepression, organic peroxides, Bxe_B2842 and 2-aminophenol

1. Introduction

Burkholderia xenovorans LB400 is the first nonpathogenic *Burkholderia* isolated and is one of the most important aerobic polychlorinated biphenyl degraders discovered thus far (Bedard et al., 1986; Chain et al., 2006; Goris et al., 2004; Maltseva, Tsoi, Quensen, Fukuda, & Tiedje, 1999; Seeger, Timmis, & Hofer, 1995; Seeger, Zielinski, Timmis, & Hofer, 1999). The *B. xenovorans* genome is 9.73 Mbp in size and contains approximately 9,000 coding sequences spanning three different chromosomes: 1, 2 and megaplasmid. The large chromosome (1) is considered the "core" chromosome because it carries the genes associated with core cellular function. The small chromosome (2) and megaplasmid carry genes associated with energy metabolism, secondary metabolism and inorganic ion and amino acid transport and metabolism. Genomic comparisons among the *Burknolderia* strains indicate that *B. xenovorans* LB400 is enriched in aromatic metabolic pathways that provide catabolic capacities for compounds such as biphenyl, 3-chlorocatechol, and 2-aminophenol (Chain et al., 2006). The genomiccontext of this strain suggests that genes of the central aromatic catabolic pathways are generally organized in operons with the genes encoding transcriptional retulators (TRs) adjacent to their regulated operons (Chain et al., 2006). In this paper, we investigate the Bxe_B2842, a TR from *B. xenovorans*, to understand its regulatory mechanism.

Sequence homology to known multiple antibiotic resistance regulator (MarR) TRs revealed that Bxe_B2842 is a member of the MarR family (Finn et al., 2010). This family of proteins comprises a wide range of TRs involved in regulating cellular processes such as metabolic pathways, stress responses, virulence and degradation or export of harmful chemicals, including phenolic compounds, antibiotics and common household detergents (Alekshun & Levy, 1999; Ariza, Cohen, Bachhawat, Levy, & Demple, 1994; Cohen, Hachler, & Levy, 1993;

Martin & Rosner, 1995; Sulavik, Gambino, & Miller, 1995). TRs in the MarR family typically have a winged helix-turn-helix (wHTH) DNA-binding motif and exist as dimmers (PDB#: 1JGS (Alekshun, Levy, Mealy, Seaton, & Head, 2001), 3VB2 and 3VOE). MarR family TRs act either to activate or repress transcription by binding to palindromic sequences within their target promoters, commonly known as DNA operator sequences (DOS) (Perera & Grove, 2010). MarR TRs can be characterized by the specific types of effectors they bind. MarR TRs that regulate metabolic pathways generally bind phenolic compounds; this binding event causes a conformational change in the TR that prevents it from binding to the DOS (Perera & Grove, 2010). Other MarR TRs regulate oxidative stress responses by interacting withreactive oxygen species (ROS) such as organic hydroperoxides. This interaction results in oxidation of cysteine residues on the TR, which causes a conformational change in the TR that attenuates its affinity for the DOS (Perera & Grove, 2010). ROS are commonly generated as by products of cellular processes. In bacteria, TRs such as OxyR, SoxR, PerR, and OhrR are responsive to elevated ROS concentrations (Fuangthong & Helmann, 2002; Helmann et al., 2003; Hidalgo, Leautaud, & Demple, 1998; Mongkolsuk & Helmann, 2002; Newberry, Fuangthong, Panmanee, Mongkolsuk, & Brennan, 2007; Zheng, Aslund, & Storz, 1998). Based on sequence homology, we postulate that the TR Bxe_B2842 is an OhrR homolog.

OhrR, the organic hydroperoxide (OHP) regulator, is a member of the MarR family of TRs and is widely found in many Gram-negative and Gram-positive bacteria. OhrR controls the expression of a gene (ohr) in *Bacillus subtilis*, *Xanthomonas campestris*, *Agrobacterium tumefaciens*, and *Streptomyces coelicolor* (Chuchue et al., 2006; Fuangthong, Atichartpongkul, Mongkolsuk, & Helmann, 2001; Oh, Shin, & Roe, 2007; Panmanee et al., 2002). The gene *ohr* encodes for a thiol peroxidase that reduces OHPs to their corresponding alcohols (Newberry et al., 2007). In Gram-negative bacteria, OhrR belongs to the two-cysteine OhrR family in which the conserved N-terminal cysteine residue is critical for its roles as a TR and OHP sensor (Atichartpongkul, Fuangthong, Vattanaviboon, & Mongkolsuk, 2010; Fuangthong & Helmann, 2002; Newberry et al., 2007; Panmanee, Vattanaviboon, Poole, & Mongkolsuk, 2006). OHP sensing by the two-cysteine OhrR family involves the initial oxidation by the OHP of the N-terminal cysteine to an unstable sulfenic acid intermediate, which then forms a disulfide bond with the conserved C-terminal cysteine (Panmanee et al., 2006). Structural evidence indicates that the oxidized OhrR undergoes a conformational change that results in dissociation from DOS (Newberry et al., 2007). Alternatively, for the one-cysteine family of OhrR, oxidation of the conserved cysteine results in the formation of a mixed disulfide bond with low molecular weight intracellular thiols (Fuangthong & Helmann, 2002; Lee, Soonsanga, & Helmann, 2007) followed by conformational change of the TR and release of the DOS.

Here, we report the characterization of *bxe_B2842*, an OhrR-typeTR found in a putative ROS-resistance operon in *B. xenovorans* composed of *osmC*, a predicted organic hydroperoxide resistance protein (*bxe_B2843*) and *phrB*, a predicted deoxyribodipyrimidine photo-lyase (*bxe_B2844*) in *Burkholderia xenovorans* (Figure 1). However, a recent publication by Cahoon and coworkers suggests that PhrB is a ROS-responsive protein involved in resistance to elevated cellular ROS and not a deoxyribodipyrimidine photo-lyase (Cahoon, Stohl, & Seifert, 2011). To date, all the reportedOhrRs have been responsive to hydroperoxides, but no studies have examined non-hydroperoxide ligands. We have utilized our recently developed frontal affinity chromatography coupled with mass spectrometry (FAC-MS) approach to TR discovery (Martí-Arbona et al., 2012) and electrophoretic mobility gel shift assays (EMSA) to identify and characterize the possible effectors of Bxe_B2842. Our investigation of Bxe_B2842 from *B. xenovorans* suggests that this TR regulates the expression of the putative ROS-resistance operon by binding to the phenolic ligand, 2-aminophenol, and that it is also responsive to OHPs.

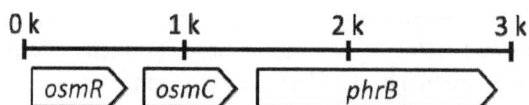

Figure 1. A putative ROS-resistance operon in *B. xenovorans* regulated by the TR *bxe_B2842* (*osmR*). The two ROS-resistance genes in the operon are *bxe_B2843* (*osmC*) and *bxe_B2844* (*phrB*)

2. Materials and Methods

2.1 Materials

Buffer components and effector molecules were purchased from Sigma Aldrich or Acros. The Exactive mass spectrometer was purchased from Thermo Scientific. DNA oligos were purchased from IDT or Invitrogen.

Restriction enzymes were purchased from New England Biolabs and polymerases from Invitrogen. Native and SDS gels were purchased from Invitrogen or Expedeon.

2.2 Cloning and Overexpression

2.2.1 GST-Tagged Protein

The gene *bxe_B2842* was cloned from the *B. xen*ovorans LB400 genomic DNA into the pGEX-KG expression vector. The *bxe_B2842* gene was amplified by conventional PCR methods (manufacturer's (Invitrogen) protocol for Platinum *Pfx* DNA polymerase) using two primers: forward 5'-GATCCATGGATGCAACGCAGCCTC-3' and reverse 5'-GATCGGATCCTCATGCCCGGACGCGG-3', containing *Nco*I and *Bam*HI sites, respectively. The resulting PCR product was digested with the respective endonucleases and ligated into the pGEX-KG vector at the *Nco*I and *Bam*HI sites. The recombinant plasmid was then transformed and the protein was expressed in Arctic Express cells. A single colony was used to inoculate a 50 mL overnight culture of LB medium containing 100 µg/mL ampicillin and 25 µg/mL gentimycin. Then, 5 mL of overnight culture was used to inoculate 1 L of LB media. The cells were grown at 30 °C for approximately 5 hours or until the cell density reached ~0.5 OD at 600 nm, induced with 1.0 mM isopropylthiogalactoside and incubated overnight at 13 °C.

2.2.2 His$_8$-Tagged Protein

The gene *bxe_B2842* was cloned from the *B. xen*ovorans LB400 genomic DNA into the pET42a(+) expression vector. The *bxe_B2842* gene was amplified by conventional PCR methods (manufacturer's (Invitrogen) protocol for Platinum *Pfx* DNA polymerase) using two primers: forward 5'-GGAATTCCATATGGACCCAGCGCCCCGCTTTGCCCTTCACGCTCGACG-3' and reverse 5'-CCGCTCGAGCCTGCGCGGCACACTCAACGATTACATGGACCGCTAG-3', containing *Nde*I and *Xho*I sites, respectively. The resulting PCR product was digested with the respective endonucleases and ligated into the pET42a(+) vector at the *Nde*I and *Xho*I sites. The C-terminal His$_8$-tagged Bxe_B2842 (Bxe_B2842-*His$_8$*) encoding recombinant plasmid was then transformed and the protein was expressed in Arctic Express cells. A single colony was used to inoculate a 50 mL overnight culture of LB medium containing 50 µg/mL kanamycin and 25 µg/mL gentimycin. Then, 5 mL of overnight culture was used to inoculate1 L of LB media. The cells were grown at 30 °C for approximately 5 hours or until the cell density reached ~0.5 OD at 600 nm, induced with 1.0 mM isopropylthiogalactoside and incubated overnight at 13 °C.

2.2.3 Protein Purification

After the overnight incubation, the cells were collected by centrifugation at 6,000 × *g* and the cell pellet was resuspended in a 20 mM Na$_2$PO$_4$, buffer containing 100 mM NaCl, 1mM dithiothretinol (DTT) and 1 mM ethylenediaminetethraacetate (EDTA) (buffer A). The resuspended cells were supplemented with 2 units/mL *DNase*I and 1 mg/mL phenylmethanesulfonyl fluoride, sonicated and centrifuged at 12,000 × *g* for 20 minutes. The cell lysate supernatant was filtered with a 0.45 µm filter and loaded onto an appropriate column (GSTrap or HisTrap column for GE Healthcare) pre-equilibrated with buffer A. The GST-Bxe_B2842 protein was eluted with a 0-50% gradient of 20 mM Tris at pH 8.0 containing 1 mM DTT) and 700 mM glutathione. The Bxe_B2842-*His$_8$* protein was eluted with a 0-55% gradient of buffer A containing 700 mM imidazole. In each case, the protein was pooled, concentrated, and loaded onto a gel-filtration column pre-equilibrated with buffer A . The tagged-Bxe_B2842 protein was collected and verified by SDS-PAGE. The protein was stored in buffer A. The *GST*- Bxe_B2842 protein was utilized for the FAC-MS and PBM experiments described below, while the Bxe_B2842-*His$_8$* protein was utilized for the EMSA experiments described below.

2.3 Frontal Affinity Chromatography - Mass Spectrometry

The frontal affinity chromatography coupled with mass spectrometry (FAC-MS) approach to TR discovery used here was similar to that described by Martí-Arbona et al. (Martí-Arbona et al., 2012). Briefly, GST-Bxe_B2842 protein (13 mg) was immobilized to a 1 ml GSTrap FF (GE Healthcare) affinity column, through which a mixture of effector-candidates was infused; the elution of these effector candidates was monitored with a mass spectrometer (Exactive Orbitrap, Thermo Scientific). The effector candidates were serine, cysteine, 2-aminophenol, homoserine, benzoic acid, *o*-phospho-L-serine, 2,5-dihydroxybenzoic acid, salicylic acid, ADP, adenine, adenosine, AMP, 2-deoxyguanosine, MTA, ATP, guanosine, hypoxanthine, xanthine, inosine, xanthosine, tryptophan, L-kynurenine, L-phenylalanine, anthranilic acid, 3-hydroxyanthranilic acid, 4-hydroxybenzoic acid, L-tyrosine, 3,4-dihydroxybenzoic acid, catechol, and chlorocatechol dissolved in the chromatography buffer (20 mM ammonium formate buffer at pH 7.3) to a final concentration of 10uM each. Compounds that were not bound by the Bxe_B2842 protein quickly eluted from the column while those for which Bxe_B2842 had affinity were retained in the column and eluted at a later time. The tighter a molecule was bound to the protein, the later it eluted. Thus, the retention time of each compound was used to distinguish between binders and non-binders. The ion intensity for each compound was normalized to its highest and plotted (elution profile) using Sigma Plot 12.3.

2.4 Electrophoretic Mobility Shift Assay (EMSA)

2.4.1 Preparation of Bxe_B2842 to Identify Possible DOS

To investigate Bxe_B2842's binding to its DNA operator sequence, EMSAs were performed using the intergenomic DNA sequences upstream from the *bxe_B2842* gene. The 201 bases upstream sequence was cloned from *B. xenovorans* LB400 genomic DNA by conventional PCR methods (manufacturer's (Invitrogen) protocol for Platinum *Pfx* DNA polymerase) with the following pairs of primers: 5'-Alexa488-CCTGCCGACCACACCGGCGGCCTGATTCC-3' /5'-GGGCAAAGCGGGGCGCTGGGTCATGG-3'   and 5'-Alexa488-CATGGACCGCTAGCCCGCCGGCTCCAGAACG-3' /5'-GCTTTGTAGAGGATGTTCTAGATGCTGTTCCTTTGTCTG-3', respectively. The Alexa488-taggedPCR product was purified using the Qiagen Quick Purification kit and stored in TB buffer until EMSAs were performed.

2.4.2 Screening for the TR/DNA/Effector Interactions

Regulation by transcriptional regulators includes two binding events. In this case, the protein Bxe_B2842 binds its DOS (first binding event) and the binding of the effector molecule to the TR/DOS complex (second binding event) causes the collapse of the binary complex and the release of the DOS. The subsections below describe the characterization of the two binding events.

2.4.2.1 TR/DOS Binary Complex Formation

The TR/DOS binding reactions were performed with 20.25 nMDOS (in separate experiments) and various concentrations of the Bxe_B2842-*His$_8$* in binding buffer containing 10 mM Tris, 200 mM KCl, 5 mM MgCl$_2$, 5% glycerol, 5.0 mM DTT at pH 7.5. After incubating binding reactions at 37°C for 1 hour, the samples were loaded onto a 7% polyacrylamide native gel for electrophoresis in a 34 mM HEPES and 43 mM imidazole running buffer that were pre-run at 150 V for 90 minutes. The electrophoretic mobility of the fluorescent Alex488-DOS was monitored with a Hitachi FMBio III fluoresence imager using a 100 micron resolution with an argon ion 488 nm laser and a 530 nm filter (25 nm bandpass, Semrock).

2.4.2.2 Effector Binding to the TR/DOS Binary Complex

Binding reactions in the presence of effectors were performed with 20.25 mM DOS, 500 nM Bxe_B2842-*His$_8$* and variable concentrations of 2-aminophenol or anthranilic acid (5-20 mM), 0-20 μM hydroperoxides (hydrogen peroxide, t-butyl hydroperoxide or cumene hydroperoxide). After incubating binding reactions at 37°C for 1 hour, the samples were loaded onto a 7% polyacrylamide native gel for electrophoresis in a 34 mM HEPES and 43 mM imidazole running buffer that were pre-run at 150 V for 90 minutes. The electrophoretic mobility of the fluorescent Alex488-DOS was monitored with a Hitachi FMBio III fluoresence imager using a 100 micron resolution with an argon ion 488 nm laser. Alexa-488 emission was detected by passage through a 530 nm filter (25 nm bandpass, Semrock).

3. Results and Discussion

3.1Bioinformatic Analysis

Bxe_B2842 is a TR belonging to the MarR family. With its 52% and 55% sequence identities (BLAST tool on NCBI website) to the characterized OHP resistance TRs (OhrR) of *B. subtilis* and *P. aeruginosa*, respectively, Bxe_B2842 mostly resembles an organic hydroperoxide resistance regulator (OhrR). Direct sequence comparisons (BLAST tool on NCI website) of Bxe_B2842 with its OhrR homologs indicate that Bxe_B2842 belongs to a two-cysteine family of OhrRs, having two mechanistically important conserved cysteines at residues 16 and 21 (Chuchue et al., 2006; Fuangthong, Atichartpongkul, Mongkolsuk, & Helmann, 2001; Oh, Shin, & Roe, 2007; Panmanee et al., 2002). According to the current model (Panmanee *et al.*, 2006), oxidation of the reactive Cys-16 would result in disulfide formation with Cys-121 and subsequent attenuation of DNA-binding. This allows RNA polymerase to bind to the promoter and initiate expression of regulated genes, in this case b*xe_B2843* and b*xe_B2844*.

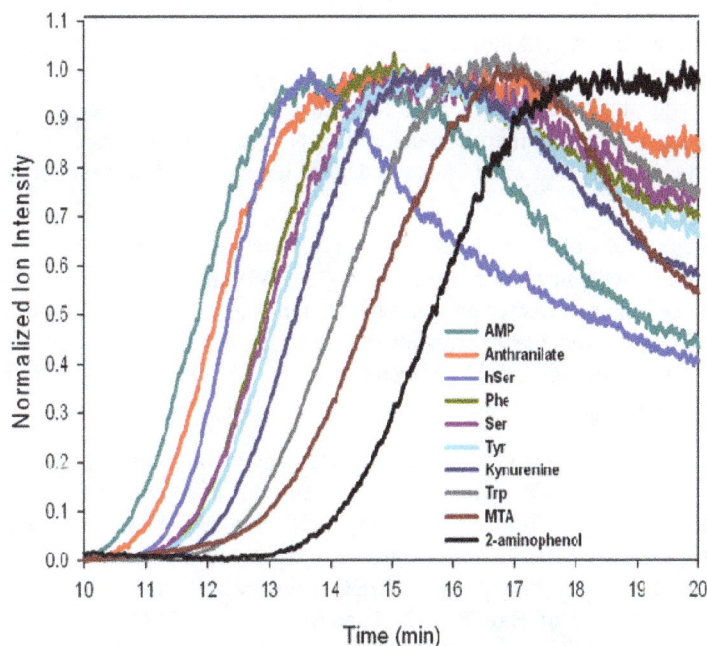

Figure 2. FAC-MS screening for possible effectors binding to GST-Bxe_B2842. The color-coded elution profiles for the individual candidate effectors (10uM dissolved in 20 mM ammonium formate, pH 7.3) are keyed in the figure itself. Compounds that did not bound to the immobilized GST-Bxe_B2842 protein quickly eluted from the column while those with higher affinity to GST-Bxe_B2842 were retained in the column and eluted at a later time. 2-Aminophenol showed the highest binding affinity of all the compounds tested

The genomic context of *bxe_B2842* shows that it is located immediately upstream of *bxe_B2843* and *bxe_B2844*. We suggest the annotation of *bxe-B2843* gene as *osmC* based on its 42% and 43% shared sequence identity with the *ohrA* and *ohrB* genes (thiol-dependent peroxidases), respectively, in *B. subtilis* and its 69% identity to the *ohr* gene in *P. aeruginosa*. Immediately downstream of the *bxe_B2843* (osmC) gene is the *bxe_B2844* gene (*phrB*), which is annotated in the NCBI database as a putative deoxyribodipyrimidine photo-lyase. The *phrB* (*bxe_B2844*) has 40% shared sequence identity with the *phrB* gene in *P. aeruginosa* PA7 and 41% with the *phrB* gene in *E. coli* K12. A recent publication by Cahoon and coworkers suggests that PhrB is a ROS-responsive protein involved in resistance to elevated cellular ROS and not a deoxyribodipyrimidine photo-lyase (Cahoon, Stohl, & Seifert, 2011). Given the genomic context of these genes, it is highly probable that the gene *bxe_B2842* is involved in the regulation of the expression of the *osmC* and *phrB* genes and is involved in the modulation of *Burkholderia*'s resistance to OHPs. Our experimental work was designed to investigate this possible function of Bxe_B2842 protein.

3.2 Discovery of Effectors by Frontal Affinity Chromatography - Mass Spectrometry (FAC-MS)

OhrR TRs bind various phenolic compounds. Structural studies of these homologs indicate a conservation of a hydrophobic effector site (Hong, Fuangthong, Helmann, & Brennan, 2005). Close examination of the genomic context of the *bxe_B2842* gene and its closest homologs in other bacteria identified a conserved cluster of annotated genes that included AMP nucleosidase, deoxyribodipyrimidine photo-lyase and homoserine kinase. A library of effector candidates was constructed that included substrates and products of these enzymes and their analogs as well as a set of commercially available phenolic compounds and organic acids. The results of the FAC-MS assay (Figure 2) showed that the immobilized GST-Bxe_B2842 was capable of binding to phenolic ligands, with 2-aminophenol being the most tightly bound of the compounds tested. The other metabolites were either not bound or weakly bound by the TR.

3.3 Identification of DNA-Binding Motif

Electrophoretic mobility shift assays (EMSA) using the intergenomic sequences of *bxe_B2842* and *osmC*, a near genomic neighbor of *bxe_B2842*, provide evidence that this TR binds to the promoter regions of these two genes in a concentration-dependent manner. We analyzed the upstream regions of *bxe_B2842* and its nine closest homologs from different *Burkholderia* species for similarity to uncover a common DOS, with a general sequence for 5'-A(A\T)T(C\A)ATTTG(C\T)(A\G)(T\C)(G/A)CAAAT(G/T)A(A\T)T-3' (palindromic operator sequence, Figure 3). Moreover, when we performed the same search for *osmC* (*bxe_B2843*) and its nine closest homologs, the same palindromic sequence was readily identified. A highly symmetrical and large consensus sequence was discovered that is present in all 20 sequences. This level of conservation and large palindromic symmetry is highly consistent withthe possibility that there is a common DNA operator sequence that is used to regulate both the *ohrR* TR gene and the downstream *osmC* gene and very likely the other member of the ROS-responsive operon *phrB*.

In the absence of an effector molecule, the EMSA result (Figure 4) demonstrated that the TR binds to the DOS sequencein a concentration-dependent manner creating a TR/DOS binary complex. This suggests that the TR acts as a repressor; when there is no effector present, the TR binds to the DOS and prevents transcription of the putative ROS-responsive operon and the expression of the OsmC protein. The common DOS suggests that Bxe_B2842 controls its own expression and that of OsmC and very likely the expression of the PhrB protein.

3.4 Characterization of the Effector and Its Effects on the TR/DOS Binary Complex

EMSA results suggested that micromolar concentrations OHPs and millimolar concentrations of 2-aminophenol are capable of disrupting the TR/DOS binary complex and of releasing the DOS. In the presence of 2-aminophenol, the protein is released from the DOS (Figure 4). Anthranilic acid was used a control because it had been identified by the FAC-MS effector screen as not being bound by the TR. As could be expected from that result, anthranilic acid did not disrupt the TR/DOS binary complex (Figure 5). This suggests that 2-aminophenol is an active effector of Bxe_B2842. Therefore, Bxe_B2842 is the first OhrR shown to have non-hydroperoxide effector-regulated activity. It is well known that 2- or 4-aminophenols are oxidized by O_2 producing ROS, which could lead to the formation of organic peroxides (Prati, Rossi, & Ravasio, 1992). It makes sense that the expression of genes involved in the detoxification of ROS (i.e., *bxe_B2843* and *bxe_B2844*) are controlled by a TR that is activated by either the presence of ROS or 2-aminophenol.

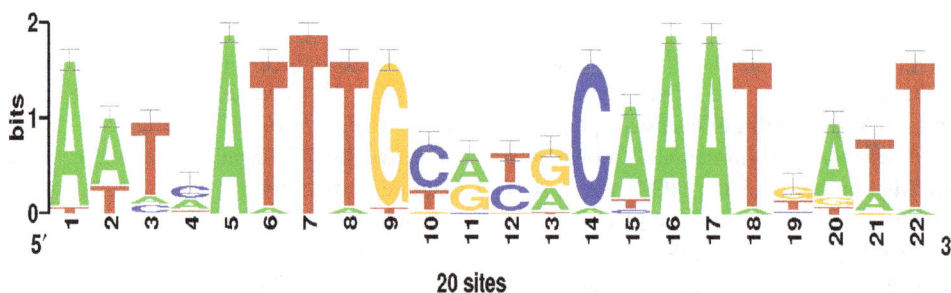

Figure 3. Consensus DNA sequence conserved in the promoter regions of *bxe_B2842* and its nine closest homologues from different *Burkholderia* species

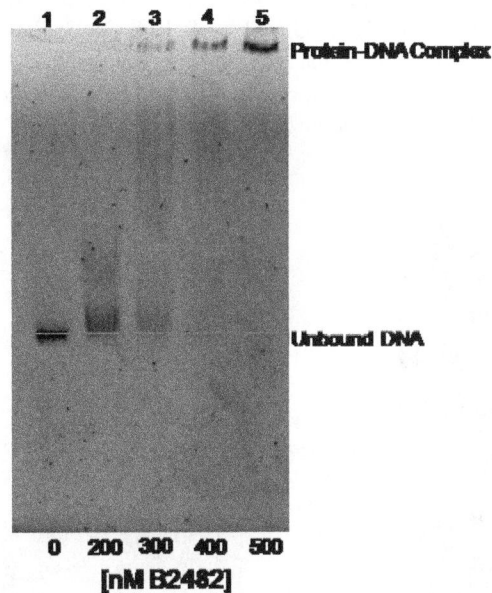

Figure 4. EMSA test for Bxe_B2842 binding to its own promoter (201 bp upstream, DOS). The concentration dependency of the DNA binding was tested by the incremental titration of the Bxe_B2842-His_8

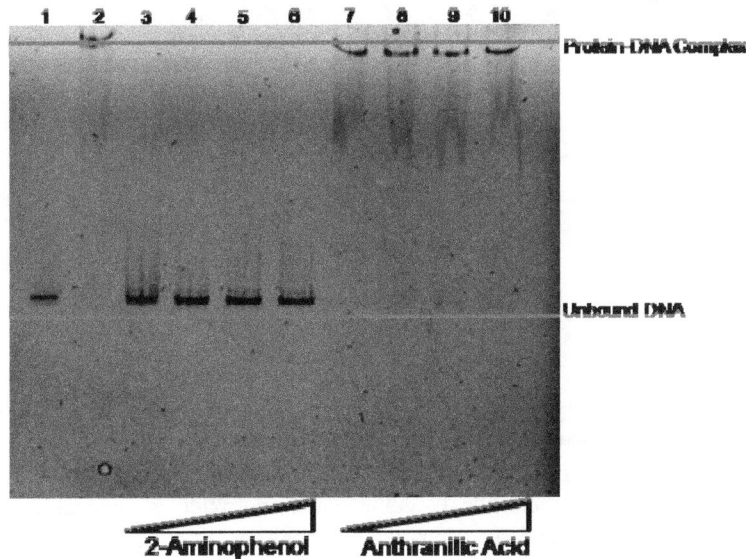

Figure 5. EMSA test for Bxe_B2842-His_8 (500 nM)/DOS (20.25 nM) with two metabolites from the FAC-MS library: anthranilate (a non-binder in the FAC-MS) and 2-aminophenol (the best binder in the FAC-MS). Lane 1 contains labeled DOS, no protein and effector. Lane 2 contains Bxe_B2842-His_8/DOS but no effector. Lanes 3 to 6 contain Bxe_B2842-His_8/DOS with incremental concentrations of 2-aminophenol ranging from 5 to 20 mM. Lanes 7 to 10 contain Bxe_B2842-His_8/DOS with incremental concentrations of 2-aminophenol ranging from 5 to 20 mM

Also consistent with this TR having non-hydroperoxide and hydroperoxide effectors are the results from EMSA with three hydroperoxides (Figure 6). H_2O_2, t-butyl hydroperoxide and cumene hydroperoxide disrupted the TR/DOS binary complex, suggesting the oxidation of the two conserved cysteines in Bxe_B2842. At low concentrations of OHP, only the TR/DOS binary complex can be detected on the gel; as the concentrations of the OHPs increase, the TR/DOS binary complex is disrupted and the free DOS can be observed. This suggests that TR oxidation causes the release of the DOS. EMSA results also indicate that the presence of hydroperoxides in the absence of protein has no effect on the DNA's electrophoretic mobility. These results are consistent with previous studies on homologous OhrRs from *B. subtilis* and *X. campestris* (Chuchue et al., 2006; Fuangthong, Atichartpongkul, Mongkolsuk, & Helmann, 2001; Oh, Shin, & Roe, 2007; Panmanee et al., 2002) and suggest that Bxe_B2842 is the regulatory proteinOhrRthat governs the transcription of the ROS-responsive operon composed of OsmC and PhrB in *B. xenovorans*. More experiments are needed to confirm the *in-vivo* regulatory effect of 2-aminophenol on *bxe*_B2842 in *B. xenovorans*.

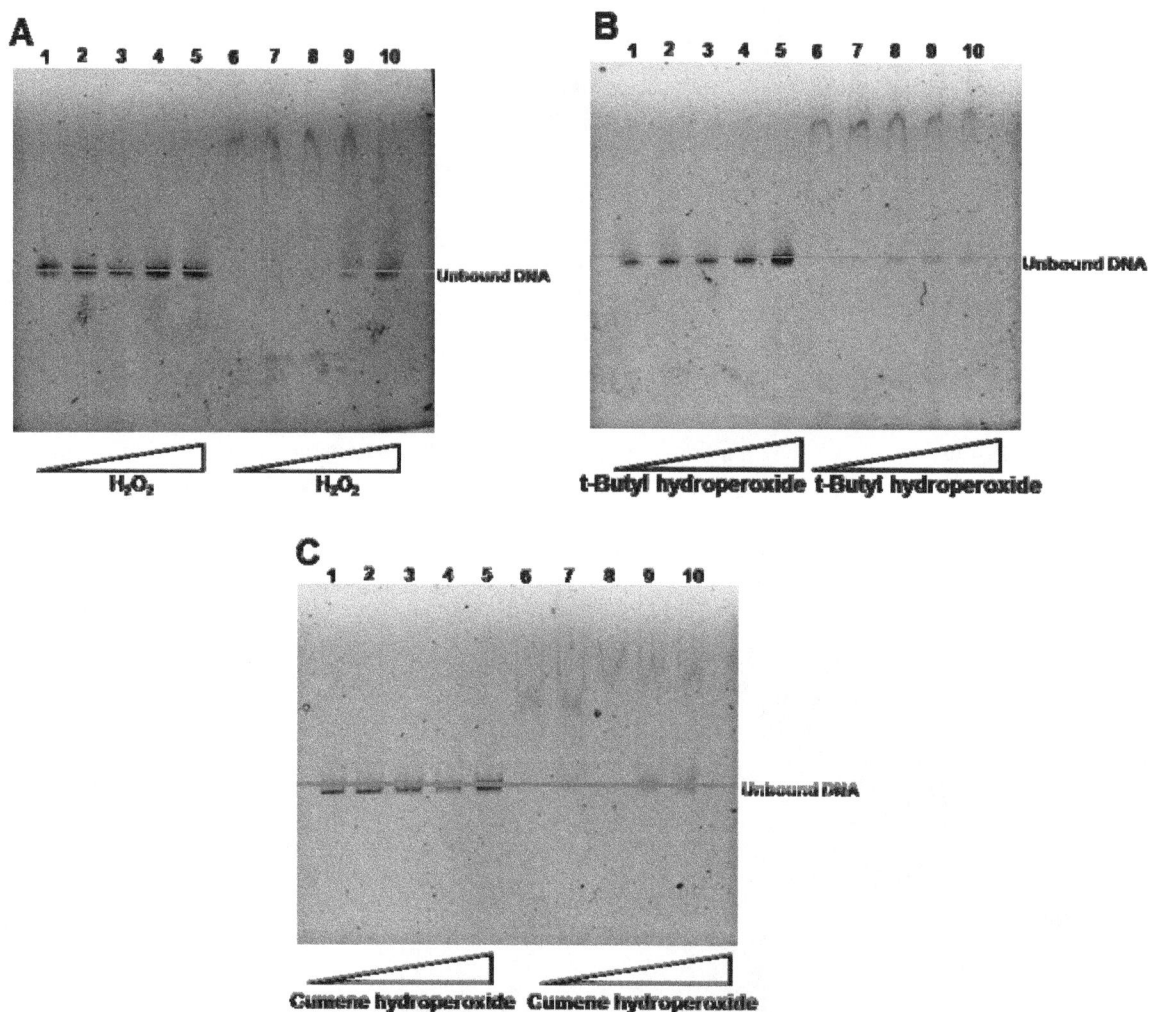

Figure 6. EMSA test for Bxe_B2842-*His*$_8$ (500 nM)/DOS (20.25 nM) in the presence of different hydroperoxides (A- hydrogen peroxide, B- t-butyl hydroperoxide, C- cumene hydroperoxide) at concentrations ranging from 0 - 20 µM. In all three gels (A-C), lanes 1 to 5 correspond to reactions containing the DOS and the hydroperoxide without the TR. Lanes 6 to 10 correspond to reactions containing Bxe_B2842-*His*$_8$/DOS and incremental concentrations of the hydroperoxides

3.5 Conclusion

Here, we identified the first organic hydroperoxide responsive TR (OhrR, Bxe_B2842) that is responsive to either hydroperoxides and the expected chemical modification or the binding of a small molecule effector that cannot modify the TR. Organic hydroperoxides are known to act by chemically oxidizing the two conserved cysteines on the C terminal of the OhrR. Oxidation of the cysteines is very unlikely for 2-aminophenol, which lacks oxidizing properties. The 2-aminophenol can thus be expected to act as a typical small molecule effector whose binding induces a conformational change and the release of the DOS. As an effector, 2-aminophenol plays a role in *B. xenovorans'* response to ROS because 2-aminophenol is an intermediate in tryptophan or nitrobenzene metabolism and can be metabolized to generate ROS. Many bacterial TRs bind an effector molecule, but the finding of two functionally related but structurally distinct types of molecules to which this TR can respond has not been previously reported.

Acknowledgements

We thank Dr. Virginia A. Unkefer for editorial improvements to this manuscript. This work was conducted in part under the auspices of the US Department of Energy and supported by the LDRD program at the Los Alamos National Laboratory (Grant No. 20090107DR).

References

Alekshun, M. N., & Levy, S. B. (1999). Alteration of the repressor activity of MarR, the negative regulator of the *Escherichia coli marRAB* locus, by multiple chemicals in vitro. *Journal of Bacteriology., 181*(15), 4669-4672.

Alekshun, M. N., Levy, S. B., Mealy, T. R., Seaton, B. A., & Head, J. F. (2001). The crystal structure of MarR, a regulator of multiple antibiotic resistance, at 2.3 A resolution. *Nature Structural Biology, 8*(8), 710-714. http://dx.doi.org/10.1038/90429

Ariza, R. R., Cohen, S. P., Bachhawat, N., Levy, S. B., & Demple, B. (1994). Repressor mutations in the *marRAB* operon that activate oxidative stress genes and multiple antibiotic resistance in *Escherichia coli. Journal of Bacteriology, 176*(1), 143-148.

Atichartpongkul, S., Fuangthong, M., Vattanaviboon, P., & Mongkolsuk, S. (2010). Analyses of the regulatory mechanism and physiological roles of Pseudomonas aeruginosa OhrR, a transcription regulator and a sensor of organic hydroperoxides. *Journal of Bacteriology, 192*(8), 2093-2101. http://dx.doi.org/10.1128/JB.01510-09

Bedard, D. L., Unterman, R., Bopp, L. H., Brennan, M. J., Haberl, M. L., & Johnson, C. (1986). Rapid assay for screening and characterizing microorganisms for the ability to degrade polychlorinated biphenyls. *Applied and Environmental Microbiology, 51*(4), 761-768.

Chain, P. S., Denef, V. J., Konstantinidis, K. T., Vergez, L. M., Agullo, L., Reyes, V. L., ... Tiedje, J. M. (2006). *Burkholderia xenovorans* LB400 harbors a multi-replicon, 9.73-Mbp genome shaped for versatility. *Proceedings of the National Academy of Sciences of the United States of America, 103*(42), 15280-15287. http://dx.doi.org/10.1073/pnas.0606924103

Chuchue, T., Tanboon, W., Prapagdee, B., Dubbs, J. M., Vattanaviboon, P., & Mongkolsuk, S. (2006). OhrR and ohr are the primary sensor/regulator and protective genes against organic hydroperoxide stress in Agrobacterium tumefaciens. *Journal of Bacteriology, 188*(3), 842-851. http://dx.doi.org/10.1128/JB.188.3.842-851.2006

Cohen, S. P., Hachler, H., & Levy, S. B. (1993). Genetic and functional analysis of the multiple antibiotic resistance (*mar*) locus in *Escherichia coli. Journal of bacteriology, 175*(5), 1484-1492.

Finn, R. D., Mistry, J., Tate, J., Coggill, P., Heger, A., Pollington, J. E., ... Bateman, A. (2010). The Pfam protein families database. *Nucleic Acids Research, 38*(Database issue), D211-222. http://dx.doi.org/10.1093/nar/gkp985

Fuangthong, M., Atichartpongkul, S., Mongkolsuk, S., & Helmann, J. D. (2001). OhrR is a repressor of ohrA, a key organic hydroperoxide resistance determinant in *Bacillus subtilis. Journal of Bacteriology, 183*(14), 4134-4141. http://dx.doi.org/10.1128/JB.183.14.4134-4141.2001

Fuangthong, M., & Helmann, J. D. (2002). The OhrR repressor senses organic hydroperoxides by reversible formation of a cysteine-sulfenic acid derivative. *Proceedings of the National Academy of Sciences of the United States of America, 99*(10), 6690-6695. http://dx.doi.org/10.1073/pnas.102483199

Goris, J., De Vos, P., Caballero-Mellado, J., Park, J., Falsen, E., Quensen, J. F., ... Vandamme, P. (2004). Classification of the biphenyl- and polychlorinated biphenyl-degrading strain LB400T and relatives as *Burkholderia xenovorans* sp. nov. *International Journal of Systematic and Evolutionary Microbiology, 54*(Part 5), 1677-1681. http://dx.doi.org/10.1099/ijs.0.63101-0

Helmann, J. D., Wu, M. F., Gaballa, A., Kobel, P. A., Morshedi, M. M., Fawcett, P., & Paddon, C. (2003). The global transcriptional response of *Bacillus subtilis* to peroxide stress is coordinated by three transcription factors. *Journal of Bacteriology, 185*(1), 243-253.

Hidalgo, E., Leautaud, V., & Demple, B. (1998). The redox-regulated SoxR protein acts from a single DNA site as a repressor and an allosteric activator. *The EMBO Journal, 17*(9), 2629-2636. http://dx.doi.org/10.1093/emboj/17.9.2629

Lee, J. W., Soonsanga, S., & Helmann, J. D. (2007). A complex thiolate switch regulates the *Bacillus subtilis* organic peroxide sensor OhrR. *Proceedings of the National Academy of Sciences of the United States of America, 104*(21), 8743-8748. http://dx.doi.org/10.1073/pnas.0702081104

Maltseva, O. V., Tsoi, T. V., Quensen, J. F., 3rd, Fukuda, M., & Tiedje, J. M. (1999). Degradation of anaerobic reductive dechlorination products of Aroclor 1242 by four aerobic bacteria. *Biodegradation, 10*(5), 363-371.

Martí-Arbona, R., Teshima, M., Anderson, P. S., Nowak-Lovato, K. L., Hong-Geller, E., Unkefer, C. J., & Unkefer, P. J. (2012). Identification of new ligands for the methionine biosynthesis transcriptional regulator (MetJ) by FAC-MS. *Journal of Molecular Microbiology and Biotechnology, 22*(4), 205-214. http://dx.doi.org/10.1159/000339717

Martin, R. G., & Rosner, J. L. (1995). Binding of purified multiple antibiotic-resistance repressor protein (MarR) to mar operator sequences. *Proceedings of the National Academy of Sciences of the United States of America, 92*(12), 5456-5460.

Mongkolsuk, S., & Helmann, J. D. (2002). Regulation of inducible peroxide stress responses. *Molecular Microbiology, 45*(1), 9-15.

Newberry, K. J., Fuangthong, M., Panmanee, W., Mongkolsuk, S., & Brennan, R. G. (2007). Structural mechanism of organic hydroperoxide induction of the transcription regulator OhrR. *Molecular Cell, 28*(4), 652-664. http://dx.doi.org/10.1016/j.molcel.2007.09.016

Oh, S. Y., Shin, J. H., & Roe, J. H. (2007). Dual role of OhrR as a repressor and an activator in response to organic hydroperoxides in Streptomyces coelicolor. *Journal of Bacteriology, 189*(17), 6284-6292. http://dx.doi.org/10.1128/JB.00632-07

Panmanee, W., Vattanaviboon, P., Eiamphungporn, W., Whangsuk, W., Sallabhan, R., & Mongkolsuk, S. (2002). OhrR, a transcription repressor that senses and responds to changes in organic peroxide levels in *Xanthomonas campestris* pv. phaseoli. *Molecular Microbiology, 45*(6), 1647-1654.

Panmanee, W., Vattanaviboon, P., Poole, L. B., & Mongkolsuk, S. (2006). Novel organic hydroperoxide-sensing and responding mechanisms for OhrR, a major bacterial sensor and regulator of organic hydroperoxide stress. *Journal of Bacteriology, 188*(4), 1389-1395. http://dx.doi.org/10.1128/JB.188.4.1389-1395.2006

Perera, I. C., & Grove, A. (2010). Molecular mechanisms of ligand-mediated attenuation of DNA binding by MarR family transcriptional regulators. *Journal of Molecular Cell Biology, 2*(5), 243-254. http://dx.doi.org/10.1093/jmcb/mjq021

Seeger, M., Timmis, K. N., & Hofer, B. (1995). Conversion of chlorobiphenyls into phenylhexadienoates and benzoates by the enzymes of the upper pathway for polychlorobiphenyl degradation encoded by the bph locus of *Pseudomonas* sp. strain LB400. *Applied and Environmental Microbiology, 61*(7), 2654-2658.

Seeger, M., Zielinski, M., Timmis, K. N., & Hofer, B. (1999). Regiospecificity of dioxygenation of di- to pentachlorobiphenyls and their degradation to chlorobenzoates by the bph-encoded catabolic pathway of *Burkholderia* sp. strain LB400. *Applied and Environmental Microbiology, 65*(8), 3614-3621.

Sulavik, M. C., Gambino, L. F., & Miller, P. F. (1995). The MarR repressor of the multiple antibiotic resistance (mar) operon in *Escherichia coli*: prototypic member of a family of bacterial regulatory proteins involved in sensing phenolic compounds. *Molecular Medicine, 1*(4), 436-446.

Zheng, M., Aslund, F., & Storz, G. (1998). Activation of the OxyR transcription factor by reversible disulfide bond formation. *Science, 279*(5357), 1718-1721.

7

Ultra Structures Assessment and Comparison of Allergenic Features of Mature and Immature Pollens of *Quercus persica* L.

Roya Zand[1]

[1] Department of biology, Islamic Azad University, Tehran north brach, Tehran, Iran

Correspondence: Roya Zand, Department of biology, Islamic Azad University, Tehran north brach, Tehran, Iran.
E-mail: Roya_zand_z@yahoo.com

Abstract

There are extensive Persian Oak forests in the west and south west of Iran. Since the pollens are one of the most plants allergenic factor, and 80 up to 90 percent of plant's allergen is pollen based, therefore in the present study the allergenic features of Persian Oak's mature and immature pollens were studied, using cello logy and anatomical methods. The pollen samples were fixed by FAA. SEM analysis showed spherical shaped pollen along with tricolpate and warty shape exine. Pollen's extract achieved using salt phosphate buffer. Electrophoresis profile of proteins showed total 16 bands in the range of 16 to 116 Kd. Mature pollen also had one distinct 52 Kd band. Allergenic test using guinea pig (350-400 Kg and 4-6 old) carried out. The blood tests, based on the numbers of eosinophils, neutrophils and immunoglobulins content of sample and treatments showed significant differences to the control. The mature pollen extract also was more allergenic than that of immature.

Keywords: vegetative organs, pollen ultra structures, allergenic features of pollen, ontogenetic features, *Quercus persica* L.

1. Introduction

Allergenic features of pollen is in the first category of hypersensitive responses. Studies have shown that IgE content is increased by pollen caused allergy (Hosseini et al., 1991). Proteins of pollen are one of allergenic factor which are embedded in cytoplasm and the pollen's coat. Minerals absorbed from environment also can activate the allergenic proteins of pollen. Interaction of pollen and air contaminators brings about releasing of allergenic aerosol of pollens. These particles penetrate into breathing system more than pollen itself (Chehregani et al., 2004). Previous studies have released that some *Leguminosea* related species have different patterns of allergenic features (Robinson et al., 2005). Persian Oak belongs to *Fagaceae*, and *Quercus* genus, by more than 500 species, is distributed throughout the word in the shrub and arboraceous forms. The fruit of Persian oak have been being used by Americans, European, Asian and African for more than thousands of years (Ozcan, 2006). This tree has extended leaves, 20 meters in height and smooth trichoms. Oaks have spirally arranged leaves, with lobate margins in many species; some have serrated leaves or entire leaves with smooth margins. The fruit is a nut called an acorn, borne in a cup-like structure known as acupule; each acorn contains one seed. Many deciduous species are marcescent, not dropping dead leaves until spring. In spring, a single oak tree produces both male flowers (in the form of catkins) and small female flowers. The live oaks are distinguished for being evergreen, but are not actually a distinct group and instead are dispersed across the genus (Ebrahimi et al., 2008). Regarding of wide distribution of Persian oak in the western forests of Iran, in the present study therefore, The allergenic feature of mature and immature oak' s pollen was assayed base on electrophoresis profiles.

2. Materials and Methods

Vegetative structures were collected near city of Khoram abad (Iran) for following anatomical assay and extraction. Buffer extract was used to prepare the proteins profile. To do this, PBS buffer (phosphate saline) by following protocol was used; 1 g of mature and immature pollen extract were separately mixed with 6 ml of PBS. The mixtures then were riled for 24 h in 4 °C on the shaker. Cold centrifuge in -4 °C then was carried out by 13000 g. Upper phase was kept in -20 for next experiments.

2.1 Animal Samples

Guinea pig in the range of 350- 500 gr were used as animal sample. The samples were isolated in certain environment (22±2 °C, 55±5 humidity, 12 light, 12 darkness photoperiodism) and fixed diet feed, for adaption to new experimental condition. The samples were randomly categorised in to three groups; group 1 were injected by mature pollen extraction, group 2 by immature pollen extract and third group were injected by salty Phosphate buffer as control sample. Injections were done once a week for 5 continues weeks by 100 μl of extraction, as a peritoneum injection (subcutaneous injection for last time injection). A week after the last injection, giving blood were done from heart of animal samples. The numbers of eosinophils using CBC, IgE using ELISA test (IU/mL) and blood sugar level (Mg/dl) were then measured for all groups. SDS-PAGE electrophoresis was used to assay the protein band profiles. 20 μl of extraction for all samples, as well as polyacrylamide gel 12% were imposed to do this. R-25 coomassie blue stain was used for dyeing the gels.

2.2 Ontogenic Studies

FAA fixator was used to fixation of flower buds of Oak. Following staining was done several weeks after fixation, by hematoxylin and eosin. Light microscope was then used for scanning of samples. Antheridium were dried and the result powder in range of 70-230 μm, was scanning by SEM.

2.3 Statistical Analysis

SPSS ver16. was used for statistical analysis. Means were compared using the duncan test at $P<0/05$, level of significance to distinguish the differences between treatments and control samples. There were three replicates for all experiments.

3. Results

3.1 Structure of Mature and Immature Antheridium

The observed results showed that antheridium consist of a layer of cutin cells, which protect the antheridium. In the immature one, mechanical layer was seen under epidrem, and under the transition cells (into the immature antheridium sac), topi layer was seen. Immature pollens embedded into the antheridiumsac sac. Epidrmal cells and mechanical layer are shown in figure 1. Transition cells were wasted in the mature antheridium, and cells of topi layer along with mature pollens were seen inside the mature antheridium sac.

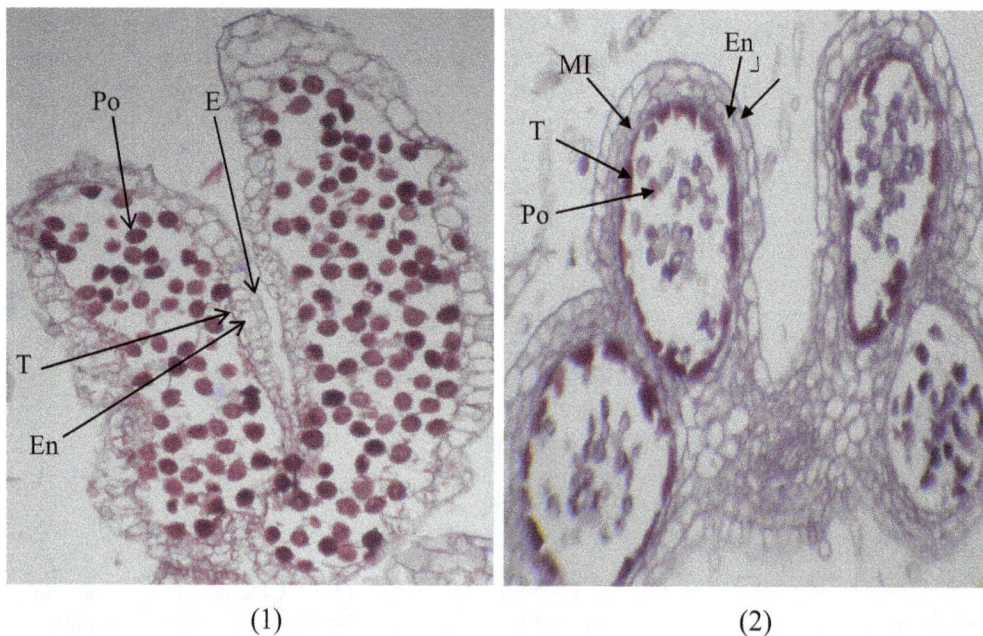

(1)   (2)

Figure 1. Mature antheridium with mature pollen (1). Immature antheridium with young microspores (2). E: Epiderm, En: Mechanical layer, MI: Transition cells layer, T: Topi layer cells, Po: Pollens

Microscopic study showed that mature pollen of Persian oak, have more oval and longer than that of immature one. The pollen were tricolpate, and the exine was warty shape (Figure 2). Mature pollens have deeper colpate, in comparison with immature one (Figure 3).

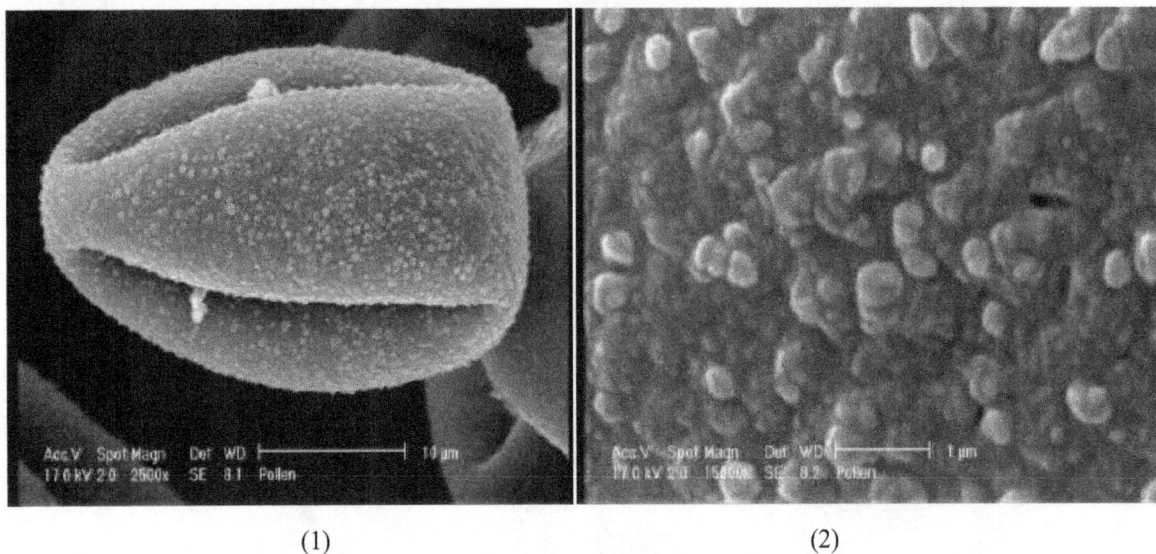

(1)   (2)

Figure 2. Microscopic image of Persian oak pollen. Mature pollen (1). Warty shape Exine (2)

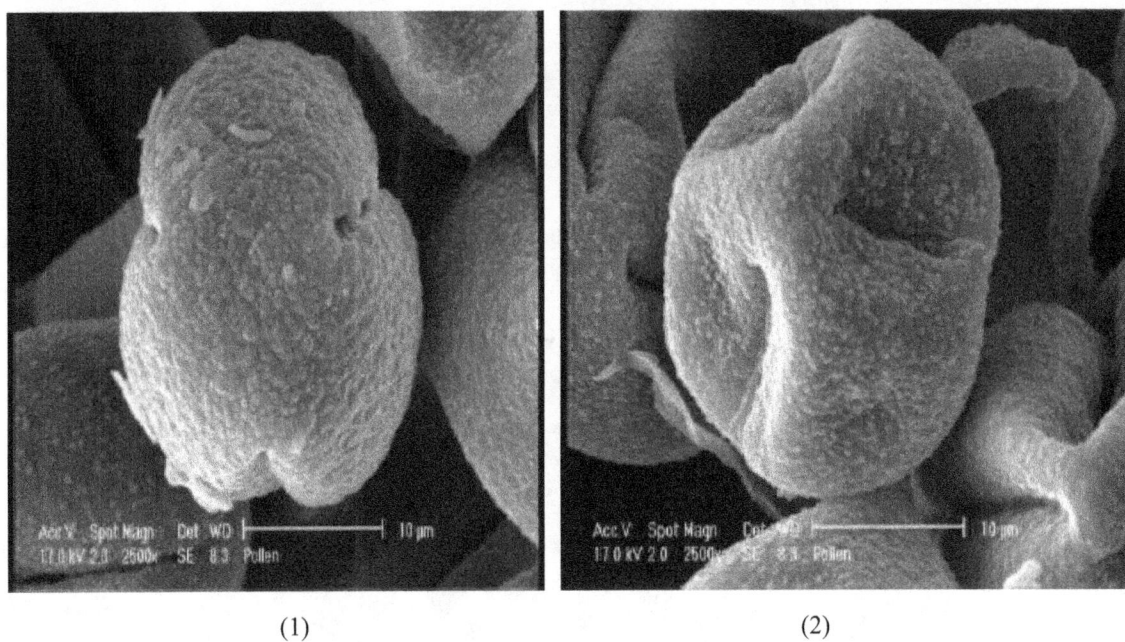

(1)   (2)

Figure 3. Colpate of (1) mature and (2) immature pollens

3.2 Electrophoresis Profile of Oak Pollen

16 proteins band in the range of 16,18,23,25,30,35,34,40,43,52,66,86,91,96 and 166 kb as well as one distinct heavier 116 kd band, were seen in the electrophoresis profiles of pollen's proteins extract. Profile of mature pollen had one more bolder 52 kd band, compared to immature one (Figure 4).

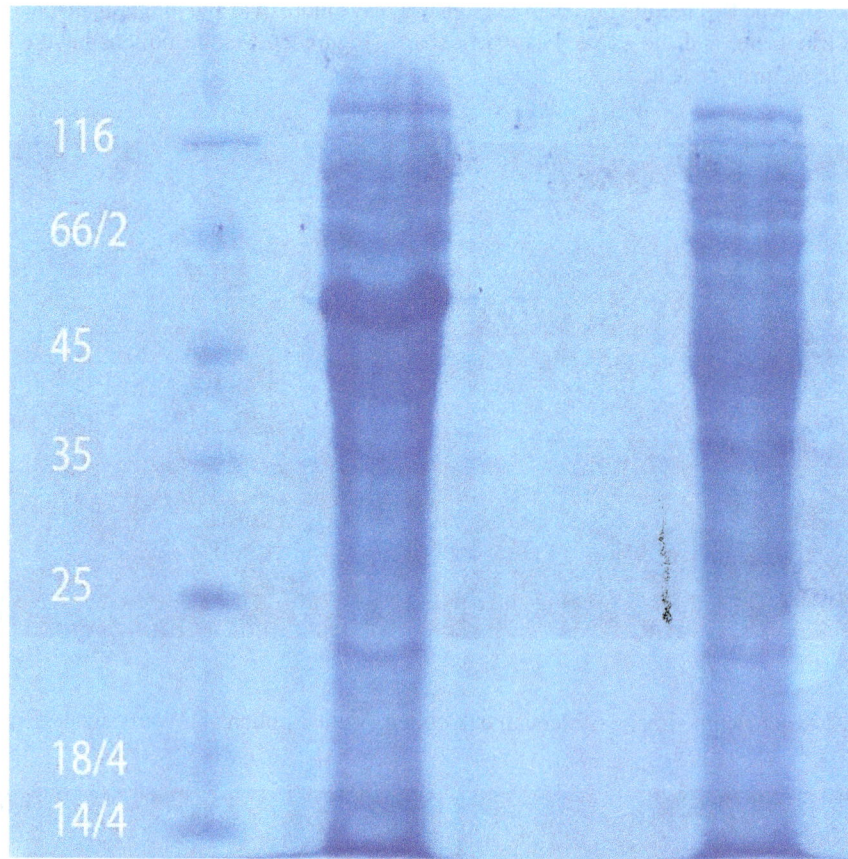

Figure 4. Electrophoresis profile of mature (left) and immature (right) pollens

3.3 Allergenic Assay

In the Figure 5, it is shown that eosinophils numbers in the blood of animal samples injected by mature pollen extract, were more than that of control and immature pollen extract injected one.

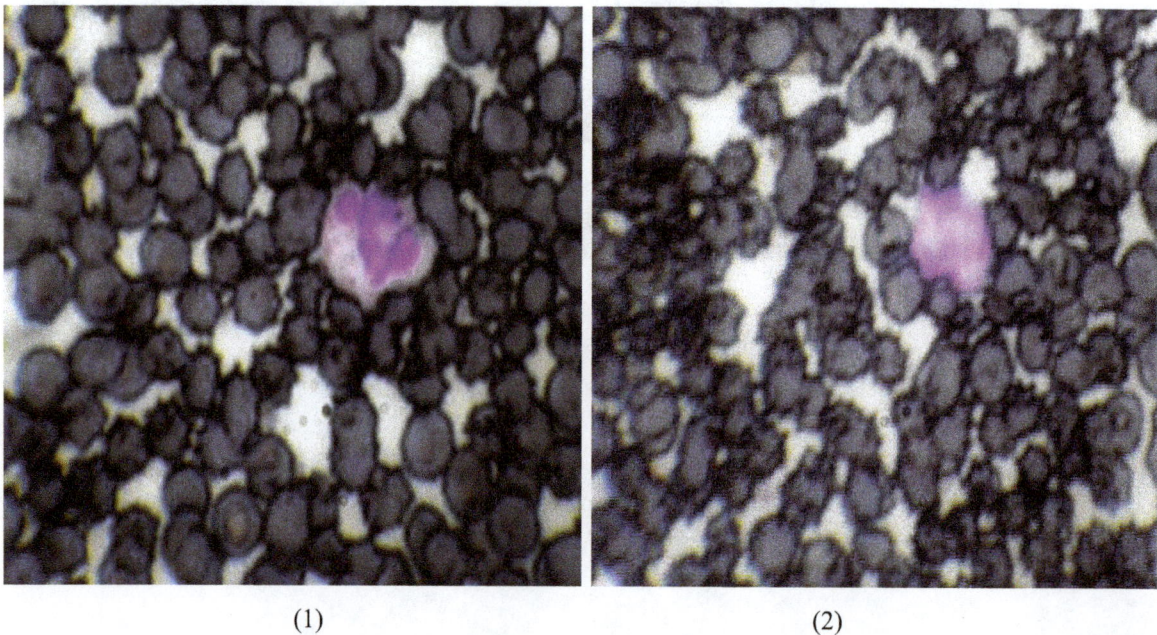

(1)   (2)

Figure 5. Eosinophils (1) and Lymphocyte (2) in the blood test of mature pollen extract injected samples

Both mature and immature pollen extract of Oak, brought about itching of eyes and Sneezing of animal samples, 30 min after dropping the extracts. The allergenic effect was more by the mature pollens (Figure 6). However, the results showed no significant difference in the skin allergenic test of control and immature pollen injection. As the figure 7 shows, mature pollen extract caused more extensive red corona (3 mm diameters) than control and immature treatment samples (about 0.3 diameters).

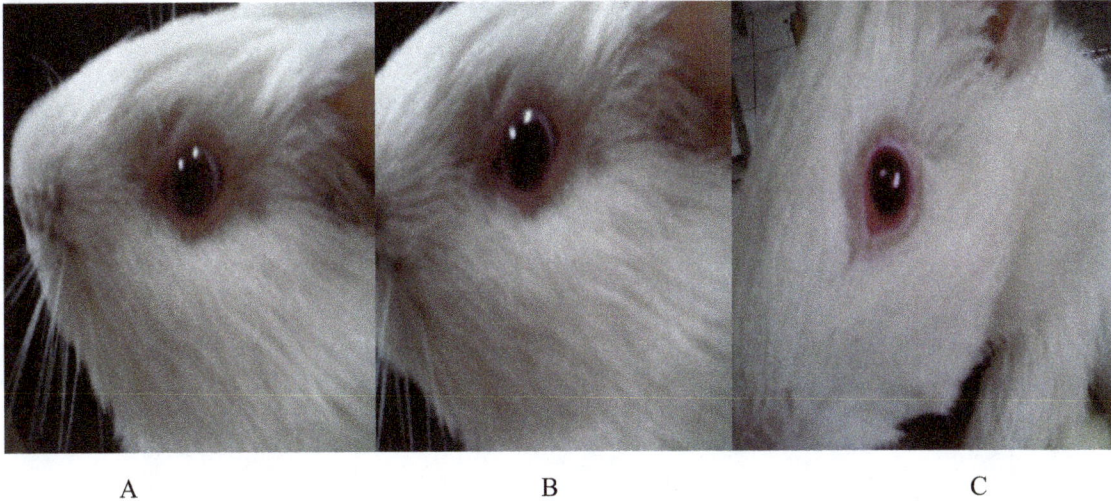

A　　　　　　　　　　B　　　　　　　　　　C

Figure 6. The eye allergenic test of (A) control, (B) immature pollen extract, and (C) mature pollen extract

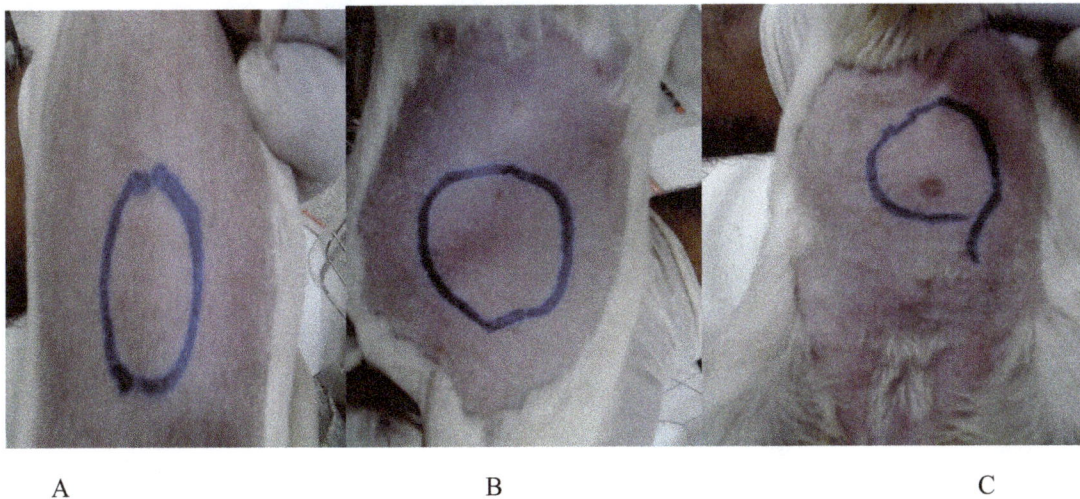

A　　　　　　　　　　B　　　　　　　　　　C

Figure 7. Skin allergenic test of (A) control, (B) immature pollen extract, and (C) mature pollen extract

3.4 Blood Tests Results

The blood sugar level of animal samples was normally 120 mg/dl. By the injection of phosphate saline buffer this level increased by 155 mg/dl. It was also raised up to 196 mg/dl and 159 mg/dl by mature and immature pollen extract injection, respectively. Eosinophils leve of control samples was 1.3%. Mature and immature pollen extract treatment increased it up to 2.7 % and 1.9 %, respectively. Basophils level of control sample was also about 1.1%, but increased by the mature (2.1%) and immature pollen extract (1.3%). Immunoglobulin level of control sample was about 4 U/ml. For mature and immature pollen extract injected samples, 10 and 8 U/ml were reported respectively (Figures 1, 2, and 3).

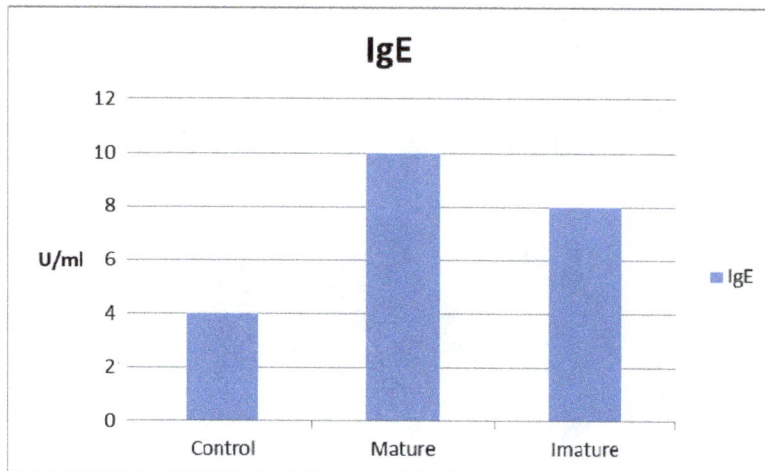

Figure 1. The serologic test; IgE levels

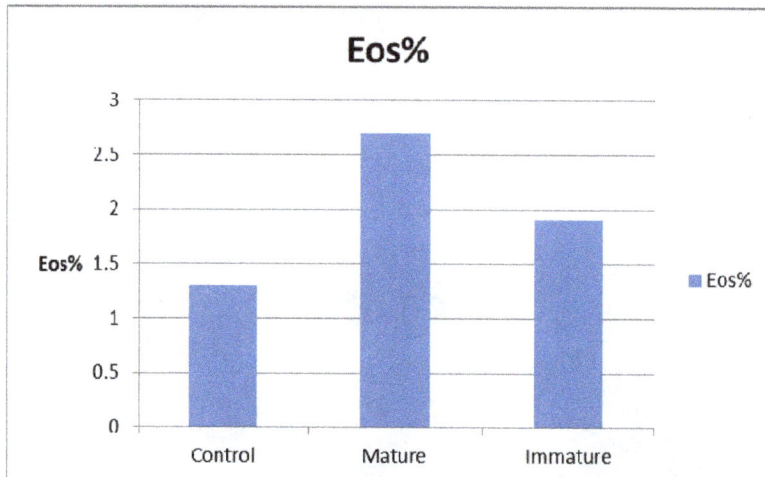

Figure 2. The serologic test; Eosinophils levels

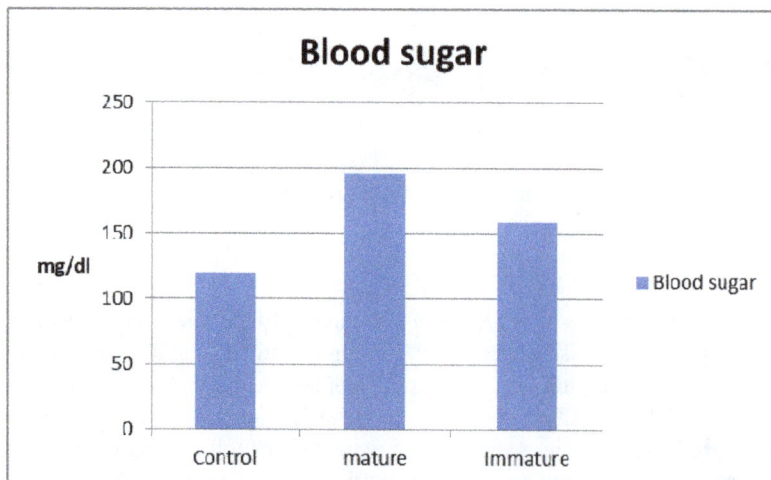

Figure 3. Comparison of blood sugar in treatments and control samples

4. Discussion

The results of present study about the general features of Persian oak which reports tricolpate pollens along with warty shape exine, was previously confirmed by Shah (2005). The nutritious cell layer that was Secretory from the beginning, are visible up until the last development phase of microspors in the margin of pollen sac. This layer is eventually changed into Amoeba- like structures. Transition layer was also wasted during the developmental process to feed the pollens (Majd et al., 1997).

Due to the lack of complete development of pollen's wall and exine and intine colpate where allergen factors are placed. The allergenic capacity of immature pollen is not as severe as mature one. Regarding of the results of skin test, it can be concluded that Persian oak pollens cause hypersensitive responses type I. Itching after 30 min of treatments also confirms the hypersensitive. Chehregari et al. (2003) and Rezanejad et al. (2003) reported the same results.

Proteins electrophoresis profile showed significant differences between mature and immature pollen, so that in mature pollens profile, one more distinct and bold band was seen. It shows synthesizing and accumulation of proteins up until the last phase of development, and also the effect of pollen size on allergenic features. Rezanejad et al. (2007) and Sing et al. (1993) confirmed the mentioned results. More IgE level in animal samples injected by mature pollen extract compared to controls, was already reported by Sing et al. (1993). Asarnoj et al. (2010), assayed the allergenic effects of Birch pollens on the human. They reported that based on the ages of samples and treatment duration, the allergenic results were different. Geroldinger et al. (2011), reported allergenic features of Birch pollens. Pérez et al. (2010) showed the effect of environmental factors on the presence of allergenic pollens of *Cupressaceae, Fraxinus ، Olea Platanus* and *Quercus*. Arbabian et al. (2011) reported more allergenic effects of Wheat pollens in the polluted aria than unpolluted one and light level of blood sugar, IgE and eosinophils. Sharif shooshtari et al. (2013) also studied the allergenic features of mature and immature pollens of *Leucanthemum*. They reported more eosinophils, and IgE content in the animal blood samples of mature pollen injected samples. Skin sensitivity was also seen in treated samples.

References

Amjad, L., & Akkafi, H. (2012). Pollen Structure of Kelussia odoratisima (Umbelliferae) from Iran. *International Journal of Scientific & Engineering Research, 3*(10).

Arbabian, S., Doustar, Y., Entezarei, M., & Nazeri, M. (2011). Effects of air pollution on allergic properties of Wheat pollens (Triticum aestivum). *Advances in Environmental Biology*, 1339-1342.

Asarnoj, A., Movérare, R., Östblom, E., Poorafshar, M., Lilja, G., Hedlin, G., ... & Wickman, M. (2010). IgE to peanut allergen components: relation to peanut symptoms and pollen sensitization in 8 - year - olds. *Allergy, 65*(9), 1189-1195.

Chalabian, F., Mansouri, M., & Sharifnia, F. (2009). The study of ultrastructure features, allergenicity and influence of air pollution on allergenicity of mature pollens in cercis siliquastrun. *Biology Journal, 4*(1), 2-8.

Chehregani, A., Majde, A., Moin, M., Gholami, M., Shariatzadeh, M. A., & Nassiri, H. (2004). Increasing allergy potency of Zinnia pollen grains in polluted areas. *Ecotoxicology and environmental safety, 58*(2), 267-272.

Chehregani, A., Majde, A., Moin, M., Gholami, M., Shariatzadeh, M. A., & Nassiri, H. (2004). Increasing allergy potency of Zinnia pollen grains in polluted areas. *Ecotoxicology and environmental safety, 58*(2), 267-272.

Ebrahimi, A., & Khiabani, M. (2008). Antimicrobial effect of Iranian oak by Disk diffusion method. *Medical Plant Seasional*, 26 -34.

Geroldinger-Simic, M., Zelniker, T., Aberer, W., Ebner, C., Egger, C., Greiderer, A., ... & Bohle, B. (2011). Birch pollen–related food allergy: Clinical aspects and the role of allergen-specific IgE and IgG 4 antibodies. *Journal of Allergy and Clinical Immunology, 127*(3), 616-622.

Majd, A., & Kiabi, S. (1997). The effect of Tehran's polluted atmosphere on ultrastructural changes and allergenicity of Cupressus arizonica pollen grains. *J Aerobiol, 13*, 407-17.

Majd, A., Chehregani, A., Moin, M., Gholami, M., Kohno, S., Nabe, T., & Shariatzade, M. A. (2004). The effects of air pollution on structures, proteins and allergenicity of pollen grains. *Aerobiologia, 20*(2), 111-118.

Majd, A., Kiabi, S. (1997). The effect of Tehrans polleution atmosphere on ultra structural changed and allergenicity of Cupressus Arizonica pollen grains. *Aerobiology*, 407-417.

Majd, A., Tajadod, G., & Ghafarzade, Z. (2013). The study of ontogenesis structures of generatrice organe and ultra structure of pollen grains in Narcissus Tazetta L. *Journal of Plant Science Research, 8*(Special Issue), 29-36.

Özcan, T. (2006). Total protein and amino acid compositions in the acorns of Turkish Quercus L. taxa. *Genetic Resources and Crop Evolution, 53*(2), 419-429.

Pérez-Badia, R., Vaquero, C., Sardinero, S., Galán, C., & García-Mozo, H. (2010). Intradiurnal variations of allergenic tree pollen in the atmosphere of Toledo [Central Spain]. *Annals of agricultural and environmental medicine, 17*(2), 269-275.

Rezanejad, F. (2007). The effect of air pollution on microsporogenesis, pollen development and soluble pollen proteins in Spartium junceum L. (Fabaceae). *Turkish Journal of Botany, 31*(3), 183-191.

Rezanejad, F. (2007). The effect of air pollution on microsporogenesis, pollen development and soluble pollen proteins in Spartium junceum L.(Fabaceae). *Turkish Journal of Botany, 31*(3), 183-191.

Rezanejad, F., & Majd, A. (2008). The effect of air pollution on pollen allergenecity in *Spartium Junceum* (Fabaceae). *Journal of Science (Teacher Training University), 7*(3-4);973-982.

Robinson, M. L. (2005). *Allergenic Plants in Southern Nevada.* The university of Nevada Reno, Nevada.

Shah, S. T., Ahmad, H. A. B. I. B., & Zamir, R. O. S. H. A. N. (2005). Pollen Morphology of three Species of Quercus (Family Fagaceae). *J Agri Soc Sci*, 1813-2235.

Shah, S. T., Ahmad, H. A. B. I. B., & Zamir, R. O. S. H. A. N. (2005). Pollen Morphology of three Species of Quercus (Family Fagaceae). *J Agri Soc Sci*, 1813-2235.

Shahali, Y., Majd, A., Pourpak, Z., Tajadod, G., Haftlang, M., & Moin, M. (2007). Comparative study of the pollen protein contents in two major varieties of Cupressus arizonica planted in Tehran. *Iranian Journal of Allergy, Asthma and Immunology, 6*(3), 123-127.

Shahali, Y., Majd, A., Pourpak, Z., Tajadod, G., Haftlang, M., & Moin, M. (2007). Comparative study of the pollen protein contents in two major varieties of Cupressus arizonica planted in Tehran. *Iranian Journal of Allergy, Asthma and Immunology, 6*(3), 123-127.

Shoushtari, M. S., Majd, A., Pourpak, Z., Shahali, Y., Moin, M., & Eslami, M. B. (2013). Differential Allergenicity of Mature and Immature Pollen Grains in Shasta Daisy (Chrysanthemum maximum Ramond). *Iranian Journal of Allergy, Asthma and Immunology, 12*(2), 99-106.

Singh, A. B., Malik, P., Parkash, D., & Gangal, S. V. (1993). Identification of specific IgE binding proteins in Castor bean (Ricinus communis) pollen obtained from different source materials. *Grana, 32*(6), 376-380.

Singh, A. B., Malik, P., Parkash, D., & Gangal, S. V. (1993). Identification of specific IgE binding proteins in Castor bean (Ricinus communis) pollen obtained from different source materials. *Grana, 32*(6), 376-380.

8

Genetic Diversity of Rhizobia Nodulating Alfalfa in Iraq as a Source of More Efficient Drought Tolerance Strains

Rana Azeez Hameed[1], Nidhal Neema Hussain[2] & Abd aljasim Muhisen Aljibouri[3]

[1] Biology Department, Al-Mustansiriyah University, Baghdad, Iraq

[2] Biology Department, Baghdad University, Baghdad, Iraq

[3] Plant Biotechnology Department, Biotechnology Research Center, Al-Nahrain University, Baghdad, Iraq

Correspondence: Rana Azeez Hameed, Biology Department, Al-Mustansiriyah University, Baghdad, Iraq. E-mail: alroomir@yahoo.com

Abstract

Sinorhizobium meliloti is a gram-negative, soil bacteria, which gain a huge importance deserved to their capability in fixing nitrogen symbiotically with an important fodder crop legume-alfalfa (*Medicago sativa*). This study aims to (i): isolate indigenous *Sinorhizobium meliloti* from different field sites in Iraq; (ii): assess the isolates tolerance to induced water shortage using polyethylene glycol-6000; (iii): evaluate genetic diversity and genetic relationships among isolates of natural population with drought tolerant abilities. Drought tolerance analysis disclose great variations between *Sinorhizobium* isolates, the highest tolerant isolates to drought were 12 from whole thirty (40%), endured from -3 up to -4 MPa(Mega pascal), at the same time the drought sensitive isolates tolerated up to -1.5 MPa, but for isolate Bs58 which tolerated up to -1 MPa. The growth decrease with the increase of drought stress. REP-PCR method was an excellent technique to distinguish between the *S.* isolates based on their tolerance to different water potential levels, the REP-PCR gave a significant band about 700bp (base pair) in size, which were present in all drought moderate and tolerant isolates, while this band was absent in all drought sensitive isolates(except one isolate). Also the ERIC-PCR method cleared the absence of two significant bands 300bp and 600bp from all drought sensitive isolates. UPGMA analysis based on rep-PCR revealed two groups branching at a similarity of 35%, first group included only one isolate Bs16, which was a drought moderate tolerant; second group contained the rest of isolates, yet the latter splits into two subgroups with 68% similarity; the first subgroup comprised all sensitive isolates with 80% similarity among them, the second subgroup included the tolerant and moderate isolates with similarity 88%.

Keywords: REP-PCR, ERIC-PCR, *Sinorhizobium*, drought

1. Introduction

Iraq is placed in arid and semi-arid regions of the world, were annual rainfall is less than 200mm, and so it goes under severe shortage of irrigation water (Tara, 2011). The effect of environment change on biota has currently gained attention. Furthermore, dry lands cover 40% of the world land surface of living. Desertification influence 70% of the world dry lands (Roy, Mazumder, & Sarma, 2009) Fabaceae family plants are often used for cultivation in degraded soil sites of arid and semi-arid regions as they can grow in dry soils that are unsuitable for most crops (Pereira, Lima, & Figueira, 2008). Alfalfa (*Medicago sativa* L.) is a deep-rooted, perennial plant belong to Fabaceae family, which is capable to utilize atmospheric nitrogen (N_2) and accumulate significant amounts of N_2 in the soil through growth (Zeng, Chen, Hu, Su, & Chen, 2007). Rhizobia is a beneficial soil bacteria that had a great agricultural value in enhance soil fertility in farming systems (Zahran, 1999). *Sinorhizobium meliloti* is able to interact with the roots of *Medicago* to form nitrogen-fixing nodules (Elboutahiri, Hami Alami &Udupa, 2010). One of the important strategy to improve the yield of arid legumes in pressurized environments should embrace a combination of stress-tolerant cultivars and stress-tolerant Rhizobia (Turner, Wright, & Siddique, 2001). Distinct phenotypic and genotypic techniques were used to classify and distinguish bacteria, although phenotypic methods are more authenticated reliable for identification and to examine genetic diversity of bacterial isolates (Gao, Terefework, Chen, & Lindstorm, 2001).

A great number of molecular methods based on polymerase chain reaction have been designed to characterize *Sinorhizobium* strains and to provide a high degree of divergence among the closely related bacterial strains. rep - PCR technique has been increasingly used to assess genetic variation of microorganisms. Entrobacteria contain families of short interspersed repetitive elements; this includes the Repetitive Extragenic Palindromic element (REP), the Entrobacterial Repetitive Intergeneric Conensus (ERIC) sequence, and the BOX elements, the role of these elements is still uncovered, yet it has been suggested that they may involve in stabilizing mRNA and binding of DNA polymerase. The REP and ERIC sequences contain a highly conserved inverted repeats, and are normally found in inter-genic regions that are transcribed, but not translated (Versalovic, Schneider, De-Bruijn &Lupski, 1994). The REP-PCR generated genomic fingerprints can be obtained not only from purified genomic DNA, but also directly from rhizobial cells derived from liquid cultures or from colonies on plates as well as from nodule tissue (De-Bruijn, 1992).

2. Materials and Methods

Sinorhizobium bacteria were isolated from Alfalfa nodules (*Medicago sativa*) plants sampled from different geographical sites in Iraq, bacteria were extracted and cultured according to Vincent (1970) procedure. The symbiotic ability of isolates were checked by PIT (Plant Infection Test) (Vincent, 1970). Thirty isolates were tested for growing on CR (CongoRed), Bromothymol blue incorporated with MS (Mannitol Salt Yeast extracted) agar media and gram stain test.

2.1 Drought Tolerance

Isolates were tested for drought tolerance by using polyethylene glycol-6000(PEG-6000 w/v) in MSY broth media at different ranges of osmotic pressure from -0.1 to -4 MPa(mega pascal), plus the control treatment(no PEG-6000)., cultures were incubated at 28±2°C on a rotary shaker in dark conditions for about seven days, then assessed growth.

2.2 DNA Extraction

The extraction of genomic DNA was done by following the protocol of wizard genomic DNA purification kit (Promega). The concentration and purity of DNA were estimated with a spectrophotometer.

2.3 rep - PCR Conditions

Amplification of DNA fragments of the selected isolates was carried out using the complementary primers to repetitive sequences in the bacterial genome:

REP1: IIIICGICGICATCIGGC

REP2: ICGICTTATCIGGCCTAC

The following thermal profiles of the reactions were used:

Initial denaturation 5 min, 94 °C

Denaturation	40sec, 94 °C	
Annealing	1min, 59 °C	32cycle
Extension	1min, 72 °C	

Final extension   10min, 72 °C

ERIC1: ATGTAAGCTCCTGGGGATTCAC

ERIC2: AAGTAAGTGACTGGGGTGAGCG

The thermal profile was:

Initial denaturation 4 min, 94 °C

Denaturation	1min, 94 °C	
Annealing	1min, 56 °C	32cycle
Extension	2min, 72 °C	

Final extension   10min, 72 °C

2.4 Statistical Analysis

Comparison of physiological trait was preformed quantitavelly on the basis of growth+ or no growth – for each isolate. As for PCR fingerprinting patterns were converted into a two-dimensional binary matrix (1, presence of a band, 0, absence of a band) and analyzed using the statistics software package (version 1.92; Past software, Ohammer, 2009) for eleven isolates of *Sinorhizobia meliloti*.

3. Results and Discussion

Studies of different researches illustrated that drought stress is considered one of the major environmental factors affecting almost most crops and casing reduction of crop yield, the population of soil bacteria decreases along the moisture stress (Hossein & Leila 2010). According to Bremer and Kramer (2000), there are two types of stress responses in microorganisms: the general stress response and specific stress response. The general stress response in normally controlled by a single or a few master regulators, which provide protection against a wide variety of environmental cues and allow cell to survive, yet it may not be enough to let the cell to grow under severe stress conditions. While under prolonged stress conditions cells employ specific stress response which utilize highly integrated networks of genetic and physiological adaptation mechanisms, and this was agreed with results conducted in this research, hence there was a great variation between the isolates, and all isolates grow well on media with induced water potential ranging from -0.1 to -1.0 MPa, while at -2.5 MPa the growth declined to only 50% reaching to 6.6% growth percent at -4.0 MPa water potential (Figure 1). Also Zahran (1999) reported that in osmotic stress a specific protein formed which was detected as new protein band in sodium dodecyl sulfate polyacrylamide gel electrophoresis profile of rhizobia.

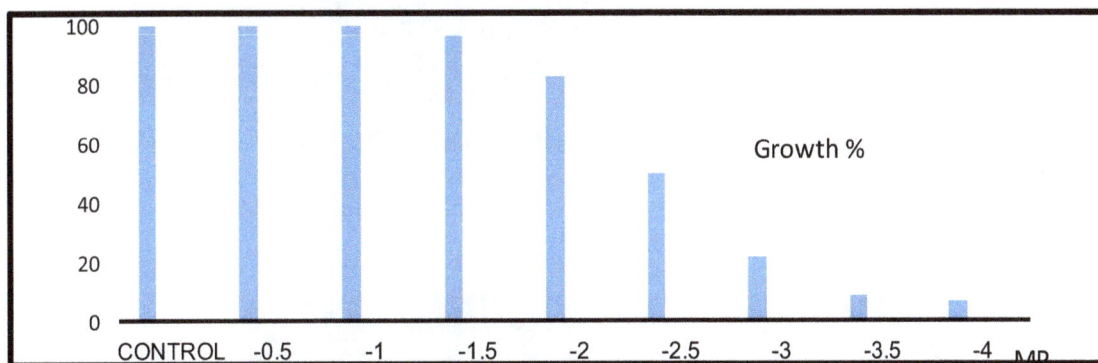

Figure 1. Effect of different water potential levels (MPa) on growth percent of *Sinrhizobium* isolates

Several reports are now accessible indicating the utility of PCR for fingerprinting of many organisms including soil proteobacteria such as rhizobia (Metha, R. Metha & Rosato., 2006); Raja, Balachandar and Sundaram (2008). REP-PCR genomic fingerprinting employ DNA primers complementary to naturally occurring and highly conserved repetitive DNA sequences, present in multiple copies in the genome of most of the Gram positive and Gram negative bacteria (Raja et al., 2008).

The Rep-PCR analysis was done using primer set REP1 and REP2 (Versalovic et al., 1994) these primers were used to amplify the repetitive sequences that present in multiple copies in gram negative bacteria genome like *Sinorhizobium* isolates consequence in (Figure 2) indicate a significant diversity between the drought sensitive and drought tolerant isolates, with the presence of 700 bp size in all drought moderate and tolerant isolates, plus Bs 58(drought sensitive isolate), while this 700 bp band was absent in the other three drought sensitive isolates.

The amplification generated a bands ranged in size about 0.3 to 1.8 kb. The majority of these bands were found in the range of 1.3 to approximately 1.8 kb, the rep-PCR technique generated highly specific, reproducible patterns that allow close strain differentiation, and corresponds with Sikora & Redzepovic (2003); Elboutahiri et al. (2010); Lisek, Paszt, Oskiera, Kulisiewicz and Malusa (2011).

The ERIC-PCR amplification was originally carried out using the primers set ERIC1 and ERIC2. The amplification generated a bands ranged about 0.3 to 1.3 kb (estimated using the DNA molecular weight marker as standard), the results demonstrated the absence of two significant bands(300 and 600 bp size each) from all the drought sensitive isolates, Figure 3.

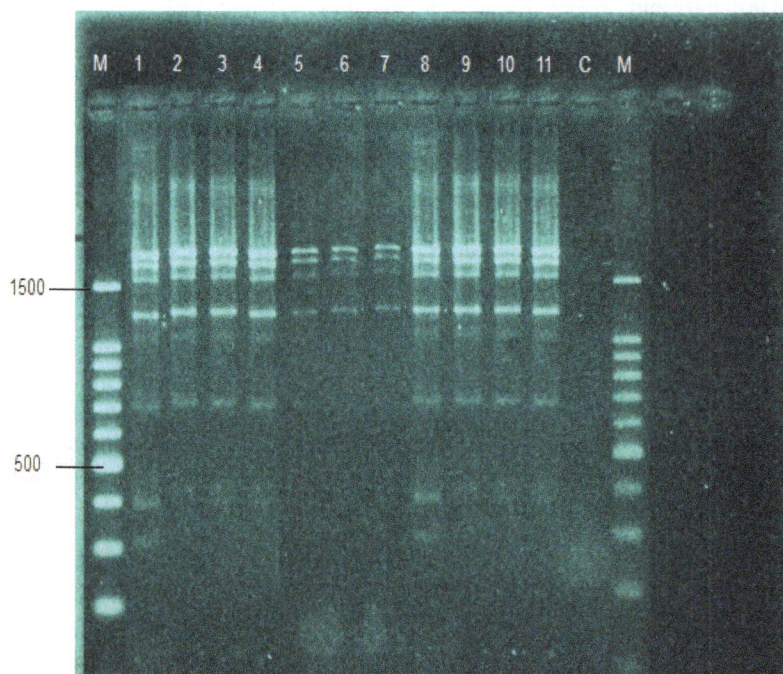

Figure 2. REP-PCR fingerprints of *Sinorhizobium meliloti* isolates generated by primers set REP1 and REP2. M=1kb DNA ladder ;1=Bs 12; 2=Bs 30; 3=Bs 38; 4=Bs 41; 5=Bs 44; 6=Bs 54; 7=Bs 55; 8=Bs 58; 9=Bs49;10= Bs 31 ;11= Bs 16 ;C=negative control and M=1kb DNA ladder

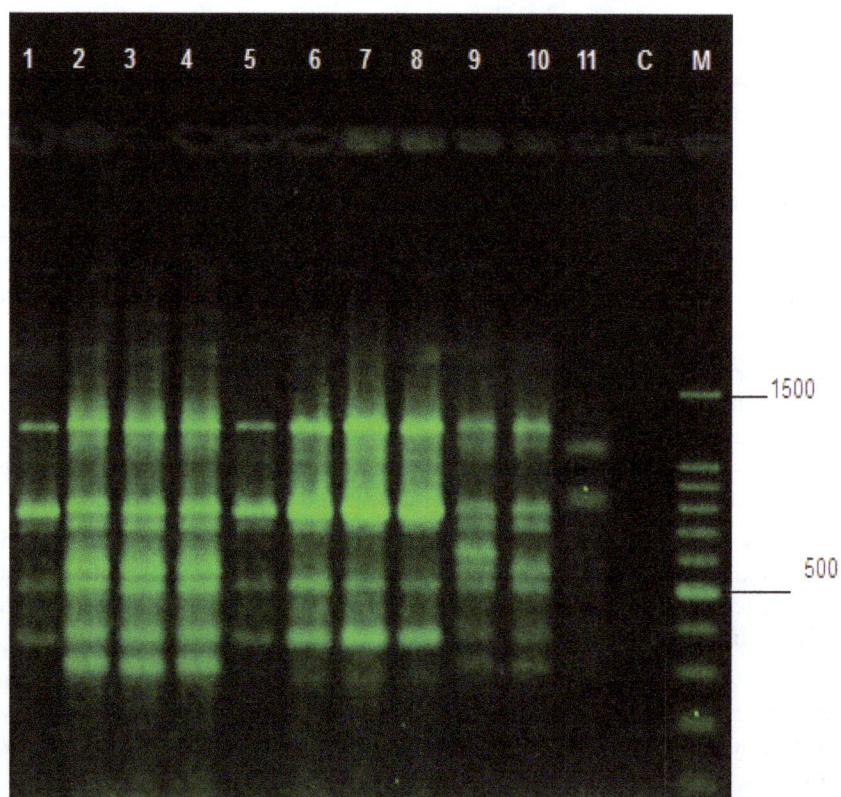

Figure 3. ERIC-PCR fingerprints of *Sinorhizobium meliloti* isolates generated by primers set ERIC1 and ERIC2.1=Bs 12; 2=Bs 30; 3=Bs 38; 4=Bs 41; 5=Bs 44; 6=Bs 54; 7=Bs 55; 8=Bs 58; 9=Bs49;10= Bs 31 ;11= Bs 16 ;C=negative control and M=1kb DNA ladder

The dendrogram in figure 4 showed the genetic similarities between *Sinorhizobium meliloti* isolates based on REP and ERIC-PCR amplification patterns, and it ranged from 35 to 100%.

The dendrogram also showed that the isolates were clustered into two groups branching at a similarity of 35% the first major group contained one isolate Bs16 which was drought and salt moderate tolerant and low /high temperature tolerant(previous work) (Hameed, Hussain, & Aljibouri, 2014).

The second major group included the rest of isolates, which splits into two subgroups with 68% similarity, the first subgroup comprised all sensitive isolates with 80% similarity between them. The second subgroup included the tolerant and moderate isolates with 88% similarity between them.

The dendrogram also showed that the isolates Bs31, 41, 30 and Bs38 were identical. The results showed that REP and ERIC-PCR indicate the efficiency and usefulness of using this technique in differentiation between *Sinorhizobium meliloti* isolates, and that the outcome obtained have was more corresponding to the results gained based on phenotypic characterization tests(previous work) (Hameed et al., 2014). These results shows that ERIC fingerprinting had a great discriminatory power which could be used for diversity analysis and species identification, and this is agreed with Aguilar, Lopez, and Riccillo (2001) and Metha et al. (2006).

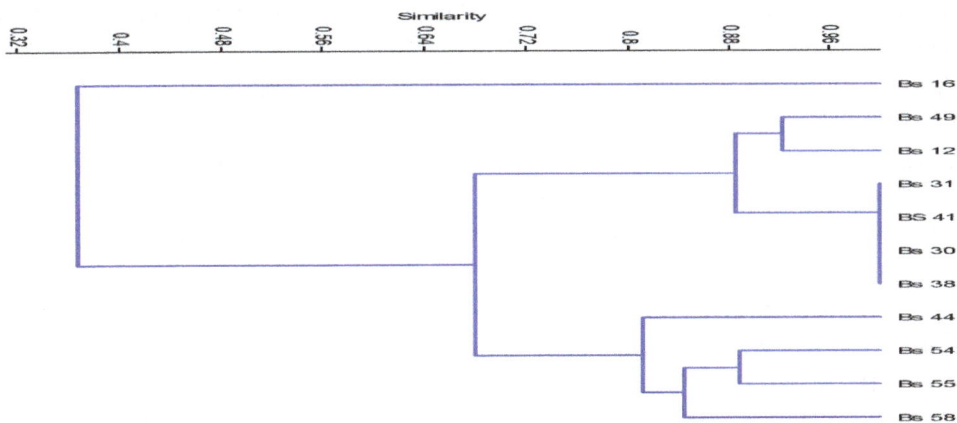

Figure 4. Dendrogram of *Sinorhizobium meliloti* isolates derived from rep-PCR fingerprints generated using REP and ERIC primers

References

Aguilar, O., Lopez, M., & Riccillo, P. (2001). The diversity of rhizobia nodulating beans in Northwest Argentina as a source of more efficient inoculant strains. *J. of Biotechnology, 91*, 181-188.

Bremer, E., & Kramer, R. (2000). Coping with osmotic challenges: osmoregulation through accumulation and release of compatible solutes in bacteria, In G. storz & R. Hengge-Aronis (ed.), *Bacterial Stress Response* (pp. 79-97).

De-Bruijin, F. (1992). Use of repetitive (repetitive extragenic palindromic and enterobacterial repetitive intergenic consenus) sequences and the polymerase chain reaction to fingerprint the genomes of *Rhizobium meliloti* isolates and other soil bacteria. *Appl. Environ. Microbiol, 58*, 2180-2187.

Elboutahiri, N., Hami Alami, I., & Udupa, S. (2010). Phenotypic and genetic diversity in *Sinorhizobium meliloti* and *S. medicae* from drought and salt affected regions of morocco, BMC Microbiology, Open access Res. Retrieved from http://www.Biomedcentral.com/1471-2180/10/15

Gao, J., Terefework, Z., Chen, W., & Lindstorm, K. (2001). Genetic diversity of rhizobia isolated from *Astragalus adsurgens* growing in different geographical regions of china. *J. Biotechnol, 91*, 155-168.

Hameed, A., Hussain N., & Aljabouri, M. (2014). Phenotypic characterization of indigenous Iraqi *Sinorhizobium meliloti* isolates for a biotic stress performance. *J. of Life Sciences, 8*(1), 1-9.

Hossein, A., & Leila, M. (2010). Assessing tolerance of rhizobial lentil symbiosis isolates to salinity and drought in dry forming condition. In *World Congress of Soil Science, Soil Solution for a Changing World, Brisbane, Australia, 13-16*.

Lisek, A., Paszt, L., Oskiera, M., Kulisiewicz, A., & Malusa, E. (2011). Use of the rep-PCR technique for differentiating isolates of rhizobacteria. *J. of Fruit and Ornamental Plant Research, 19*(1), 5-12.

Metha, A., Metha, R., & Rosato, B. (2006). ERIC and REP-PCR amplify non-repetitive fragments from the genome of *Drechslera avehae* and *Stemphylium solani, 5ᵗʰcongress of European Microbiologists, 211*(1), 51-55.

Pereira, S., Lima, A., & Figueira, E. (2008). *Rhizobium leguminosarium* isolated from agricultural ecosystems subjected to different climatic influences: The relation between genetic diversity salt tolerance and nodulation efficiency. In T.-X. Liu (Ed.), *Soil Ecology Research Developments* (pp. 247-263). Nova Science Publisher, Inc., New York.

Raja, P., Balachandar, D., & Sundaram, S. (2008). PCR fingerprinting for identification and discrimination of plant-associated facultative methylobacteria. *Indian journal of Biotechnology, 7*, 508-514.

Roy, R., Mazumder, P., & Sarma, G. (2009). Prolin, protein, catalase and root traits as indices of drought resistance in bold grained rice (*Oryza sativa*) genotypes. *African J. of Biotechnology, 8*(23), 652-658.

Sikora, S., & Redzepovic, S. (2003). Genotypic characterization of indigenous Soybean rhizobia by PCR-RFLP of 16S-Rdna, rep-PCR and RAPD analysis. *Food technol. Biotechnol., 41*(1) 61-67.

Tara, M. (2016). Iraq country pasture/forage resource profile online link. Retrieved from http://www.fao.org/ay/doc/Iraq.html

Turner, N., Wright, G., & Siddique, K. (2001). Adaptation of grain legumes(pulses) to water limited environments. *Advances in Agronomy, 71*, 193-231.

Versalovic, J., Schneider, M., De-Bruijin, F., & Lupski, J. (1994). Genomic fingerprinting of bacteria using repetitive sequence based on polymerase chain reaction methods. *Mol. Cellular Biol., 5*, 25-40.

Vincent, J. (1970). Amanual for the practical study of the root-nodule bacteria. *IBP Handbook No. 15*. Blackwell, Oxford, UK.

Zahran, H. (1999). Rhizobium-legume symbiosis and nitrogen fixation under sever conditions and in arid climate. *Microbial Mol. Biol. Rev., 63*, 968-989.

Zeng, Z., Chen, W., Hu, Y., Su, X., & Chen, D. (2007). Screening of highly effective *Sinorhizobium meliloti* strains for "rector" alfalfa and testing of its competitive nodulating ability in the field. *Pedosphere, 17*, 219-228.

9

Effect of a Mushroom (*Coriolus versicolor*) Based Probiotic on the Expression of Toll-like Receptors and Signal Transduction in Goat Neutrophils

Kingsley Ekwemalor[1], Emmanuel Asiamah[1] & Mulumebet Worku [2]

[1] Department of Energy and Environmental Systems, North Carolina Agricultural and Technical State University, USA

[2] Department of Animal Sciences, North Carolina Agricultural and Technical State University, USA

Correspondence: Mulumebet Worku, Department of Animal Science, North Carolina Agricultural and Technical State University, 1601 E Market Street, 27411, Greensboro, USA. E-mail: Worku @ncat.edu

Abstract

Neutrophils recognize and destroy pathogens through activation of the Toll like receptor (TLR) system as part of the inflammatory response of innate immunity. The expression and modulation of genes in the TLR signaling pathway in caprine blood neutrophils was investigated. Following initial screening for infection, goats (N=15) were assigned to three groups of five (n=5) individuals. Goats were drenched daily with 10 mL of powdered CorPet (Mycology labs Inc) soaked in hot (treatment I) or cold (treatment II) sterile filtered endotoxin free water, for a 4-week period. A control group of five age-matched goats received sterile water (treatment III). Blood was collected weekly and analyzed for packed cell volume and white blood cell differential counts. At weeks 1 and 4 neutrophils were isolated, using differential centrifugation and hypotonic lysis of red blood cells. The concentration and purity of total RNA isolated using Trizol was determined on a Nanodrop spectrophotometer. The RETROscript kit was used to synthesize cDNA. The expression of 84 genes in the human TLR signaling pathway RT^2 PCR Array was evaluated using real time PCR and the Livak method. The house keeping gene GAPDH was used to normalize the data. At week 1 untreated goats expressed 48 genes in the pathway. Goat neutrophils expressed 10 TLRs. Mushroom extracts modulated expression of and signaling by TLR. These results will help in the definition of the role of TLR expression in neutrophils and its contribution to goat innate immunity. Further this may aid in the design of therapeutics for goat health.

Keywords: goat, mushroom, neutrophil, toll-like receptors

1. Introduction

Goats live in a wide variety of microbe-rich environments. It is crucial to have a sensitive innate defense mechanism which relies in part by recognizing conserved molecules that are unique to some classes of potential pathogens. The innate immune system is based principally on physical and chemical barriers to infection, as well as on different cell types recognizing invading pathogens and activating antimicrobial immune responses (Basset et al., 2003).

They are key innate immune effector cells that provide early defense against invading microorganisms (Prince et al., 2011). They initiate antimicrobial and proinflammatory functions. They transit rapidly to sites of infection, where they limit infection and allow recruitment and activation of other immune cells through the release of inflammatory mediators and antimicrobial products, resulting in pathogen clearance and ultimately, in the initiation of an adaptive response (Yamashiro et al., 2001). Primary sensing of pathogen associated molecular patterns (PAMPs) to alert the innate immune system is achieved by an array of germ-like encoded receptors known as pattern recognition receptors (PRRs) (Tirurugaan et al., 2010). One of the important PRRs that play a key role in innate immunity is the type 1 transmembrane proteins called Toll-like Receptors (TLRs). Toll-like Receptors recognize microbial markers namely proteins, carbohydrates, lipids, nucleic acids and/or their combinations in an efficient, non-self-reactive manner to initiate a complex signaling cascade and activate a wide variety of transcription factors and inflammatory cytokines (Akira & Takeda, 2004). These cell surface molecules also activate complement, phagocytosis, inflammation and apoptosis in response to pathogen detection (West et al., 2006) finally culminating in the initiation of adaptive immunity through the induction of pro-inflammatory mediators (Janeway & Medzhitov, 2002).

Appropriate recognition of the invading pathogen is fundamental for the prompt and proper activation of the immune response. Their targets include bacteria, fungi, protozoa, viruses, virally infected cells and tumor cells. This function is facilitated by the expression of TLR family members by neutrophils, allowing the recognition of an extensive repertoire of PAMPs and thus triggering the response of invading pathogens (Prince et al., 2011). Pathogen associated molecular pattern such as beta-glucans, which form the main cell wall skeleton in mushrooms and as fungi, are recognized immediately by PRR (Kumagai & Akira, 2010), such as TLR2, dectin-1 and CR3. Mushrooms activate B-lymphocytes and macrophages through TLR, modulating the immune system and inducing the production of cytokines (Liao et al., 2004).

Mushrooms are known for their nutritional and medicinal value and also for the diversity of bioactive compounds they contain. The mushroom *Coriolus versicolor* (CV) has been reported to have an effect by boosting suppressed immune function, extending the survival rate and improving quality of life of cancer patients (LY Eliza et al., 2014). Various products derived from this mushroom and claimed to have medicinal value are commercially available. The active ingredients in *CV* are polysaccharides, in particular the polysaccharide krestin (PSK) and polysaccharide-peptide (PSP) (Chan & Yeung, 2006). They exert their therapeutic effects by modulating the host's immune response. Both preclinical and clinical evidences have demonstrated that extracts from *CV* display a wide array of biological activities, including stimulatory effects on the immune system and inhibition of cancer growth (Zhou et al., 2007). The aqueous extract of mushrooms has been found to be effective in activating T and B lymphocytes, macrophages, natural killer cells, and lymphocyte-activated killer cells, as well as promoting the production of antibodies and various cytokines, such as IL-2 and IL-6, and tumor necrotic factor (TNF) *in vivo* (Lull et al., 2005, Rowan et al., 2003*)*. CV can remedy intestinal disorders, suppress microbial infection the immune response (Cui & Chisti, 2003) and improve immune function by increasing neutrophil count (Tsang et al., 2003). The objective of this study was to determine the effect of a mushroom based probiotic on the expression of genes in the TLR signaling pathway in Caprine blood neutrophils and to evaluate their modulation.

2. Materials and Methods

2.1 Animals

Fifteen clinically healthy female SpanishXBoer goats from the goat herd at the North Carolina Agricultural and Technical University Small Ruminant Research Unit were used in this study. Animals were clinically healthy and not under any treatment. Initial sampling was carried out to determine the health of the animals. The study was approved by the Institutional Animal Care and Use Committee.

2.2 Preparation of Mushroom Extracts

CorPet (CV) powder was purchased from Mycology Research Labs Ltd (United Kingdom). It contains *Coriolus versicolor*, Microcrystalline cellulose (bulking agent), Silica (anti-caking agent), vegetable and Magnesium. Hot extracts (treatment I) was prepared by weighing 25 g of CorPet powder in 250 ml of sterile endotoxin free water and heating to 100^0C with stirring for 20 minutes. Twenty-five (25) grams of CorPet was stirred in 250 ml of sterile endotoxin free distilled water and served as cold extract (treatment II). The extracts were left to cool and then stored at 4^0C until it was used. Distilled water (treatment III) served as control.

2.3 CorPet Drench Administration

Ten (10) mL of the hot and cold extracts were given to each goat (5 goats per group) daily. Ten (10) mL of distilled water was administered to the control group. Extracts were administered daily for 30 days using a 10 ml syringe.

2.4 Neutrophil Isolation

Peripheral blood (15 ml) was collected from the jugular vein into vacutainer tubes containing anticoagulant weekly and analyzed for packed cell volume, white blood cell differential count and live cell count using TC20 (Biorad). Neutrophils were isolated from blood samples by differential centrifugation and hypotonic lyses of the red blood cells according to the modified procedure of Carlson and Kaneko (1973). The cell pellet was resuspended in 5 ml of sterile phosphate-buffered saline; pH 7.4. All reagents used in this study were prepared using sterile endotoxin free water. Endotoxin assay was performed as described by Adjei-Fremah et al. (2016a). Viable cells were counted on a TC10 (Biorad) cell counter using Trypan blue dye exclusion technique. Isolated neutrophils were adjusted to a concentration of 1×10^7 viable cells/ml in PBS and used for RNA extraction. The white blood cell differential counts were determined as described by Schaim et al. (1975). White blood cell differential counts were performed using an Olympus B 201 microscope using a 100x magnification.

2.5 Isolation of RNA

Total RNA, was isolated using Tri-reagent (Molecular Research Centre, Inc. Cincinnati, OH) following extraction procedure previously described by Asiamah et al. (2016). The quantity and quality of RNA was measured with the ND-1000 UV/VIS Nanodrop spectrophotometer (260 nm and 260/280 nm respectively). RNase free water was used as a blank.

2.6 Real Time PCR

Reversed transcription was performed using Oligo (dT) primers with 2 ug of the total RNA from each treatment group using a Complementary DNA (cDNA) RETRO script Kit (Ambion Inc., Austin, TX) following the manufacturer's instructions. The cDNA products were measured for purity and concentration using the Nanodrop spectrophotometer (NanoDrop Technologies). Quantification was performed in the CFX96™ Biorad Real-Time PCR detection system with the addition of the dye SYBR Green using the Qiagen Human TLR RT-PCR array (Qiagen, Valencia, CA) containing specific primer sets for 84 relevant TLR pathway genes, 5 housekeeping genes, and 2 negative controls (Table 1). Gene expression was normalized to GAPDH (housekeeping genes) to determine the fold change in gene expression between test and control samples by using the $2^{-\Delta\Delta Ct}$ method (Livak & Schmittgen, 2001).

Table 1. Functional gene grouping Qiagen human TLR RT-PCR

Toll-Like Receptors: CD180 (LY64), SIGIRR, TLR1, TLR2, TLR3, TLR4, TLR5, TLR6, TLR7, TLR8, TLR9, TLR10.
Pathogen-Specific Responses:
Bacterial: CCL2 (MCP-1), CD14, CD180 (LY64), FOS, HRAS, IL10, IL12A, IL1B, IL6, IL8, IRAK1, HMGB1, HSPA1A (HSP70 1A), JUN, LTA (TNFB), LY86 (MD-1), LY96, NFKBIA (IKBA/MAD3), PTGS2 (COX2), RELA, RIPK2, TLR2, TLR4, TLR6, TNFRSF1A, TICAM1 (TRIF).
Viral: EIF2AK2 (PRKR), IFNB1, IFNG, IL12A, IL6, IRF3, PRKRA, RELA, TBK1, TLR3, TLR7, TLR8, TNF, TICAM1 (TRIF).
Fungal/Parasitic: CLEC4E, HRAS, HSPA1A (HSP70 1A), IL8, TLR2, TIRAP.
TLR Signaling:
Negative Regulation: SARM1, SIGIRR, TOLLIP.
TICAM1 (TRIF)-Dependent (MYD88-Independent): IRF3, MAP3K7 (TAK1), TAB1, NR2C2, PELI1, TBK1, TICAM2, TLR3, TLR4, TRAF6, TICAM1 (TRIF).
MYD88-Dependent: IRAK1, IRAK2, MAP3K7 (TAK1), TAB1, MYD88, NR2C2, TIRAP, TLR1, TLR10, TLR2, TLR4, TLR5, TLR6, TLR7, TLR8, TLR9, TRAF6.
Downstream Pathways and Target Genes:
NFκB Pathway: BTK, CASP8, CHUK (IKKa), ECSIT (SITPEC), FADD, IKBKB, IL10, IL1B, IRAK1, IRAK2, IRF3, LY96, MAP3K1 (MEKK), MAP3K7, MAP4K4, NFKB1, NFKB2, NFKBIA (IKBA/MAD3), NFKBIL1, NFRKB, PPARA, REL, RELA, TNF, TNFRSF1A, UBE2N, UBE2V1.
JNK/p38 Pathway: ELK1, FOS, IL1B, JUN, MAP2K3 (MEK3), MAP2K4 (JNKK1), MAP3K1 (MEKK), MAP3K7, MAPK8 (JNK1), MAPK8IP3, TNF.
JAK/STAT Pathway: CCL2 (MCP-1), CSF2 (GM-CSF), IFNG, IL12A, IL2, IL6.
Interferon Regulatory Factor (IRF) Pathway: CXCL10 (INP10), IFNA1, IFNB1, IFNG, IRF1, IRF3, TBK1.
Cytokine-Mediated Signaling Pathway: CCL2 (MCP-1), CSF3 (GCSF), IL1A, IL1B, IL6, IRAK1, IRAK2, RELA, SIGIRR, TNF, TNFRSF1A.
Regulation of Adaptive Immunity: CD80, CD86, HSPD1, IFNG, IL10, IL12A, IL1B, IL2, MAP3K7, TRAF6.
Adaptors & TLR Interacting Proteins: BTK, CD14, HMGB1, HRAS, HSPA1A (HSP70 1A), HSPD1, LY86 (MD-1), LY96 (MD-2), MAPK8IP3, MYD88, PELI1, RIPK2, SARM1, TICAM1 (TRIF), TICAM2 (TRAM), TIRAP, TOLLIP.
Effectors: CASP8 (FLICE), EIF2AK2 (PRKR), FADD, IRAK1, IRAK2, MAP3K7 (TAK1), TAB1, NR2C2, PPARA, PRKRA, ECSIT (SITPEC), TRAF6, UBE2N, UBE2V1.

2.7 Statistical Analysis

Statistical analysis was conducted using the statistical analysis software SAS (SAS Institute Inc., Cary, NC). Analysis of variance was performed to evaluate the significance of treatment differences on RNA concentration $p < 0.05$.

3. Results

This study evaluated the expression and modulation of genes in the TLR signaling pathway in goat neutrophils. There was no significant difference ($p>0.05$) in the percentages of neutrophils, lymphocytes, monocytes, basophils and eosinophils between treatments and control groups (Figure 1). There was no difference in RNA concentration between treatment and control groups ($p<0.42$). There was an observable decrease in RNA concentration in samples from treated goats compared to controls at week 1 and week 3 compared to an increase at week 2 and 4 (Figure 2).

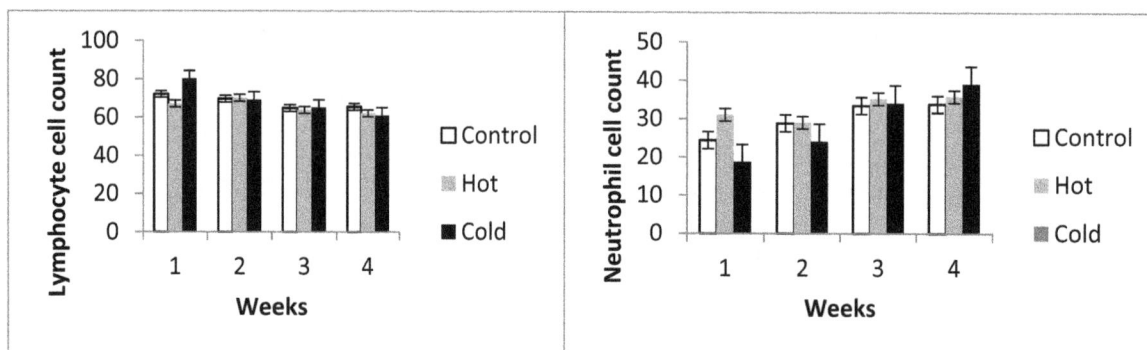

Figure 1. Effect of CorPet extract on white blood cell count

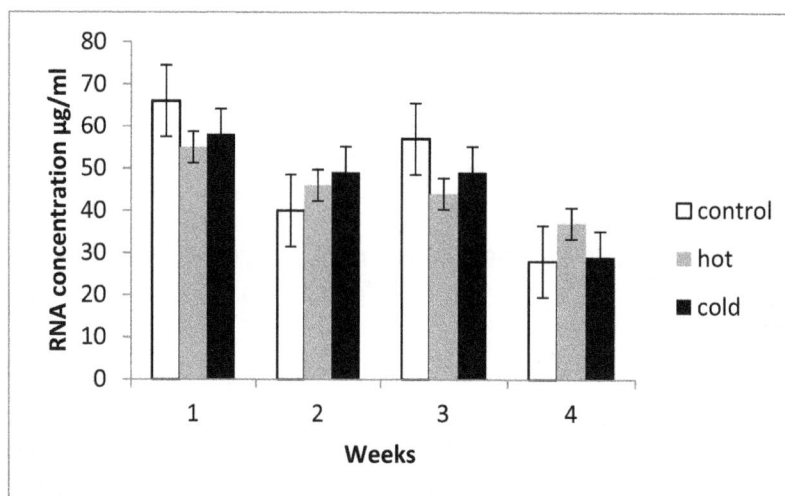

Figure 2. Effect of CorPet on RNA concentration over a week period

Genes on the human TLR array associated with fungal and parasitic response, toll-like receptor signaling, cytokine signaling and downstream signaling of toll-like receptors (Table 1) were expressed in goat peripheral blood neutrophils. At the beginning of the study 48 genes were expressed in untreated goats. Treatment with CV modulated the expression of these genes. After 4 weeks, hot extract modulated the expression of 40 genes; cold extract modulated the expression of 6 genes while goats from the control group expressed 8 genes (**Table 2**).

Table 2. Summary of number of genes expressed.

Treatment	Number of genes expressed	
	Week 1	Week 4
Control	14	7
Hot	40	84
Cold	47	6

3.1 Effect of Treatment on 84 Genes in TLR Signaling Pathway

In the control group animals 14 genes were expressed in week 1. These genes include CASP8, CCL2, IFNB1, IL1B, IRF3, MAP3K47, MAPK8, NFRKB, SARMI, TBK1, TLR2, TLR8, TLRF6 and UEE2VI. After 4 weeks 7 genes were expressed which included CLEC4E, CXCL10, ELK1, NFKB1L1, TLR 8, UBE2N and UBE2V1. Toll-like receptor 2 and UBE2VI were the genes expressed in week 1 and week 4. Both genes were down-regulated. These genes are not associated with fungi detection via PRR.

In the hot extract treatment group 48 genes were expressed in week 1. Following treatment, after 4 weeks all 84 genes were expressed in the TLR signaling pathway. Functional grouping of genes associated with fungal and toll-like receptors are listed in Table 3.

Table 3. Functional grouping of differentially expressed genes for the hot treatment group after 4 weeks

Functional group	Gene
Fungal/parasitic	CLEC4E, HRAS, HSPA1A (HSP70 1A), IL8, TLR2, TIRAP.
Toll-like receptors	CD180 (LY64), SIGIRR, TLR1, TLR2, TLR3, TLR4, TLR5, TLR6, TLR7, TLR8, TLR9, TLR10.

Genes with a high fold change include MAPK8, NFKB2, PPARA, TOLLIP, TNFRSF1A, IRAK1 and CSF3 (Table 4).

Table 4. Differentially expressed genes and fold change in expression of animals drenched with hot extract treatment

Gene	Function	Fold change
MAPK8	Involved in a wide variety of cellular processes such as proliferation, differentiation, transcription regulation and development	425
NFKB2	Central activator of genes involved in inflammation and immune function	2
PPARA	Affects the expression of target genes involved in cell proliferation, cell differentiation and in immune and inflammation responses	596
TOLLIP	Protein that interacts with several Toll-like receptor (TLR) signaling cascade components	328
IRAK1	Responsible for IL1-induced upregulation of the transcription factor NF-kappa B	107

In the cold treatment group 48 genes were expressed in week 1. After 4 weeks 6 genes were expressed. These genes include: MAK2K3, CXCL10, ELK1, FOS, MYD88 and UBE2V1. Genes that were common in both week one and week 4 (CXCL10, MAK2K3 and MYD88) were down regulated.

3.2 Effect of Treatment on Toll-Like Receptors

In goat neutrophils all 10 TLR were expressed. In the control group 2 TLR were expressed at week 1. After 4 weeks only 1 TLR was expressed. In the hot extract group, 3 TLR were expressed at week 1. After 4 weeks the hot extract induced the expression of all 10 toll-like receptors. In the cold extract group 6 TLR were expressed in week 1 but after 4 weeks there was no expression of TLR observed (Table 5).

Table 5. Effect of treatment on TLR expression.

Treatment	week	Toll-like Receptor
Control	0	TLR2 , TLR8
Control	4	TLR8
Hot	0	TLR4, TLR5, TLR7
Hot	4	TLR1, TLR2, TLR3, TLR4, TLR5, TLR6, TLR7, TLR8, TLR9, TLR10
Cold	0	TLR4, TLR5,TLR6, TLR7, TLR9, TLR10
Cold	4	-

There was variability of TLR expressed among animals which can be attributed to genetic variability. Also there was variation of TLR expressed in the hot and cold treatment groups which can also be attributed to extract preparation method. Treatment with CV impacted the expression of genes important to TLR-mediated signal transduction in goat blood. Thus extract preparation had an effect on gene expression.

4. Discussion

Probiotics are nutritional supplements containing potentially beneficial microorganisms which confer beneficial health effects in variety of conditions and diseases in animals. In this study, the effect of a mushroom (*Coriolus versicolor*) based probiotics on the expression and modulation of genes in the TLR signal transduction pathway was evaluated in goat neutrophils. Hot extract treatment of CV had an effect on the expression of 6 TLRs. Ten TLR's have been identified in goats and expression has been reported in different tissue (Dhanasekaran et al., 2014, Tirumurugaan et al., 2010, Worku et al., 2016), however there are no reports regarding the expression of TLR in goat neutrophils. The TLRs recognize conserved PAMPs that are unique to microorganisms and are absent from higher eukaryotes (Han et al., 2003). The various TLRs exhibit different patterns of expression. The TLR1 family (TLR1, 2, 6 and 10) is involved in the recognition of gram-positive and gram-negative bacteria and heterodimers of TLR1 or TLR6 with TLR2 are crucial for the identification of several PAMPs (Kwong et al., 2011). Batbayar et al., (2011) reported that expression of TLR2, TLR4 and TLR6 was increased by β-glucan from *Ganoderma lucidum* which corresponds to the results obtained in this study. Results have shown that fungi binds to fungal ligand recognition regions on TLR1, 2, 4, 5, 6 and 10 which are found on the cell surface (Sasai & Yamamototo, 2013). This report corresponds to the expression of these TLRs in our study with the exception of TLR 3, 7, 8 and 9. Result showed that there was variation in expression of TLR among animals. Goyal et al., (Goyal et al., 2012) reported a nucleotide polymorphism observed in TLR7 gene of 24 goats representing 12 different breeds. Also in another study conducted by Zhou and Hickford (2008), they detected 5 nucleotide polymorphisms of TLR4 using 374 New Zealand goats of different breeds. These results correspond to the variation in expression of TLR in our study and can be exploited for association with disease susceptibility in goats.

Binding of ligands to TLRs triggers at least two important cell signaling pathways. One pathway involves MyD88, an adaptor protein shared by most of the TLRs that leads to the activation of the transcription factor NF-κB resulting in the release of pro-inflammatory cytokines (Raja et al., 2011, Adjei-Fremah et al., 2016b). Administration of CV modulated the expression of genes involved in multiple signaling pathways such as TLR-mediated signaling induction pathway, nuclear factor κB (NF- κB), jun N-terminal kinase (JNK) and p38 mitogen-activated protein kinase (MAPK), janus kinase/signal transducers and activators of transcription (JAK-STAT), interferon regulatory factor (IFN) and cytokine mediated signaling pathways. Studies have shown that NF-κB plays a crucial role in immune responses and is an important transcriptional regulator of inflammatory cytokine genes. Yang et al. (2015) reported that polysaccharide from the CV activates mouse B cells through the MAPK and NF-κB signaling pathways. Also studies by Liu et al., (Liu et al., 2016) reported that extracts from shiitake mushroom (*Lentinula edodes*) induced the production of G-CSF, GM-CSF, M-CSF by activating the MAPKT/NF- κB signaling pathway in bone marrow cells. This corresponds with our results with the expression of genes involved in the MAPK and NF-κB signaling pathways. Previous study in our lab also shows that CV induced the secretion of G-CSF and GM-CSF (Ekwemalor, 2015).

Innate immunity can be stimulated by the activation of pattern recognition receptors. TLR's recognize pathogen-associated molecular patterns (PAMPs) that are expressed on infectious agents, and mediate the production of cytokines necessary for the development of effective immunity. Mushrooms which have β-glucans as their major component appear to have an important role in the innate immune response to fungal pathogens and in initiating a protective adaptive response (Brown & Gordon, 2003). Several receptors for β-glucans acting as PRRs have been identified to date include Dectin-1, complement receptor 3 and TLR (Kang et al., 2013). Studies have suggested that polysaccharides from *Ganoderma lucidum* (Reishi) and *Phellinus linteus* extracts can act through TLR pathways to induce inflammatory responses in mouse cells (Prince et al., 2011).

Our results show a systemic effect of oral administration of CorPet probiotics which had an effect on TLR signal transduction in neutrophil in goat peripheral blood. Thus evidence is presented for activation of caprine neutrophils through dietary supplements such as probiotics. Specifically the expression and modulation of all ten TLR in goat neutrophils is presented. The method of extract (hot or cold) preparation may impact the PAMPs detected by neutrophils and provides support for opportunities for differential of TLR by pathogens and therapeutics in goat neutrophils. The possible difference in extract preparation between hot and cold extracts might be as a result of heating. Studies have shown differences in the effect of aqueous preparation as hot or cold formulation (Jung et al., 2014), but obvious reason needs to be clarified through further studies.

5. Conclusion

We report the expression and modulation of genes in the TLR signal transduction pathway in goat neutrophils. In light of the critical role of TLR in controlling bacterial infections the expression and modulation of these genes may be critical to goat health and modulation of innate immunity using mushroom based probiotics. The expression of 10 TLRs in goat neutrophils has implications for further understanding of innate immunity in goats and efforts to modulate it through novel therapeutics. Our findings illustrate a repertoire of TLRs associated with inflammatory activation in goats, and they encourage further exploration.

Acknowledgements

We appreciate the assistance of Gary Summers and Hamid Ismail during sample collection and Tiffany Martin, Jordan Page and Allie McMahan for their assistance in this project. Funding support was from USDA project NC.X-271-5-13-120-1, Modulation of Receptor Cross-Talk for cattle, Sheep and Goat Innate Immunity.

References

Adjei-Fremah, S., Asiamah, E. K., Ekwemalor, K., Jackai, L., Schimmel, K., & Worku, M. (2016). Modulation of Bovine Wnt Signaling Pathway Genes by Cowpea Phenolic Extract. *Journal of Agricultural Science, 8*(3), 21. http://dx.doi.org/10.5539/jas.v8n3p21

Adjei-Fremah, S., Ekwemalor, K., Asiamah, E., Ismail, H., & Worku, M. (2016). Transcriptional profiling of the effect of lipopolysaccharide (LPS) pretreatment in blood from probiotics-treated dairy cows. *Genomics Data, 10*, 15-18. http://dx.doi.org/10.1016/j.gdata.2016.08.016

Akira, S., & Takeda, K. (2004). Toll-like receptor signalling. *Nature reviews immunology, 4*(7), 499-511. http://dx.doi.org/10.1038/nri1391

Asiamah, E. K., Adjei-Fremah, S., Osei, B., Ekwemalor, K., & Worku, M. (2016). An Extract of Sericea Lespedeza Modulates Production of Inflammatory Markers in Pathogen Associated Molecular Pattern (PAMP) Activated Ruminant Blood. *Journal of Agricultural Science, 8*(9), 1. http://dx.doi.org/10.5539/jas.v8n9p1

Basset, C., Holton, J., O'Mahony, R., & Roitt, I. (2003). Innate immunity and pathogen–host interaction. *Vaccine, 21*, S12-S23. http://dx.doi.org/10.1016/S0264-410X(03)00195-6

Batbayar, S., Kim, M. J., & Kim, H. W. (2011). Medicinal Mushroom Lingzhi or Reishi, Ganoderma lucidum (W. Curt.: Fr.) P. Karst., β-Glucan Induces Toll-like Receptors and Fails to Induce Inflammatory Cytokines in NF-κB Inhibitor-Treated Macrophages. *International journal of medicinal mushrooms, 13*(3). http://dx.doi.org/10.1615/IntJMedMushr.v13.i3.10

Brown, G. D., & Gordon, S. (2003). Fungal β-glucans and mammalian immunity. *Immunity, 19*(3), 311-315. http://dx.doi.org/10.1016/S1074-7613(03)00233-4

Carlson, G. P., & Kaneko, J. J. (1973). Isolation of leukocytes from bovine peripheral blood. *Experimental Biology and Medicine, 142*(3), 853-856. http://dx.doi.org/10.3181/00379727-142-37131

Chan, S. L., & Yeung, J. H. (2006). Effects of polysaccharide peptide (PSP) from Coriolus versicolor on the pharmacokinetics of cyclophosphamide in the rat and cytotoxicity in HepG2 cells. *Food and Chemical Toxicology, 44*(5), 689-694. http://dx.doi.org/10.1016/j.fct.2005.10.001

Cui, J., & Chisti, Y. (2003). Polysaccharopeptides of Coriolus versicolor: physiological activity, uses, and production. *Biotechnology advances, 21*(2), 109-122. http://dx.doi.org/10.1016/S0734-9750(03)00002-8

Dhanasekaran, S., Biswas, M., Vignesh, A. R., Ramya, R., Raj, G. D., Tirumurugaan, K. G., ... & Subbiah, E. (2014). Toll-like receptor responses to Peste des petits ruminants virus in goats and water buffalo. *PloS one, 9*(11), e111609. http://dx.doi.org/10.1371/journal.pone.0111609

Ekwemalor, K. (2015). *The Effect of a Mushroom (Coriolus versicolor) Based Probiotic on Innate immunity in Goats Naturally Infected with Gastrointestinal Parasites* (Doctoral dissertation, North Carolina Agricultural and Technical State University).Goyal, S., Dubey, P. K., Tripathy, K., Mahajan, R., Pan, S., Dixit, S. P., & Kataria, R. S. (2012). Detection of polymorphism and sequence characterization of Toll-like receptor 7 gene of Indian goat revealing close relationship between ruminant species. *Animal biotechnology, 23*(3), 194-203. http://dx.doi.org/10.1080/ 10495398.2012.684417

Han, S. B., Yoon, Y. D., Ahn, H. J., Lee, H. S., Lee, C. W., Yoon, W. K., ... Kim, H. M. (2003). Toll-like receptor-mediated activation of B cells and macrophages by polysaccharide isolated from cell culture of

Acanthopanax senticosus. *International immunopharmacology, 3*(9), 1301-1312. http://dx.doi.org/10.1016/S1567-5769(03) 00118-8

Janeway Jr, C. A., & Medzhitov, R. (2002). Innate immune recognition.*Annual review of immunology, 20*(1), 197-216. http://dx.doi.org/10.1146/annurev.immunol.20.083001.084359

Jung, I. L. (2014). Soluble extract from Moringa oleifera leaves with a new anticancer activity. *PloS one, 9*(4), e95492. http://dx.doi.org/10.1371/journal.pone.0095492

Kang, S. C., Koo, H. J., Park, S., Lim, J. D., Kim, Y. J., Kim, T., ... & Sohn, E. H. (2013). Effects of β-glucans from Coriolus versicolor on macrophage phagocytosis are related to the Akt and CK2/Ikaros. *International journal of biological macromolecules, 57*, 9-16. http://dx.doi.org/10.1016/j.ijbiomac.2013.03.017

Kumagai, Y., & Akira, S. (2010). Identification and functions of pattern-recognition receptors. *Journal of Allergy and Clinical Immunology, 125*(5), 985-992. http://dx.doi.org/10.1016/j.jaci.2010.01.058

Kwong, L. S., Parsons, R., Patterson, R., Coffey, T. J., Thonur, L., Chang, J. S., ... & Hope, J. C. (2011). Characterisation of antibodies to bovine toll-like receptor (TLR)-2 and cross-reactivity with ovine TLR2. *Veterinary immunology and immunopathology, 139*(2), 313-318. http://dx.doi.org/10.1016/j.vetimm.2010.10.014

Liao, S. C., Cheng, Y. C., Wang, Y. C., Wang, C. W., Yang, S. M., Yu, C. K., ... & Shieh, J. M. (2004). IL-19 induced Th2 cytokines and was up-regulated in asthma patients. *The Journal of Immunology, 173*(11), 6712-6718. http://dx.doi.org/10.4049/jimmunol.173.11.6712

Liu, Q., Dong, L., Li, H., Yuan, J., Peng, Y., & Dai, S. (2016). Lentinan mitigates therarubicin-induced myelosuppression by activating bone marrow-derived macrophages in an MAPK/NF-κB-dependent manner. *Oncology reports, 36*(1), 315-323. http://dx.doi.org/10.3892/or.2016.4769

Livak, K. J., & Schmittgen, T. D. (2001). Analysis of relative gene expression data using real-time quantitative PCR and the 2− ΔΔCT method. *Methods, 25*(4), 402-408. http://dx.doi.org/10.1006/meth.2001.1262

Lull, C., Wichers, H. J., & Savelkoul, H. F. (2005). Antiinflammatory and immunomodulating properties of fungal metabolites. *Mediators of inflammation, 2005*(2), 63-80. http://dx.doi.org/10.1155/MI.2005.63

LY Eliza, W., K Fai, C., & P Chung, L. (2012). Efficacy of Yun Zhi (Coriolus versicolor) on survival in cancer patients: systematic review and meta-analysis. *Recent patents on inflammation & allergy drug discovery, 6*(1), 78-87. http://dx.doi.org/10.2174/187221312798889310

Prince, L. R., Whyte, M. K., Sabroe, I., & Parker, L. C. (2011). The role of TLRs in neutrophil activation. *Current opinion in pharmacology, 11*(4), 397-403. http://dx.doi.org/10.1016/j.coph.2011.06.007

Raja, A., Vignesh, A. R., Mary, B. A., Tirumurugaan, K. G., Raj, G. D., Kataria, R., ... & Kumanan, K. (2011). Sequence analysis of Toll-like receptor genes 1–10 of goat (Capra hircus). *Veterinary immunology and immunopathology, 140*(3), 252-258. http://dx.doi.org/10.1016/j.vetimm.2011.01.007

Rowan, N. J., Smith, J. E., & Sullivan, R. (2003). Immunomodulatory activities of mushroom glucans and polysaccharide–protein complexes in animals and humans (a review). *International Journal of Medicinal Mushrooms, 5*(2). http://dx.doi.org/10.1615/InterJMedicMush.v5.i2.10

Sasai, M., & Yamamoto, M. (2013). Pathogen recognition receptors: ligands and signaling pathways by Toll-like receptors. *International reviews of immunology, 32*(2), 116-133. http://dx.doi.org/10.3109/08830185.2013.774391

Schaim, O. W., Jain, N. C., & Carrol, E. J. (1975). *Veterinary Haematology.* Lea Febiger Philadelphia USA.

Tirumurugaan, K. G., Dhanasekaran, S., Raj, G. D., Raja, A., Kumanan, K., & Ramaswamy, V. (2010). Differential expression of toll-like receptor mRNA in selected tissues of goat (Capra hircus). *Veterinary immunology and immunopathology, 133*(2), 296-301. http://dx.doi.org/10.1016/j.vetimm.2009.08.015

Tsang, K. W., Lam, C. L., Yan, C., Mak, J. C., Ooi, G. C., Ho, J. C., ... & Lam, W. K. (2003). Coriolus versicolor polysaccharide peptide slows progression of advanced non-small cell lung cancer. *Respiratory medicine, 97*(6), 618-624. http://dx.doi.org/10.1053/rmed.2003.1490

West, A. P., Koblansky, A. A., & Ghosh, S. (2006). Recognition and signaling by toll-like receptors. *Annu. Rev. Cell Dev. Biol., 22*, 409-437. http://dx.doi.org/10.1146/annurev.cellbio.21.122303.115827

Worku, M., Abdalla, A., Adjei-Fremah, S., & Ismail, H. (2016). The Impact of Diet on Expression of Genes Involved in Innate Immunity in Goat Blood. *Journal of Agricultural Science, 8*(3), 1. http://dx.doi.org/10.5539/jas.v8n3p1

Yamashiro, S., Kamohara, H., Wang, J. M., Yang, D., Gong, W. H., & Yoshimura, T. (2001). Phenotypic and functional change of cytokine-activated neutrophils: inflammatory neutrophils are heterogeneous and enhance adaptive immune responses. *Journal of leukocyte biology, 69*(5), 698-704. http://dx.doi.org/10.1159/000071555

Yang, S. F., Zhuang, T. F., Si, Y. M., Qi, K. Y., & Zhao, J. (2015). Coriolus versicolor mushroom polysaccharides exert immunoregulatory effects on mouse B cells via membrane Ig and TLR-4 to activate the MAPK and NF-κB signaling pathways. *Molecular immunology, 64*(1), 144-151. http://dx.doi.org/10.1016/j.molimm. 2014.11.007

Zhou, H., & Hickford, J. G. H. (2008). Allelic polymorphism of the caprine calpastatin (CAST) gene identified by PCR–SSCP. *Meat science, 79*(2), 403-405. http://dx.doi.org/10.1016/j.meatsci.2007.10.015

Zhou, L., Ivanov, I. I., Spolski, R., Min, R., Shenderov, K., Egawa, T., & Littman, D. R. (2007). IL-6 programs TH-17 cell differentiation by promoting sequential engagement of the IL-21 and IL-23 pathways. *Nature immunology, 8*(9), 967-974. http://dx.doi.org/10.1038/ni1488

10

Gene Expression Patterns in Functionally Different Cochlear Compartments of the Newborn Rat

Johann Gross[1], Heidi Olze[1] & Birgit Mazurek[1]

[1] Molecular Biology Research Laboratory, Department of Otorhinolaryngology, Charité Universitätsmedizin Berlin, Campus Charité Mitte, Berlin, Germany

Correspondence: Johann Gross, Molecular Biology Research Laboratory, Department of Otorhinolaryngology, Charité Universitätsmedizin Berlin, Campus Charité Mitte, Berlin, Germany. E-mail: johann.gross@charite.de; johann.gross@arcor.de

Abstract

In an experimental model of organotypic cultures of the stria vascularis (SV), the organ of Corti (OC) and the modiolus (MOD), we compared the expression levels and injury/hypoxia induced response of 36 genes associated with the cells´ energy-producing and energy-consuming processes, using the microarray technique. A decrease of expression was observed for most of the voltage-dependent K^+- and Ca^{++}- channels as an effective mechanism to lower energetic demands. We identified two gene networks of transcripts that are differentially expressed across the three regions. One cluster is associated with the transcription factor hypoxia-inducing factor (*Hif-1a*) and the second one with the caspase and calpain cell death genes *Casp3, Capn1, Capn2* and *Capns1*. The *Hif-1a* gene subset consists of genes belonging to the glucose metabolism (glucose transporter *Slc2a1*, glycolytic enzymes *Gapdh, Hk1* and *Eno2*), the Na^+/K^+ homeostasis (ATPase *Atp1a1*) and the glutamate pathway (NMDA receptor associated protein 1 *Grina*, glutamate transporter *Slc1a1, Slc1a3*). The *Slc2a1, Gapdh, Hk1, Slc1a3, Grina* and *Atp1a1* transcripts are also members of the cell death subset indicating a role they have to play in the differential regional cell death rates. The newly identified genes *Grina* and calnexin (*Canx*) may play specific and yet unknown roles in regulating cell death induced by injury and hypoxia in the inner ear. We assume that the differential regional response occurs on the basis of endogenous gene regulatory mechanisms and may be important to maintaining the cochlea's function following damage from trauma and hypoxia.

Keywords: cell death, gene expression, hypoxia, injury, inner ear, microarray

1. Introduction

The cochlea consists of three main complex structures, each serving a specific function: the organ of Corti (OC), the modiolus (MOD) and the stria vascularis (SV). The OC with its inner and outer hair cells transforms the mechanical signal into an electrical one via hair cell depolarization and signal amplification. The specific function of the SV is to produce and maintain the ionic composition of the endolymph, a very specific fluid with high concentrations of potassium and low concentrations of sodium. The MOD, the conically shaped central axis in the cochlea, contains the spiral ganglion neurons (SGNs). These bipolar neuronal cells transmit the electrical signals from the hair cells to the cochlear nuclei in the brainstem.

Organotypic cultures of the SV, the OC and the MOD were used to experimentally study the differential gene expression of these regions to injury stress and hypoxia (Gross et al., 2007). In freshly prepared tissue, about 2-10 % of all nuclei were found to be stained by propidium iodide, indicating cell damage during preparation of the cochlear tissue. After 24h in culture, the number of necrotic cells in the MOD region increased from 8 % to 25-35 %), whereas the number of such cells remained unchanged in the OC and SV regions (Gross et al., 2008). Gene expression markers indicate that two basic pathogenetic mechanisms are involved in this experimental model: mechanically induced inflammation and hypoxia. The expression of genes involved in apoptosis and necrosis (*Casp3, Capn1, Capn2* and *Capns1*), reactive oxygen species metabolism (*Sod3, Nos2*), inflammation (*Ccl20*) as well as selected transcription factors (*Hif-1a, Jun,* and *Bmyc*) have to play a key role in the differential regional response (Mazurek et al., 2011; Gross, Olze, & Mazurek, 2014).

Other pathways implicated in injury and hypoxia include processes involved in energy production and regulation of ion homeostasis which could become critical for cell survival and regeneration (Michiels, 2004). The energy balance of a cell under physiological and pathological conditions depends on its ATP production and consumption. The energy demand of the inner ear is high and can be compared to that of brain tissue. To meet the energy demand under conditions of inflammation and hypoxia, anaerobic ATP supply is triggered (Pasteur effect; Boutilier & St-Pierre, 2000). ATP production via glycolysis is associated with increased glucose transport and consumption (Frezza et al., 2011).

Maintaining ion homeostasis to allow de- and repolarization of the cells belongs to the processes that require high amounts of energy (Rolfe & Brown, 1997; Buttgereit & Brand, 1995). In the cochlea, the maintenance of ion homeostasis, K^+-cycling and its role in the endocochlear potential are coupled to Na,K-ATPase (Wangemann, 2002). K^+-cycling is particularly important, as in response to the stimulated stereocilia, endolymphatic K^+ flows into the sensory hair cells via the apical transduction channel and is released from the hair cells into the perilymph via basolateral K^+ channels. K^+ may be taken up by fibrocytes in the spiral ligament and transported from cell to cell via gap junctions into strial intermediate cells which secrete it to the endolymph.

Calcium entry and the maintenance of the multiple segregated transduction pathways is controlled by a combination of calcium channels, Ca^{++}-ATPases and buffering mechanisms. Elevated levels of intracellular calcium under hypoxia or injury stress are the result of a massive influx of extracellular calcium through activated channels or the release of calcium from intracellular stores like the endoplasmatic reticulum or the mitochondria (Brini & Carafoli, 2009; Lang, Vallon, Knipper, & Wangemann, 2007). Main routes for calcium influx are voltage-gated Ca^{++} channels, purinergic receptors and ionotropic glutamate receptors (Martinez-Sanchez et al., 2004). For active efflux of intracellular Ca^{++}, the main routes have found to be its export via the plasma membrane calcium ATPase (PMCA) and the transport via the Na^+/Ca^{++} exchanger. Another mechanism to decrease cytosolic Ca^{++} concentrations is uptake into intracellular stores via the sarco- and endo-plasmic reticulum calcium ATPase (SERCA). The calcium buffering mechanisms consist of several Ca^{++} binding proteins.

The aim of the present study is to analyze the basal and injury-induced expression of genes associated with the energy-producing and energy-consuming processes, i.e., of genes associated with the glucose metabolism, the regulation of Na^+-/K^+- and Ca^{++}-homeostasis, including the glutamate pathway. A microarray study was used as the guide that directed us to selecting a total of 36 genes associated with energy production and energy consumption, including the glutamate pathway (Mazurek et al., 2006). The use of the neurobiological array RN-U34 offers the possibility of identifying several transcripts in the inner ear that had not been described previously.

2. Materials and Methods

2.1 Explant Cultures

The cochleae from 3 to 5-day old Wistar rats were dissected into OC, MOD and SV (Sobkowicz, Loftus, & Slapnick, 1993). Details of the preparation of the fragments, the culture conditions and the testing of the viability of the explants were reported previously (Gross et al., 2007). Briefly, the fragments were incubated in four-well tissue culture dishes in Dulbecco`s Modified Eagle Medium/F12Nutrient (1:1) Mixtures (Gibco, Karlsruhe, Germany) supplemented with 10 % fetal bovine serum. Fragments of one ear were kept in culture under normoxic conditions, fragments of the second ear were exposed to moderate hypoxia (oxygen partial pressure inside the culture medium was 10-20 mm Hg) for 5 h, starting three hours after plating. The number of dead cells was determined in freshly prepared tissue (controls) and after 24 h in culture using the live/dead viability test by propidium iodide (PI) and calcein AM staining (Gross et al., 2008).

2.2 cDNA Microarray Analysis

The cDNA microarray analysis was carried out using the Affymetrix Rat Neurobiology U34 Array (RN-U34; Affymetrix, Santa Clara, USA). The complete data sets from this study have been deposited to the Gene Expression Omnibus (GEO) database according to the MIAME standard and can be accessed by ID GSE5446. Each of the total RNA samples of the MOD, OC and SV used in the microarray study originated from 6 animals. Altogether, 16 RNA preparations arising from three independent series were analyzed within one year: four samples from freshly prepared tissue (OC1, OC2, MOD, SV) and 12 experimental samples from cultures of OC, MOD and SV under normoxic (n = 2) and hypoxic conditions (n = 2). Further details of the RNA isolation and quantification and the cDNA microarray analysis were previously reported (Gross et al., 2007).

2.3 Statistical Analysis

Intensity of expression was classified on the basis of the histogram of normalized log2 signals and resulted in a normal distribution (data not shown). Values at or above the 75th percentile of the cumulative intensities are considered to be high level expression (> log2 = 12.55, = 6000 relative units; bold in Tables 1-3) and values below the 25th percentile to be low level expression (< log2 = 10.43, = 1380 relative units; italics in Tables 1-3); values between them are considered to be moderate (normal typeface in Tables 1-3). In this study, gene expression was not found to differ significantly between normoxic and hypoxic environments, neither for the numbers of PI-stained nuclei nor for the expression of HIF-1a mRNA or for that of other genes, with the data having been combined to result in four samples per region. Obviously, the hypoxia conditions we used were too mild to induce specific expression changes. Overlapping gene expression patterns induced by hypoxia and mechanical injury may also contribute to this observation. Three features of the gene expression are presented: (i) The absolute expression levels, classified as low, moderate or high using the log2 data. (ii) The fold change of the expression level was calculated as the ratio between the expression intensity of the 24h cultures and the expression intensity of freshly prepared tissue. The mean coefficient of the variation was 11.8 ± 7.6 % (n = 36) for signal intensity and 17.6 ± 8.1 % (n = 108; Tables 1-3) for the expression change. To test the significance levels of the fold changes we used the paired t-test or the Wilcoxon paired test (Tables 1-3). (iii) We used the Pearson´s correlation analysis with the Bonferroni post hoc test to identify co-expression changes among selected transcripts across the three regions.

3. Results

3.1 Glucose Transporter and Glycolytic Enzymes

The chip comprises the transcripts of glucose transporters *Slc2a1* (Glut1) and *Slc2a3* (Glut3) and the glycolytic enzymes *Gapdh* (glyceraldehyde-3-phosphate dehydrogenase), *Hk1* (hexokinase 1) and *Eno2* (neuron-specific enolase; Table 1). As expected, the *Slc2a1* and the *Gapdh* transcripts show high expression levels in all regions, whereas *Hk1* and *Eno2* belong to the subset of genes with low or moderate expression levels. In culture, with the exception of *Eno2* in the MOD region, all transcripts increase significantly across the three regions.

Table 1. Expression of genes associated with glucose transport and glycolysis in the organotypic cultures of the modiolus, the organ of Corti and the stria vascularis

Gene	Expression			Fold change			Name/function
	MOD	OC	SV	MOD	OC	SV	
Slc2a1 (S68135)*[1]	**12109**	**8328**	**19903**	2.4	7.9[#1]	3.6	Glucose transp. 1 (Glut1)
Slc2a3 (D13962)*[2]	1764	1542	*1132*	1.3	2.1	3.4	Glucose transp. 3 (Glut3)
Gapdh (X02231.1)*[3]	**15383**	**24992**	**24879**	7.2	5.4	5.0	Gap-Dehydrogenase
Hk1 (J04526.1)*[4]	*1012*	1556	1866	2.2[#2]	1.6	1.3	Hexokinase 1
Eno2 (X07729)	5069	4818	1235	0.9	1.9[#3]	5.0[#4]	Enolase 2, gamma

Note. Expression intensity (relative units, RU) was categorized on the basis of the histogram of the normalized log2 signals (see materials and methods). Fold change (Wilcoxon paired test, n = 12): *[1]T(12) = 0.00, p = 0.002; *[2]T(12) = 0.00, p = 0.002; *[3]T(12) = 2.00, p = 0.004; *[4]T(12) = 2.00, p = 0.004. Glut - glucose transporter, Gap-glyceraldehyde-3-phosphate. [#]Significance of expression changes (paired t-test, n = 4): [#1]p<0.000 vs MOD, p<0.006 vs SV, [#2]p<0.02 vs SV, [#3]p<0.004 vs MOD, [#4]p<0.000 vs MOD, 0.006 vs OC.

3.2 Na^+/K^+-Homeostasis

Several hundred genes known to encode ion channel proteins are involved in the regulation of Na^+/K^+-homeostasis (Gabashvili, Sokolowski, Morton, & Giersch, 2007). The chip identified the transcript for one sodium and six potassium channels (Table 2), most of them previously not identified in the inner ear. These channels show low to moderate basal expression levels and do not change (*Scn3a, Kcnk1, Kcnq3* and *Kcns3*) or decrease (*Pias3, Alg10* and *Kcnh2*). The basal expression levels and the fold changes in culture are similar in all regions.

Sodium-potassium ATPase is of importance not only for the neuronal resting potential but also for re-establishing the ion homeostasis in the inner ear following injury and hypoxia (Johar, Priya, & Wong-Riley, 2012; Wangemann, 2002). The chip comprises two isoforms of the alpha subunit and two isoforms of the beta subunit (Table 2). The basal expression levels of the various isoforms are unevenly distributed in the three regions and, as expected, relatively high in the OC and SV regions. The most remarkable findings in the present work are the increase of the *Atp1a1* subunit and the decrease of the *Atp1a3, Atp1b2* and the *Atp1b3* subunits.

3.3 Calcium Homeostasis

Calcium entry and the maintenance of the multiple segregated transduction pathways is controlled by a combination of voltage-dependent calcium channels, purinergic and ionotropic glutamate receptors, the Ca^{++}-ATPases and several buffering mechanisms (Table 3).

3.3.1 Calcium Channels

The chip contains information for several voltage-dependent Ca^{++}-channels which respond to cell membrane potential changes and have a role to play in changes of the local intracellular Ca^{++} homeostasis and the formation of the nerve impulse (*Cacna1d* /D38101, *Cacna1g*/AF027984, *Cacna2d1*/M8662, *Cacnb3*/M88751). These transcripts show low to moderate basal expression levels and decrease in culture 0.4-0.7-fold (data not shown). The expression of the voltage-dependent calcium channel *Cacn1c* and of the purinergic channel *P2rx* remained unchanged in all regions.

Table 2. Expression of genes associated with Na^+/K^+-transport in the organotypic cultures of the modiolus, the organ of Corti and the stria vascularis

Gene	Expression			Fold change			Name/function
	MOD	OC	SV	MOD	OC	SV	
Scn3a (Y00766)	*1206*	*1159*	*897*	0.8	0.8	0.7	VG Na^+channel, type III, alpha
Kcnk1 (AF022819)	*780*	3514	2542	1.2	1.2	1.2	K^+-channel, SF K, M1 (Twik)
Kcnq3 (AF091247)	*1280*	*1333*	*1084*	1.1	0.8	0.8	K^+-VGCh, SF KQT-like, M3
Kcns3 (Y17607)	*404*	2211	1433	1.0	0.7	1.0	K^+- VGCh, SF S, M3
Pias3 (AF032872)*[1]	4281	3549	3688	0.5	0.7	0.5	Binding protein to Kv-channels
Alg10 (U78090)*[2]	3299	3180	2446	0.6	0.8	0.5	Regulatory component
Kcnh2 (U75210)*[3]	4063	3846	2847	0.6	0.6	0.7	K^+-VGCh, SF H, M2 (ERG1)
Atp1a1 (M74494)*[4]	1498	**6001**	**15083**	**8.3**	**4.1**	**2.9**	ATPase, Na^+/K^+ transp., alpha1
Atp1a3 (M28648)*[5]	2475	5905	*1207*	0.2	0.4	0.3	ATPase, Na^+/K^+ transp., alpha3
Atp1b2 (J04629)*[6]	3914	**6888**	**2628**	0.5	0.4	0.3	ATPase, Na^+/K^+ transp., beta2
Atp1b3 (D84450)*[7]	**16726**	**8315**	**14474**	0.7	0.9	0.8	ATPase, Na^+/K^+ transp., beta3

Note. Expression see legend to Table 1. Abbr.: VGCh-Voltage gated channel; SF-subfamily; M-Member; V-voltage; Pias3 - Protein inhibitor of activated STAT, 3; Alg10 - Asparagine-linked glycosylation 10, regulatory component of non-inactivating K^+ channels, voltage-gated K^+ channel binding protein, involved in modulating the expression of Kv2 channels; transp.- transporting. Fold change (Wilcoxon paired test, n = 12): [1-7]$T(12) = 0.00, p = 0.002$.

3.3.2 Glutamate Pathway

The following transcripts involved in regulating the activity of the glutamate pathway were detected by the chip: three sequences associated with the ionotropic glutamate receptors (*Grina, Grin2B,* and *Grik5*), the glutamate receptor interacting protein 2 (*Grip2*), two glutamate transporters (*Slc1a1* and *Slc1a3*) and the glutamate-ammonia ligase (*Glul*). These transcripts have not been previously characterized in the inner ear. *Grina* encodes for a glutamate-binding subunit of an NMDA receptor-associated complex protein (NMDARA1), also called glutamate-binding protein (Kumar, Tilakaratne, Johnson, Allen, & Michaelis, 1991; Nielsen et al., 2011). It is characterized by a high basal expression level in all regions; in culture, its expression level remained unchanged in the OC and the SV and tended to increase in the MOD region. *Grin2b* encodes for the ionotropic NMDA2B receptor (NR2B), *Grik5* for the ionotropic kainate 5 receptor and *Grip2* for the glutamate interacting protein 2, which binds and affects AMPA receptors. These transcripts are moderately expressed and decrease in culture in all regions (Table 3).

Slc1a1 and *Slc1a3*, two glutamate transporters, are members of the excitatory amino acid transporter (EAAT) family of high-affinity sodium-dependent glutamate carriers encoded by the genes of the SLC1 family. *Slc1a1* encodes for the solute carrier family 1 (the neuronal/epithelial high affinity glutamate transporter, also known as EAAT3, EAAC1; Chen, Kujawa, & Sewell, 2010b). *Slc1a1* shows a low basal expression, and in culture, its expression tends to increase in the SV. *Slc1a3* encodes for the solute carrier family 1 member 3 (glial high affinity glutamate transporter, also known as EAAT1, GLAST). This transporter shows a high basal expression level; in culture, its expression decreases clearly in all regions. *Glul* encodes for the glutamate-ammonia ligase (also known as glutamine synthetase) which converts glutamate to glutamine; this transcript shows a high basal expression level, and in culture, its expression tends to increase in the OC.

Table 3. Expression of genes associated with Ca^{++} homeostasis in the modiolus, the organ of Corti and the stria vascularis

Gene	Expression			Fold change			Name/function
	MOD	OC	SV	MOD	OC	SV	
Cacna1c (M59786)	2396	*1292*	1845	0.9	1.1	0.8	VCa-ch, L-type,a-1C
P2rx2 (AF020756)	4585	**10920**	**9326**	0.7	1.3	1.1	Purinerg. rec. P2X, ch2
P2rx4 (U47031)	2416	1882	2194	0.9	0.9	1.1	Purinerg. rec. P2X, ch4
Grina (S61973)	**10282**	**21127**	**9227**	1.7$^{\#1}$	0.8	0.9	GR-NMDA-ass.prot.1
Grin2b (U11419)*[1]	2864	2414	2292	0.8	0.7	0.7	GR-NMDA2B
Grip2 (AF090113)*[2]	2794	2400	2052	0.7	0.7	0.7	GR-interact.prot.2
Grik5 (Z11581)*[3]	2567	3917	3010	0.6	0.6	0.8	GR-kainate5
Slc1a1 (D63772)	1658	*1060*	*827*	0.8	1.1	2.4$^{\#2}$	Eaat3
Slc1a3 (S59158)*[4]	**12059**	**23828**	**9227**	0.2$^{\#3}$	0.6	0.5	Eaat1
Glul (M91652)*[5]	**19899**	**17553**	**36544**	1.3	1.8$^{\#4}$	1.1	Glul
Atp2b1 (L04739)	1538	1486	*1120*	0.8	0.7	0.9	Pmca1b
Atp2b2 (J03754)*[6]	*785*	2190	*716*	0.1	0.5	0.3	Pmca2
Calm2 (M17069)*[7]	**37541**	**37770**	**33732**	0.5	0.7	0.7	Calmodulin 2
Calm3 (X14265)	**6740**	5556	5074	0.8	1.2	1.0	Calmodulin 3
Canx (L18889)*[8]	5700	**7439**	5467	2.2	2.2	2.6	Calnexin
Ppp3ca (D90035)*[9]	**6015**	**6021**	5400	0.9	0.8	0.7	Calcineurin

Note. Expression see legend to Table 1. Abbr.: VCa-ch – Voltage dependent calcium channel; Purinerg. rec.- purinergic receptor, ligand-ion channel; GR-NMDA-ass.prot.1 - glutamate receptor, N-methyl D-aspartate-associated protein 1; NMDA2B - glutamate receptor, ionotrop, N-methyl D-aspartate 2B; GR-interact.prot.2- GR-interacting protein 2; Eaat3 -neuronal/epithelial high affinity glutamate transporter, system Xag, member 1; Eaat1 - glial high affinity glutamate transporter, member 3; Glul - glutamate-ammonia ligase. Pmca – plasma membrane calcium ATPase. Fold change (Wilcoxon paired test, n = 12): ^{*1,2}T(12) = 0.00, p = 0.002; ^{*3}T(12) = 1.00, p = 0.003; $^{*4-9}$T(12) = 0.00, p = 0.002; $^{\#1}$p<0.005 vs OC and 0.001 vs SV; $^{\#2}$p<0.022 vs MOD; $^{\#3}$p<0.001vs OC; $^{\#4}$p<0.02 vs SV (n = 4).

3.3.3 Calcium ATPases and Calcium Binding Proteins

Ca^{++}-ATPases contribute largely to re-establishing Ca^{++}- ion homeostasis after unregulated Ca^{++}- influx, which is an energy demanding process (Buttgereit & Brand, 1995). The chip comprises two transcripts for plasma membrane calcium ATPase (PMCA; *Atp2b1, Atp2b2;* Table 3) and two for smooth endoplasmic reticulum calcium ATPase (SERCA; *Atp2a2, Atp2a3*). *Atp2b1* encodes for PMCA1, an enzyme with a housekeeping function. *Atp2b2* encodes for PMCA2, an enzyme with special functions in maintaining Ca^{2+} homeostasis in hair cells (Brini & Carafoli, 2009). Unlike PMCAs, SERCAs accumulate Ca^{++} into vesicles of the endoplasmic reticulum at the expense of ATP hydrolysis. The SERCA transcripts *Atp2a2* (J04739) and *Atp2a3* (M30581) show moderate expression levels and decrease significantly in all regions (0.5 - 0.6 fold, data not shown).

The present array data showed high to moderate expression levels for the Ca^{++}-binding proteins calmodulin (CaM) *Calm2 and Calm3,* for calnexin (*Canx*) and for calcineurin (*Ppp3ca*; Table 3). In culture, *Calm2* and *Ppp3ca* expression decreased in all regions, whereas *Canx* increased. *Calm3* remained unchanged. These features are in line with the finding that calcium-binding proteins constitute a high portion of the total cellular protein in all mammalian cells and are involved in protecting from calcium overload.

3.4 Co-Expression Analysis

Hypoxia inducible factor (HIF) is a key transcription factor regulating adaptation to hypoxia and tissue injury and it plays an important part in cell survival (Semenza, 2001). To characterize possible associations between the expression changes of *Hif-1a* and the transcripts analyzed in this study, we correlated the expression changes across the three regions. We observed that *Hif-1a* expression has correlations to nine transcripts associated with the metabolism of glucose (*Slc2a1, Slc2a3, Gapdh, Hk1* and *Eno2*), Na$^+$/K$^+$ homeostasis (*Atp1a1*) and the glutamate pathway (*Slc1a1, Slc1a3, Grina;* Figure 1A). A more detailed analysis of the data sets shows that the significance between two genes includes a different regional response (Figure 1B-E). The glucose metabolism associated genes belong to the classical target genes of HIF-1alpha (Greijer & van der Wall, 2004; Marin-Hernandez, Gallardo-Perez, Ralph, Rodriguez-Enriquez, & Moreno-Sanchez, 2009; Yu et al., 2012). Remarkably, the higher *Hif-1a* expression in the MOD region is associated with a relative decrease of expression levels of *Slc1a1/a3* and a relative increase of the *Grina* transcript.

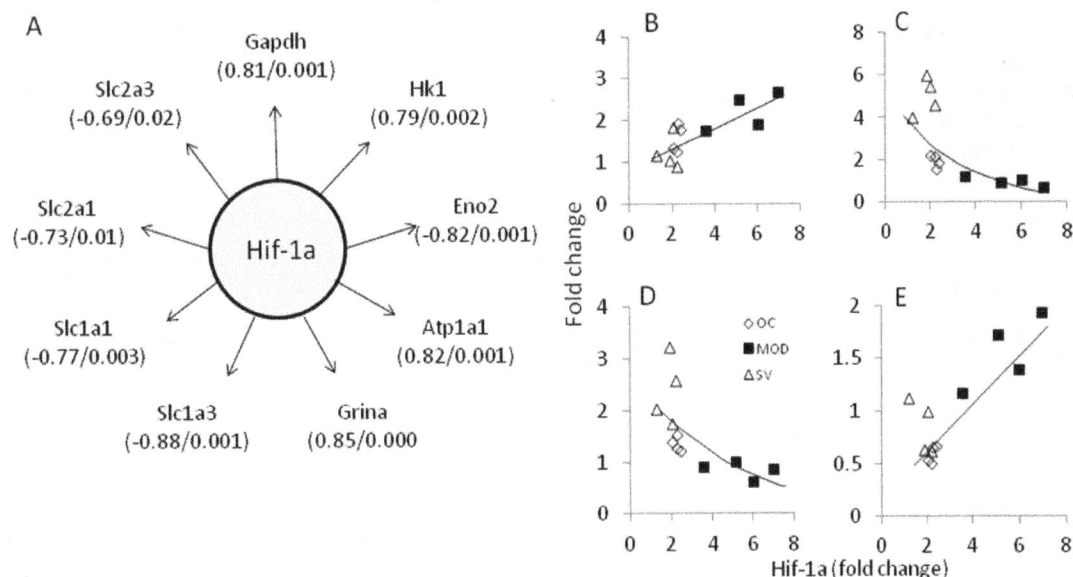

Figure 1. Relationship between *Hif-1a* expression and transcripts belonging to the glucose metabolism, the Na$^+$, K$^+$ homeostasis and the glutamate pathway

Note. (A) Diagram illustrating the cluster of transcripts associated with *Hif-1a* expression. Numbers indicate correlation coefficients *r* and significance levels *p*, n = 12. Marginal significance was observed for *Slc2a1* and *Slc2a3*. (B-E) Examples of scatter plots illustrating the correlations between *Hif-1a* and selected transcripts of Figure 1A. B - *Hk1*; C – *Eno2;* D - *Slc1a1;* E - *Grina*. The best fit curve to the *Eno2* and *Slc1a1* data is an exponential function.

To identify co-expression changes across the three regions in relation to cell death, we correlated the changes of these transcripts with molecules known to be mediators in apoptotic and necrotic cell death (Gross, Olze, & Mazurek, 2014). The present study shows that the *Slc2a1, Slc1a3, Grina, Atp1a1, Gapdh* and *Hk1* transcripts correlate closely with the cell death subunits (Figure 2A). These correlations are based on different responses in the MOD region compared to OC and SV (Figure 2B-E).

4. Discussion

4.1 Up-Regulated Transcripts

The up- and down-regulation of transcript levels appear plausible and efficient in terms of energy expenditure of the underlying processes. In the present study, we observe an up-regulation of transcripts involved in energy production and protective mechanisms. It is known that an increased mRNA synthesis which may contribute to an increased transcript levels is very energy-demanding (Simpson, Carruthers, & Vannucci, 2007). The parallel increase of the glucose transporter transcripts *Slc2a1* and *Slc2a3*, in particular in the OC and SV regions, and the up-regulation of *Gapdh* and *Hk1* may primarily contribute to increasing energy production (Edamatsu, Kondo, & Ando, 2011; Marin-Hernandez, Gallardo-Perez, Ralph, Rodriguez-Enriquez, & Moreno-Sanchez, 2009). The differential expression of *Hk1* and *Eno2* may have additional functions. The mitochondrial-bound isoform HK1 may interact with the membrane permeability transition (MPT) pore through the voltage-dependent anion channel (VDAC) which inhibits the cytochrome c release induced by the pro-apoptotic proteins Bax and Bid (Azoulay-Zohar, Israelson, Abu-Hamad, & Shoshan-Barmatz, 2004; Marin-Hernandez, Gallardo-Perez, Ralph,

Rodriguez-Enriquez, & Moreno-Sanchez, 2009). *Eno2* encodes for the neuron-specific enolase (NSE) which is the gamma-gamma enolase isoenzyme. The unchanged expression level in the MOD and the increase of *Eno2* in the OC and the SV may have a role to play in the cells' adaptation to stress (Yan et al., 2011). The expression increase of *Atp1a1* in all regions underlies the important functional role of Na,K-ATPase for cell survival, as studied using specific inhibitors (Johar, Priya, & Wong-Riley, 2012; Fu, Ding, Jiang, & Salvi, 2012). The calcium-binding protein *Canx* is important as it assumes the role of a chaperone in order to transport newly synthesized proteins from the endoplasmatic reticulum to the outer cellular membrane (Zuppini et al., 2002). Previous work showed that *Grina* is a member of the transmembrane BAX inhibitor motif (TMBIM3) known as an anti-apoptotic protein that controls apoptosis through the modulation of ER calcium homeostasis (Rojas-Rivera et al., 2012). *Grina's* high basal expression level and its increased expression in the MOD may well be involved in cell protection (Goswami et al., 2012). What is of interest here is the specific increase of *Slc1a1* (glutamate transporter) in the SV and of *Glul* (glutamate-ammonia ligase) in the OC. There is *in vitro* evidence of a polarized brain-to-blood transport of glutamate by endothelial cells co-cultured with astrocytes (Helms, Madelung, Waagepetersen, Nielsen, & Brodin, 2012). GLUL converts glutamate to glutamine and may contribute to the elimination of toxic glutamate (Takumi et al., 1997).

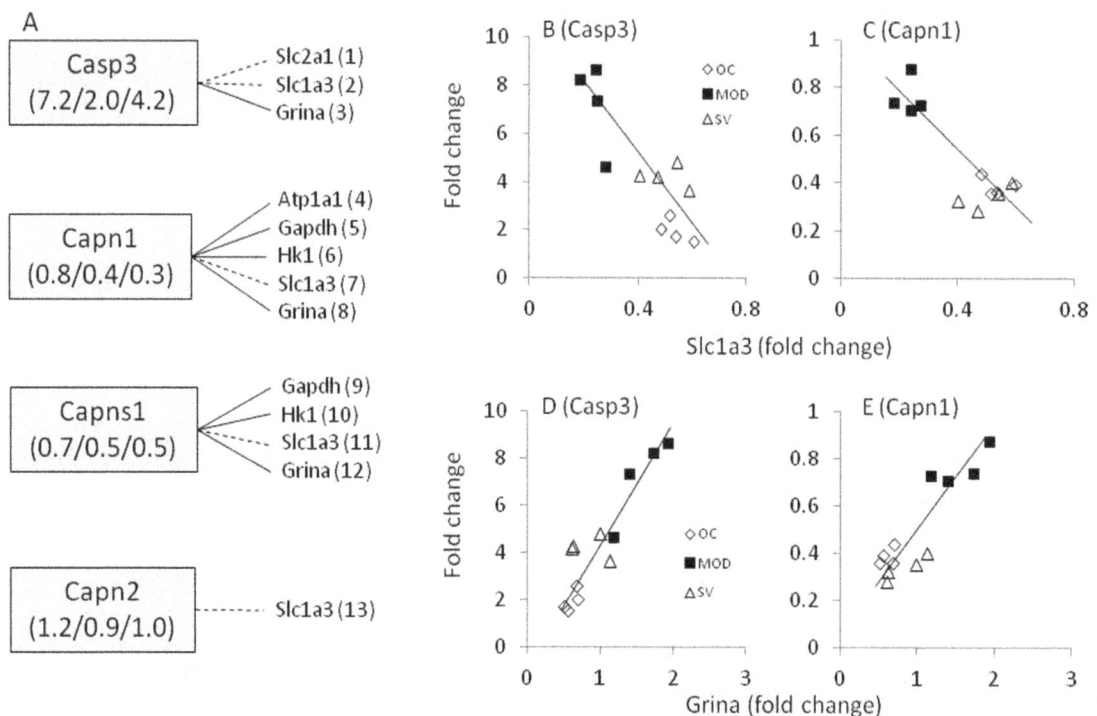

Figure 2. Members of the cell death cluster

Note. (A) Diagram illustrating the cluster of transcripts involved in apoptosis and necrosis. Lines 1-13 indicate significant correlations with correlation coefficients in the range $r = 0.78 – 0.92$ and significance levels in the range $p < 0.000 – 0.003$. Broken lines indicate negative correlations. Numbers within the rectangle indicate the fold change in MOD/OC/SV (Gross, Olze, & Mazurek, 2014). (B-E) Examples of scatter plots illustrating the correlations between transcripts of the glutamate system and cell death transcripts. Correlation coefficients and significance levels (r/p): B,- 0.87/0.000; C, -0.85/0.000; D, 0.92/0.000; E, 0.90/0.000.

4.2 Down-Regulated Transcripts

An important way of adapting to the energetic deficit following tissue injury and hypoxia is to decrease major ATP consuming functions. Several subsets of transcripts are characterized by a clear expression decrease, among them the K⁺ and Ca⁺⁺ - channels, the Atpase subunits *Atp1a3*, *Atp1b2*, the two SERCA transcripts, three glutamate receptors and the glutamate transporter subunit *Slc1a3*. Certain hypoxia-tolerant lower vertebrate species resort to decreasing ion channels as a mechanism to lower their energetic demands (Boutilier & St-Pierre, 2000). Because macromolecule turnover and ion-motive ATPases are major ATP consumers it comes as no surprise in the present study that the mRNA levels of ion channels are found to be down-regulated. Suppression of ion channel densities probably associated with lower cell membrane permeability decreases the energetic costs of maintaining electrochemical gradients (so-called 'channel arrest'; Hochachka, Buck, Doll, & Land, 1996). Activities that are essential to the maintenance of life should be able to function at lower energy charge values (Atkinson, according to Buttgereit & Brand, 1995; Wieser & Krumschnabel, 2001). Whereas the changes of most of the transcripts may be advantageous for cell survival, the down-regulation of the glial high affinity glutamate transporter *Slc1a3* appears to exert a rather damaging effect, in particular in the MOD region, because this strong expression decrease of *Slc1a3* may contribute to elevated extracellular glutamate concentrations and cell damage in the MOD region (Bianchi, Bardelli, Chiu, & Bussolati, 2014; Gegelashvili & Schousboe, 1997).

4.3 Transcripts Without Significant Changes

Several K^+ channels, the purinergic ion channels, the voltage-dependent calcium channel *Cacn1c* and calmodulin *Calm3* belong to the subset of transcripts without significant changes. The unchanged expression of these channels may allow speculations to be made about the importance of these particular transcripts to cell survival (Rolfe & Brown, 1997). The *Cacn1c* subunit is part of the Cav1.2 channel that plays an important role in synaptic-activity-dependent gene expression and may be important for regenerative processes. With the exception of the *Cacnb3* data, no data are available for these channels in the inner ear (Kuhn et al., 2009). The high expression levels of the calcium-binding proteins *Calm3* and of the purinergic ion channel *P2rx2* in the OC and the SV made us assume that the corresponding proteins have a role to play in cell survival or cell repair. Calcineurin plays an important role in refilling Ca^{2+} stores of the endoplasmatic reticulum and maintaining optimum conditions for protein processing and folding (Bollo et al., 2010).

4.4 Hif-1a Associated Genes

Observations to the extent that the energy balance, glucose uptake by the cells, ion homeostasis and glutamatergic neurotransmission are interlinked, support the assumption of a membership of these genes in a functional gene network, with *HIF-1a* being an important regulator of efficient adaptation across the three regions (Greene & Greenamyre, 1996; Rodriguez-Rodriguez, Almeida, & Bolanos, 2013). For example, glutamate transport depends on the ATPase to remove Na^+ from the cytoplasm, and the activity of Na,K-ATPase is dependent on ATP production (Casey, Pakay, Guppy, & Arthur, 2002). Our data suggest that the differential expression changes of *Atp1a1*, the two glutamate transporters and of *Grina* optimize the adaption process of the glutamate system during repair and regeneration.

The co-expression changes between *Hif-1a* and several genes on the transcript levels could be explained by unique features observed for HIF-1 expression in recent years. Several authors showed a crucial role for Hif-1a mRNA turnover to exist in HIF-1 signaling, and a regulatory role of mRNA turnover as a modulator of HIF-1a function, independent of the oxygen tension (Fahling et al., 2012; Schodel, Mole, & Ratcliffe, 2013; Eltzschig & Carmeliet, 2011). Little is known about the factors that regulate expression changes of *Grina* and of the glutamate transporters. Endothelins, a family of peptides up-regulated in the injured brain, negatively regulate glial glutamate transporter expression (Rozyczka, Figiel, & Engele, 2004). Studies in astrocytes indicate that the suppression of *Slc1a3* is Ca^{++}-dependent (Liu, Yang, & Tzeng, 2008). By inhibiting the down-regulation of *Slc1a3* and buffering the glutamate homeostasis, taurine protects retinal cells *in vitro* under hypoxic conditions (Chen et al., 2010a).

Other than for these genes, we observed no significant correlations between *Hif-1a* and Ca^{++}-ATPases. The expression and activity of proteins involved in Ca^{2+} regulation are subject to the autoregulatory principle, which means that they are regulated by the Ca^{2+} signal itself (Brini & Carafoli, 2009). Other investigations have shown that the level of expression of proteins associated with Ca^{++} homeostasis is regulated by transcription (recently reviewed by Ritchie, Zhou, & Soboloff, 2011).

4.5 Cell Death Associated Genes

The present work suggests that *Slc2a1, Slc1a3, Grina, Atp1a1, Gapdh* and *Hk1* may be involved in the differential cell death rate. The type of cell death immediately after the preparation of the cultures can be categorized as accidental necrotic cell death (ANCD), whereas the cell death measured 24h after the damaging event corresponds most probably to secondary necrotic cell death (SNCD) or late apoptotic cell death (Krysko et al., 2011). The observed genetic responses may be important in terms of secondary cell-death prevention. Preventing the glutamate transporter *Slc1a3* from strongly decreasing in the MOD region or the increase of *Grina* can be assumed to be of special importance to the survival of the SGNs following injury and hypoxia. Up to now, it is unclear which factors show the highest regulatory strength to induce cell death or to maintain survival.

4.6 Conclusions

This study documents previously undescribed genetic features of the different compartments of the inner ear. Both gene clusters, the *Hif-1a* (Figure 1) and the cell death cluster (Figure 2), are based on the co-expression changes of genes across OC, MOD and SV and are the result of a differential, tissue-specific response to injury and hypoxia. The different response occurs on the basis of endogenous gene regulatory mechanisms developed in the course of evolution. The differential response appears important to maintain the cochlea's function following damage from environmental factors. For example, the decrease of ion channels in all regions or the differential response of *Hif-1a* and other transcription factors may be crucial for inner ear function.

The basal expression and the injury/hypoxia-induced patterns may contribute to the region-related difference in the cell death rates immediately after injury and after 24 h in culture. The newly identified genes *Grina* and *Canx* may play specific and yet unknown roles in regulating cell death induced by injury and hypoxia in the inner ear.

However, caution is indicated in the interpretation of these data for several reasons. First, the noise of microarray might influence some of the experimental results. However, the significant correlations between microarray data and quantitative RT-PCR values observed in several studies, justify our approach. Second, the data are not complete in the sense that only some members of pathways have been experimentally determined. Third, we are aware that the response of immature tissue is different from that of mature tissue. Nevertheless, responses are very similar to that in mature tissue (Kennedy, 2012). Fourth, the analysis quantified mRNA levels, but the data do not indicate whether subsequent proteins are generated and where they are located. Beyond the up and down-regulation of transcript expression, the question arises of whether such changes are functional. However, many similarities in the response of the corresponding proteins, even in *in vivo* studies, lead to the conclusion that these observations are far more important than only for the present model.

Acknowledgements

We would like to thank the University Hospital Charité for support. It gives us great pleasure to thank Johannes Wendt for his generous help in critically reading and correcting this article.

References

Azoulay-Zohar, H., Israelson, A., Abu-Hamad, S., & Shoshan-Barmatz, V. (2004). In self-defence: hexokinase promotes voltage-dependent anion channel closure and prevents mitochondria-mediated apoptotic cell death. *Biochem. J., 377*, 347-355. http://dx.doi.org/10.1042/BJ20031465

Bianchi, M. G., Bardelli, D., Chiu, M., & Bussolati, O. (2014). Changes in the expression of the glutamate transporter EAAT3/EAAC1 in health and disease. *Cell Mol. Life Sci., 71*, 2001-2015. http://dx.doi.org/10.1007/s00018-013-1484-0

Bollo, M., Paredes, R. M., Holstein, D., Zheleznova, N., Camacho, P., & Lechleiter, J. D. (2010). Calcineurin interacts with PERK and dephosphorylates calnexin to relieve ER stress in mammals and frogs. *PLoS. ONE., 5*, e11925. http://dx.doi.org/10.1371/journal.pone.0011925

Boutilier, R. G., & St-Pierre, J. (2000). Surviving hypoxia without really dying. *Comp Biochem. Physiol A Mol. Integr. Physiol., 126*, 481-490. http://dx.doi.org/10.1016/S1095-6433(00)00234-8

Brini, M., & Carafoli, E. (2009). Calcium pumps in health and disease. *Physiol Rev., 89*, 1341-1378. http://dx.doi.org/10.1152/physrev.00032.2008

Buttgereit, F., & Brand, M. D. (1995). A hierarchy of ATP-consuming processes in mammalian cells. *Biochem. J., 312*(Pt 1), 163-167.

Casey, T. M., Pakay, J. L., Guppy, M., & Arthur, P. G. (2002). Hypoxia causes downregulation of protein and RNA synthesis in noncontracting Mammalian cardiomyocytes. *Circ. Res., 90*, 777-783. http://dx.doi.org/10.1161/01.RES.0000015592.95986.03

Chen, F., Mi, M., Zhang, Q., Wei, N., Chen, K., Xu, H., ... Chang, H. (2010a). Taurine buffers glutamate homeostasis in retinal cells in vitro under hypoxic conditions. *Ophthalmic Res., 44*, 105-112. http://dx.doi.org/10.1159/000312818

Chen, Z., Kujawa, S. G., & Sewell, W. F. (2010b). Functional roles of high-affinity glutamate transporters in cochlear afferent synaptic transmission in the mouse. *J. Neurophysiol., 103*, 2581-2586. http://dx.doi.org/10.1152/jn.00018.2010

Edamatsu, M., Kondo, Y., & Ando, M. (2011). Multiple expression of glucose transporters in the lateral wall of the cochlear duct studied by quantitative real-time PCR assay. *Neurosci. Lett., 490*, 72-77. ttp://dx.doi.org/10.1016/j.neulet.2010.12.029

Eltzschig, H. K., & Carmeliet, P. (2011). Hypoxia and inflammation. *N. Engl. J Med., 364*, 656-665. http://dx.doi.org/10.1056/NEJMra0910283

Fahling, M., Persson, A. B., Klinger, B., Benko, E., Steege, A., Kasim, M., ... Mrowka, R. (2012). Multilevel regulation of HIF-1 signaling by TTP. *Mol. Biol. Cell., 23*, 4129-4141. http://dx.doi.org/10.1091/mbc.E11-11-0949

Frezza, C., Zheng, L., Tennant, D. A., Papkovsky, D. B., Hedley, B. A., Kalna, G., ... Gottlieb, E. (2011). Metabolic profiling of hypoxic cells revealed a catabolic signature required for cell survival. *PLoS. ONE., 6*, e24411. http://dx.doi.org/10.1371/journal.pone.0024411

Fu, Y., Ding, D., Jiang, H., & Salvi, R. (2012). Ouabain-induced cochlear degeneration in rat. *Neurotox. Res., 22*, 158-169. http://dx.doi.org/10.1007/s12640-012-9320-0

Gabashvili, I. S., Sokolowski, B. H., Morton, C. C., & Giersch, A. B. (2007). Ion channel gene expression in the inner ear. *J Assoc. Res. Otolaryngol., 8*, 305-328. http://dx.doi.org/10.1007/s10162-007-0082-y

Gegelashvili, G., & Schousboe, A. (1997). High affinity glutamate transporters: regulation of expression and activity. *Mol Pharmacol., 52*, 6-15.

Goswami, D. B., Jernigan, C. S., Chandran, A., Iyo, A. H., May, W. L., Austin, M. C., ... Karolewicz, B. (2012). Gene expression analysis of novel genes in the prefrontal cortex of major depressive disorder subjects. Prog. Neuropsychopharmacol. *Biol. Psychiatry., 43C*, 126-133. http://dx.doi.org/10.1016/j.pnpbp.2012.12.010

Greene, J. G., & Greenamyre, J. T. (1996). Bioenergetics and glutamate excitotoxicity. *Prog. Neurobiol., 48*, 613-634. http://dx.doi.org/10.1016/0301-0082(96)00006-8

Greijer, A. E., & van der Wall, W. E. (2004). The role of hypoxia inducible factor 1 (HIF-1) in hypoxia induced apoptosis. *J. Clin. Pathol., 57*, 1009-1014. http://dx.doi.org/10.1136/jcp.2003.015032

Gross, J., Machulik, A., Amarjargal, N., Moller, R., Ungethum, U., Kuban, R. J., ... Mazurek, B. (2007). Expression of apoptosis related genes in the organ of Corti, modiolus and stria vascularis of newborn rats. *Brain Res., 1162*, 56-68. http://dx.doi.org/10.1016/j.brainres.2007.05.061

Gross, J., Machulik, A., Moller, R., Fuchs, J., Amarjargal, N., Ungethuem, U., ... Mazurek, B. (2008). mRNA expression of members of the IGF system in the organ of Corti,the modiolus and the stria vascularis of newborn rats. *Growth Factors., 26*, 180-191. http://dx.doi.org/10.1080/08977190802194317

Gross, J., Olze, H., & Mazurek, B. (2014). Differential Expression of Transcription Factors and Inflammation-, ROS-, and Cell Death-Related Genes in Organotypic Cultures in the Modiolus, the Organ of Corti and the Stria Vascularis of Newborn Rats. *Cell Mol. Neurobiol., 34*, 523-538. http://dx.doi.org/10.1007/s10571-014-0036-y

Helms, H. C., Madelung, R., Waagepetersen, H. S., Nielsen, C. U., & Brodin, B. (2012). In vitro evidence for the brain glutamate efflux hypothesis: brain endothelial cells cocultured with astrocytes display a polarized brain-to-blood transport of glutamate. *Glia., 60*, 882-893. http://dx.doi.org/10.1002/glia.22321

Hochachka, P. W., Buck, L. T., Doll, C. J., & Land, S. C. (1996). Unifying theory of hypoxia tolerance: molecular/metabolic defense and rescue mechanisms for surviving oxygen lack. *Proc. Natl. Acad. Sci. U. S. A., 93*, 9493-9498. http://dx.doi.org/10.1073/pnas.93.18.9493

Johar, K., Priya, A., & Wong-Riley, M. T. (2012). Regulation of Na+/K+-ATPase by nuclear respiratory factor 1: Implication in the tight coupling of neuronal activity, energy generation, and energy consumption. *J Biol. Chem., 287*, 40381-40390. http://dx.doi.org/10.1074/jbc.M112.414573

Kennedy, H. J. (2012). New developments in understanding the mechanisms and function of spontaneous electrical activity in the developing mammalian auditory system. *J Assoc. Res. Otolaryngol., 13*, 437-445. http://dx.doi.org/10.1007/s10162-012-0325-4

Kuhn, S., Knirsch, M., Ruttiger, L., Kasperek, S., Winter, H., Freichel, M., ... Engel, J. (2009). Ba2+ currents in inner and outer hair cells of mice lacking the voltage-dependent Ca2+ channel subunits beta3 or beta4. *Channels (Austin.)., 3*, 366-376. http://dx.doi.org/10.4161/chan.3.5.9774

Kumar, K. N., Tilakaratne, N., Johnson, P. S., Allen, A. E., & Michaelis, E. K. (1991). Cloning of cDNA for the glutamate-binding subunit of an NMDA receptor complex. *Nature., 354*, 70-73. http://dx.doi.org/10.1038/354070a0

Krysko, D. V., Agostinis, P., Krysko, O., Garg, A. D., Bachert, C., Lambrecht, B. N., & Vandenabeele, P. (2011). Emerging role of damage-associated molecular patterns derived from mitochondria in inflammation. *Trends Immunol., 32*, 157-164. http://dx.doi.org/10.1016/j.it.2011.01.005

Lang, F., Vallon, V., Knipper, M., & Wangemann, P. (2007). Functional significance of channels and transporters expressed in the inner ear and kidney. *Am J Physiol Cell Physiol., 293*, C1187-C1208. http://dx.doi.org/10.1152/ajpcell.00024.2007

Liu, Y. P., Yang, C. S., & Tzeng, S. F. (2008). Inhibitory regulation of glutamate aspartate transporter (GLAST) expression in astrocytes by cadmium-induced calcium influx. *J Neurochem., 105*, 137-150. http://dx.doi.org/10.1111/j.1471-4159.2007.05118.x

Marin-Hernandez, A., Gallardo-Perez, J. C., Ralph, S. J., Rodriguez-Enriquez, S., & Moreno-Sanchez, R. (2009). HIF-1alpha modulates energy metabolism in cancer cells by inducing over-expression of specific glycolytic isoforms. *Mini. Rev. Med. Chem., 9*, 1084-1101. http://dx.doi.org/10.2174/138955709788922610

Martinez-Sanchez, M., Striggow, F., Schroder, U. H., Kahlert, S., Reymann, K. G., & Reiser, G. (2004). Na(+) and Ca(2+) homeostasis pathways, cell death and protection after oxygen-glucose-deprivation in organotypic hippocampal slice cultures. *Neuroscience., 128*, 729-740. http://dx.doi.org/10.1016/j.neuroscience.2004.06.074

Mazurek, B., Amarjargal, N., Haupt, H., Fuchs, J., Olze, H., Machulik, A., & Gross, J. (2011). Expression of genes implicated in oxidative stress in the cochlea of newborn rats. *Hear. Res., 277*, 54-60. http://dx.doi.org/10.1016/j.heares.2011.03.011

Mazurek, B., Machulik, A., Amarjargal, N., Kuban, R. J., Ungethuem, U., Fuchs, J., ... Gross, J. (2006). Gene expression of organ of Corti (OC), modiolus (MOD) and stria vascularis (SV) of newborn rats. Gene Expression Omnibus (GEO) website (http://www.ncbi.nlm.nih.gov/geo/). ID GSE5446.

Michiels, C. (2004). Physiological and pathological responses to hypoxia. *Am J Pathol., 164*, 1875-1882. http://dx.doi.org/10.1016/S0002-9440(10)63747-9

Nielsen, J. A., Chambers, M. A., Romm, E., Lee, L. Y., Berndt, J. A., & Hudson, L. D. (2011). Mouse transmembrane BAX inhibitor motif 3 (Tmbim3) encodes a 38 kDa transmembrane protein expressed in the central nervous system. *Mol. Cell Biochem., 357*, 73-81. http://dx.doi.org/10.1007/s11010-011-0877-3

Ritchie, M. F., Zhou, Y., & Soboloff, J. (2011). Transcriptional mechanisms regulating Ca(2+) homeostasis. *Cell Calcium., 49*, 314-321. http://dx.doi.org/10.1016/j.ceca.2010.10.001

Rodriguez-Rodriguez, P., Almeida, A., & Bolanos, J. P. (2013). Brain energy metabolism in glutamate-receptor activation and excitotoxicity: role for APC/C-Cdh1 in the balance glycolysis/pentose phosphate pathway. Neurochem. *Int., 62*, 750-756. http://dx.doi.org/10.1016/j.neuint.2013.02.005

Rojas-Rivera, D., Armisen, R., Colombo, A., Martinez, G., Eguiguren, A. L., Diaz, A., ... Hetz, C. (2012). TMBIM3/GRINA is a novel unfolded protein response (UPR) target gene that controls apoptosis through the modulation of ER calcium homeostasis. *Cell Death. Differ., 19*, 1013-1026. http://dx.doi.org/10.1038/cdd.2011.189

Rolfe, D. F., & Brown, G. C. (1997). Cellular energy utilization and molecular origin of standard metabolic rate in mammals. *Physiol Rev., 77*, 731-758.

Rozyczka, J., Figiel, M., & Engele, J. (2004). Endothelins negatively regulate glial glutamate transporter expression. *Brain Pathol., 14*, 406-414. http://dx.doi.org/10.1111/j.1750-3639.2004.tb00084.x

Schodel, J., Mole, D. R., & Ratcliffe, P. J. (2013). Pan-genomic binding of hypoxia-inducible transcription factors. *Biol. Chem., 394*, 507-517. http://dx.doi.org/10.1515/hsz-2012-0351

Semenza, G. L. (2001). Hypoxia-inducible factor 1: control of oxygen homeostasis in health and disease. *Pediatr. Res., 49*, 614-617. http://dx.doi.org/10.1203/00006450-200105000-00002

Simpson, I. A., Carruthers, A., & Vannucci, S. J. (2007). Supply and demand in cerebral energy metabolism: the role of nutrient transporters. *J Cereb. Blood Flow Metab., 27*, 1766-1791. http://dx.doi.org/10.1038/sj.jcbfm.9600521

Sobkowicz, H. M., Loftus, J. M., & Slapnick, S. M. (1993). Tissue culture of the organ of Corti. *Acta Otolaryngol. Suppl (Stockholm)., 502*, 3-36.

Takumi, Y., Matsubara, A., Danbolt, N. C., Laake, J. H., Storm-Mathisen, J., Usami, S., ... Ottersen, O. P. (1997). Discrete cellular and subcellular localization of glutamine synthetase and the glutamate transporter GLAST in the rat vestibular end organ. *Neuroscience., 79*, 1137-1144. http://dx.doi.org/10.1016/S0306-4522(97)00025-0

Wangemann, P. (2002). K+ cycling and the endocochlear potential. *Hear. Res., 165*, 1-9. http://dx.doi.org/10.1016/S0378-5955(02)00279-4

Wieser, W., & Krumschnabel, G. (2001). Hierarchies of ATP-consuming processes: direct compared with indirect measurements, and comparative aspects. *Biochem. J., 355*, 389-395. http://dx.doi.org/10.1042/0264-6021:3550389

Yan, T., Skaftnesmo, K. O., Leiss, L., Sleire, L., Wang, J., Li, X., & Enger, P. O. (2011). Neuronal markers are expressed in human gliomas and NSE knockdown sensitizes glioblastoma cells to radiotherapy and temozolomide. *BMC. Cancer., 11*, 524. http://dx.doi.org/10.1186/1471-2407-11-524

Yu, J., Li, J., Zhang, S., Xu, X., Zheng, M., Jiang, G., & Li, F. (2012). IGF-1 induces hypoxia-inducible factor 1alpha-mediated GLUT3 expression through PI3K/Akt/mTOR dependent pathways in PC12 cells. *Brain Res., 1430*, 18-24. http://dx.doi.org/10.1016/j.brainres.2011.10.046

Zuppini, A., Groenendyk, J., Cormack, L. A., Shore, G., Opas, M., Bleackley, R. C., & Michalak, M. (2002). Calnexin deficiency and endoplasmic reticulum stress-induced apoptosis. *Biochemistry., 41*, 2850-2858. http://dx.doi.org/10.1021/bi015967+

11

Characterization of Structure, Divergence and Regulation Patterns of Plant Promoters

Yingchun Liu[1][*], Jiaming Yin[1][*], Meili Xiao[1], Annaliese S. Mason[3], Caihua Gao[1], Honglei Liu[1], Jiana Li[1] & Donghui Fu[2]

[1] Engineering Research Center of South Upland Agriculture of Ministry of Education, College of Agronomy and Biotechnology, Southwest University, Chongqing, China

[2] Key Laboratory of Crop Physiology, Ecology and Genetic Breeding, Ministry of Education, Jiangxi Agricultural University, Nanchang, China

[3] School of Agriculture and Food Sciences and ARC Centre for Integrative Legume Research, The University of Queensland, Brisbane, Australia

[*] These authors contributed equally

Correspondence: Donghui Fu, Key Laboratory of Crop Physiology, Ecology and Genetic Breeding, Ministry of Education, Jiangxi Agricultural University, Nanchang 330045, China. E-mail: fudhui@163.com

Abstract

Plant promoters have attracted increasing attention because of their irreplaceable role in modulating the spatio-temporal expression of genes interacting with transcription factors (TFs). Despite their importance, the basic characteristics of plant promoters are not well understood. In order to determine sequence diversity within promoter regions, evolutionary divergence of promoters between plant species, and the general structural characteristics of promoter sequences, we downloaded and analyzed 3922 plant promoter sequences from a wide range of plant species. The average plant promoter GC content was lower in dicotyledons than in monocotyledons, which might suggest different evolutionary pressures for promoter sequences between the two clades. Approximately 3.3% of plant promoters harbored minisatellite sequences, and 15.4% of plant promoters harbored microsatellite sequences (also called simple sequence repeats). Very few transposable elements were detected within the plant promoters. The most common transcription factor binding site (TFBS) motif was AGAGAGAGA, followed by TTAGGGTTT and then GCCGCC. Transcribed gene regions with promoters containing the corresponding TFBSs were predicted to be most commonly involved in metabolic processes, biological regulation, and stimulus response in plants. These results reveal some basic structural characteristics of plant promoters and clarify the evolutionary forces shaping plant promoters. This data might facilitate cloning of plant promoter sequences and aid in our understanding of gene spatio-temporal expression patterns in plants.

Keywords: transcription factor binding sites, minisatellite, microsatellite, transposable elements, functional annotation, GC content, evolutionary forces

1. Introduction

Promoters are sections of DNA sequence that lie upstream of the transcribed sequences and regulate their expression (Hernandez-Garcia et al., 2010). Promoters contain binding sites for transcription factors (TFs), and interact with these TFs to modulate gene expression. RNA polymerase initiates transcription at promoter sequences, and hence binding of RNA polymerase by TFs within promoter sequences regulates spatio-temporal expression of the downstream transcribed sequence (Camp et al., 2003; Halfon & Zhu, 2009; Freeman et al., 2011). Therefore, promoters are critical for priming or halting gene expression (Wolf et al., 2010; Mastroeni et al., 2011), especially in stress signaling and transcriptional activation during pathogen infection (Hwang et al., 2009; Pandey & Somssich, 2009). To date, numerous promoters have been identified in animals (Romania et al., 2011), plants (Wang et al., 2011), viruses (Smith et al., 2011b), and microorganisms (Cooper et al., 2011).

Promoters may be classified into two types according to the degree of matching between the regulatory protein and the transcription start site (TSS): Peak promoters and broad promoters. Peak promoters initiate the process of transcription in a narrow genomic region, while broad promoters switch on transcription in a wide genomic region (Nozaki et al., 2011). Cap-analysis gene expression data can be used to indentify those two types of promoters (Carninci et al., 2006). These peak promoters generally contain TATA-boxes (except in mammals) and regulate tissue-specific transcripts in eukaryotes (Hoskins et al., 2011). For most promoters, gene transcription starts from broad regions that are usually associated with CpG islands. These broad promoters have a wide distribution of TSSs, usually over a 100-bp region, and start sites that are preferentially comprised of pyrimidine/purine dinucleotides (Carninci et al., 2006).

Promoters can be divided into prokaryotic- and eukaryotic-type promoters, which differ mainly in promoter motifs. A typical promoter sequence is thought to comprise certain motifs positioned at specific sites upstream of TSS. Two hexameric motifs centered at or near the -10 and -35 positions relative to the TSS are observed in a prokaryotic promoter, whilst a TATA box, a CCAAT box, and a GC box are usually observed in eukaryotic promoters (Bansal & Kanhere, 2005). These three types of boxes play a major role in precise initiation of transcription (Molina & Grotewold, 2005). Nevertheless, not every eukaryotic gene promoter has all three motifs (Anish et al., 2009). In addition, some novel motifs in promoter sequences, e.g. AGTTAGG (Abdullah et al., 2010), G-quadruplex (Chowdhury et al., 2010), and TATGAAAAGAATATGAGAA motifs (Wu & Huang, 2004), have been identified. Other promoter motifs, such as GATA (Obara et al., 2005) and AAAAT (Van Oers et al., 2007), are not conserved but are essential for some promoter functionality. Overall, eukaryotic promoters display more complex structures and regulation patterns than prokaryotic promoters (Bansal & Kanhere, 2005).

Promoters undergo mutations such as nucleotide substitutions, small insertions and deletions in a similar fashion to transcribed sequences (Seliverstov et al., 2009). The evolution and conservation of promoters has been scrutinized through comparative genomics studies in mammals. Previous studies include comparisons between humans and chimpanzees (Deyneko et al., 2010), and between rats, mice, rhesus monkeys, and humans for promoters of hepatic lipase genes (Van Deursen et al., 2007). GC-rich monotone gradients have been observed in eukaryotes while AT-rich monotone gradients have been observed in bacteria, along with strand biases (Calistri et al., 2011).

Each gene can have several promoters that control its spatio-temporal expression. Although promoters are important in investigating patterns of gene expression and for transgenic work, promoters are cloned far less often than transcribed gene sequences. A total of 3922 plant promoters in the Plant Promoter Database (PlantProm DB; http://linux1.softberry.com/berry.phtml) have been collected to date. Knowledge of the basic structural and evolutionary characteristics of plant promoters remain unknown, making plant promoter sequences hard to identify. To facilitate better characterization of plant promoter sequences, the 3922 available plant promoter sequences were downloaded and analyzed. Basic promoter characteristics were dissected, presence of special motifs, minisatellite sequences, microsatellite sequences, and transposable elements (TEs). We present the results of this analysis, and propose mechanisms for promoter divergence and evolution.

2. Materials and Methods

2.1 Acquisition of Plant Promoter Sequences

All plant promoter sequences from monocotyledons and dicotyledons (the latter mainly from *Arabidopsis thaliana*) were downloaded from the PlantProm DB (Release 2009.02; http://linux1.softberry.com); an annotated, non-redundant collection of proximal promoter sequences (Shahmuradov et al., 2003). These promoters could potentially be recognized by RNA polymerase II and contained experimentally determined TSSs from diverse plant species (Solovyev et al., 2003). The PlantProm DB contains both the predicted TSSs and the experimentally verified promoter TSSs, identified using approaches such as full-length cDNA/5'ESTs mapping, cap-analysis gene expression, and serial analysis of gene expression.

2.2 Detection of Microsatellite Sequences

Microsatellite sequences (also called simple sequence repeats; SSRs) are tandem repeat sequences with repeated unit lengths of 1-10 bp, present in most organisms (Morgante et al., 2002). The software SSR Locator (Da Maia et al., 2008) was used to mine SSRs with mono-, di-, tri-, tetra-, penta-, hexa-, hepta-, octa-, nova-, and decanucleotide motifs which contained a minimum of 10, 5, 4, 3, 2, 2, 2, 2, 2, and 2 repeats, respectively; only SSR sequences with a total length ≥ 20 bp were assigned as true SSRs (Gao et al., 2011), and subject to analysis.

2.3 Detection of Minisatellite Sequences

Minisatellites, a type of tandem repeat sequence, consist of a short series of 11-100 bp repeat units. Tandem Repeats Finder 4.04 (http://tandem.bu.edu/trf/trf.download.html) developed by Gary Benson of the Bioinformatics Program at Boston University, was used to detect minisatellite sequences (Martin, 2006). Default parameters were used: Alignment parameters were match = 2, mismatch = 7, indel = 7, the minimum alignment score to report a repeat was 50 and the maximum period size was 100 bp.

2.4 Detection of Transposable Elements

There are two classes of transposable elements (TEs): DNA transposons and retrotransposons (Zhang et al., 2004). The Long Terminal Repeat (LTR)-Finder 1.05 (http://tlife.fudan.edu.cn/ltr_finder/) was used to detect full-length LTR retrotranspsons in genome sequences. The parameters of minimal LTR length, minimal distance between LTRs, and the output threshold score were set to 50, 100, and 3.0, respectively (Gao et al., 2012). The RepeatMasker 3.0SE-AB program (www.repeatmasker.org) was used to detect all types of transposons using the abblast (formerly known as WUBlast) search engine with *A. thaliana* set as the reference species. Since LTR-type retrotransposons detected by the LTR-Finder tool with default parameters exhibit intact retrotransposon sequence characteristics, LTR-Finder predictions were used instead of LTR retrotransposon predictions from RepeatMasker.

2.5 Prediction of Transcription Factor Binding Sites (TFBSs)

The online software NSITE-PL (http://linux1.softberry.com) with default parameters was used to predict transcription factor binding sites by recognition of regulatory motifs of plant promoters.

2.6 Functional Annotation by Blast2Go

The sequences of the transcribed gene regions with promoters containing TFBSs were in downloaded in a batch from NCBI (http://www.ncbi.nlm.nih.gov/sites/batchentrez). Blast2Go V2.6.0 (Conesa et al., 2005) (http://www.blast2go.org), a functional annotation prediction tool for unknown sequences, was used with default parameters to predict the putative functions of the transcribed gene regions with promoters containing TFBSs. Functional annotations of these genes were carried out for cellular component, biological process and molecular function.

2.7 Alignment of Plant Promoter Sequences

All plant promoters underwent all-by-all BlastN analysis using the basic local alignment search tool (BLAST) (http://www.ncbi.nlm.nih.gov/blast) (Cameron & Williams, 2007) with an E value of less than e^{-10}. The alignment results were imported into Cytoscape V2.7.0 (an open source platform for complex network analysis and visualization) (http://www.cytoscape.org) to classify different groups using the 'import network from table' function.

2.8 Phylogenic Dendrogram of Plant Promoter Sequences

The Molecular Evolutionary Genetics Analysis (MEGA; http://www.megasoftware.net) 4.0 software was used to draw the phylogenetic dendrogram of different plant promoter sequence groups using the maximum composite likelihood (MCL) model with the bootstrap value set as 1000 (Kumar et al., 2007).

3. Results

3.1 Plant Promoter Sequence Sets

A total of 3922 plant promoter sequences were downloaded from the PlantProm DB: 98 from monocotyledons and 3824 from dicotyledons. Monocotyledon sequences comprised 36 plant promoters from *Zea*, 32 from *Hordeum*, and 19 from *Triticum*. Dicotyledon sequences comprised 3537 plant promoters from *Arabidopsis*, 49 from *Nicotiana*, 46 from *Solanum*, 31 from *Glycine* and 31 from *Pisum*. Another 130 plant promoters were acquired from other genera, including *Phaseolus* (13), *Brassica* (9), and *Avena* (4).

3.2 Distribution of GC Content of Plant Promoters

GC content was calculated for each plant promoter sequence. The GC content of plant promoters ranged from 13.1% to 72.6%, with an average of 34.6%. The GC content of dicotyledon promoters ranged from 13.1% to 58.6% with an average of 34.1%, whilst the GC content of monocotyledon promoters ranged from 33.0% to 72.6% with an average of 50.5%. The centre of the GC content distribution for most dicotyledon promoters was from 30% to 40%, median 34.26%, whereas the centre of the GC content distribution of most dicotyledon promoters ranged from 50% to 60%, median 51% (Figure 1).

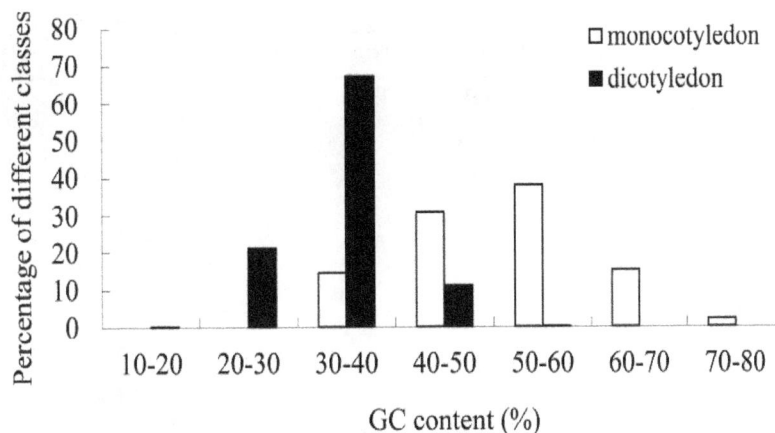

Figure 1. Proportion of plant promoter sequences with different ranges of GC content in different classes: monocotyledon and dicotyledon

3.3 Basic Characteristics of Plant Promoters

3.3.1 Detection of Microsatellites

Approximately 15% of the analyzed plant promoters (605 out of 3922) contained one or more microsatellites. Of these, 93% (563 out of 605) contained a single microsatellite, 6.5% (39 out of 605) contained two microsatellites, and 0.5% (3 out of 605) contained three microsatellites. Microsatellites with monomer motifs were by far the most common microsatellite type in the promoters (74.92%). Dimeric and trimeric microsatellite motifs were the next most common and accounted for, respectively, 15.39% and 6.14% of promoter-containing microsatellites (Table 1).

Microsatellites with monomer motifs were almost all A/T types (486 out of 487; 99.79%), with a single C monomer motif. A-motifs comprised the majority of the microsatellites with monomer repeats (345 out of 487; 70.84%) and T-motifs the minority (141 out of 487, 28.95%). AG/CT and GA/TC microsatellites comprised 71% of microsatellites with dimer motifs (Table 1).

3.3.2 Detection of Minisatellite Sequences

Approximately 2.24% of promoters (88 out of 3922) contained minisatellite sequences. No minisatellite sequences were found in monocotyledons. The length of the repeat unit ranged from 11 to 116 bp with an average of 24 bp, and the average number minisatellite repeats was 2.3, ranging from 1.9 to 3.8.

3.3.3 Analysis of TEs

No intact LTR retrotransposons were detected using LTR-finder. RepeatMasker detected 50 interspersed repeats, 6 truncated retrotransposons, and 34 DNA transposons (0.04%, 0.34%, and 0.08% of all promoters, respectively; Table 2). The most common TE types were MuDR-IS905 (0.13%), followed by hobo-Activator (0.12%), L1/CIN4 (0.02%), and Ty1/Copia (0.02%).

Table 1. Type and distribution of microsatellites in the collected plant promoter sequences

Group	Type	Type 1[a]		Type 2[b]		Subtotal	The subtotal/ The group total [%]	Overall [%]
		Number	Percentage[%]	Number	Percentage [%]			
Monomers	A/T	345	70.99	141	29.01	486	99.79	74.77
	C	1	100.00	-	-	1	0.21	0.15
Dimers	GA/TC	5	10.64	42	89.36	47	47.00	7.23
	AG/CT	8	33.33	16	66.67	24	24.00	3.69
	TA	12	100.00	-	-	12	12.00	1.85
	AT	10	100.00	-	-	10	10.00	1.54
	AC	5	100.00	-	-	5	5.00	0.77
	CA	2	100.00	-	-	2	2.00	0.31
Trimers	AGA/TCT	4	40.00	6	60.00	10	25.00	1.54
	AAG/CTT	2	22.22	7	77.78	9	22.50	1.38
	GAA/TTC	5	62.50	3	37.50	8	20.00	1.23
	AAC	3	100.00	-	-	3	7.50	0.46
	ACA	3	100.00	-	-	3	7.50	0.46
	ATC/GAT	1	50.00	1	50.00	2	5.00	0.31
	CCA	2	100.00	-	-	2	5.00	0.31
	ATT	1	100.00			1	2.50	0.15
	GTC	1	100.00			1	2.50	0.15
	TCG	1	100.00	-	-	1	2.50	0.15
Other		23				23		3.46

[a] the left hand side motif

[b] the right hand side motif (reverse complement of [a]).

The percentage of Type 1 and Type 2 motifs was derived by the number of Type 1 or Type 2 motifs divided by the subtotal.

Table 2. Predictions of presence of different types of transposable elements (TEs) in plant promoter sequences

TEs	Number of TEs	Average length of TE harbored in promoter sequence[bp]	Percentage of plant promoter sequences containing TEs in all promoters [%]
DNA transposons	34	97.2	0.34
Retroelements	6	57.8	0.04
Unclassified	10	83.3	0.08

3.3.4 Analysis of TFBSs

We used the online software NSITE-PL to predict 31259 TFBS motifs from 3922 plant promoter sequences. On average, one promoter contained eight TFBS motifs. Motif lengths ranged from 4 to 51 bp (predominantly ≤ 30 bp; 99.9%) with an average length of 11 bp.

TFBS with 10-bp motifs comprised the highest proportion of TFBS in the promoters (25.5%), followed by TFBS with 12-bp motifs (14.2%), then TFBS with 9-bp motifs (14.18%) (Figure 2). TFBS motif length mostly ranged from 6 to 17 bp (97.8% of all promoters). Up to 50% of the motifs were classified into 545 motif types, demonstrating that some key TFBS motifs are widely distributed in promoters. Most TFBS motifs possessed the characteristics of simple repeat sequences.

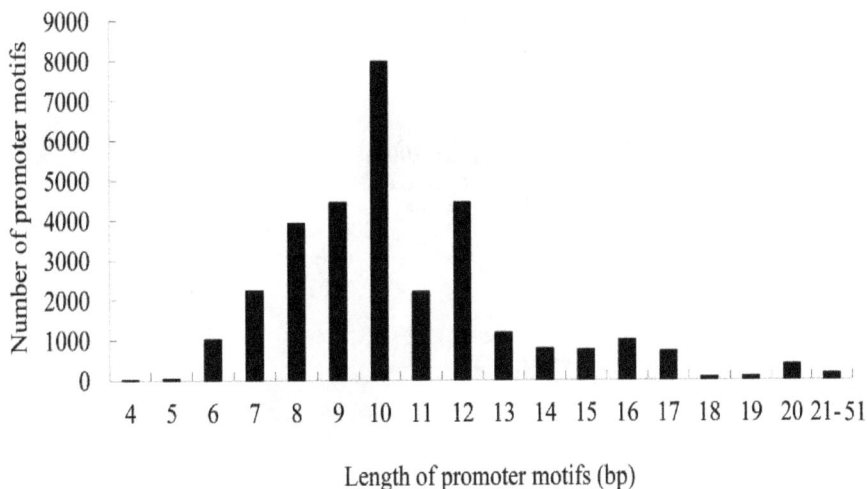

Figure 2. Distribution of motif lengths of transcription factor binding site (TFBS) in plant promoters. Promoter TFBS motifs were predicted using the software NSITE-PL to process plant promoter sequences

The TFBS motif with the highest frequency was AGAGAGAGA (1.6%; 495 out of 31259), which has previously been suggested to be a regulatory element for light responsive photo-transduction regulation in plants (Parida et al., 2009). The second most common TFBS motif was TTAGGGTTT (1.3%; 392 out of 31259); this motif has been shown to interact directly with MYB2-box-like elements in the promoters of osmotic, drought, and ABA-induced genes (Yun et al., 2010). The next most common TFBS motif was GCCGCC (1.1%; 336 out of 31259), involved in the cell cycle, jasmonic acid (JA) responsiveness and sugar signaling (Hu et al., 2011) (Figure 3). The three most common motifs comprised 4.0% of the total motif types, with the remaining motifs present at lower frequencies. The G+C content of TFBS motifs varied from 0.0% to 100.0%, with an average of 43.35%. Motifs with G+C content ranging from 0.0% to 50.0% accounted for 74.7% of all motifs, suggesting that critical promoter motifs exist in AT-rich regions.

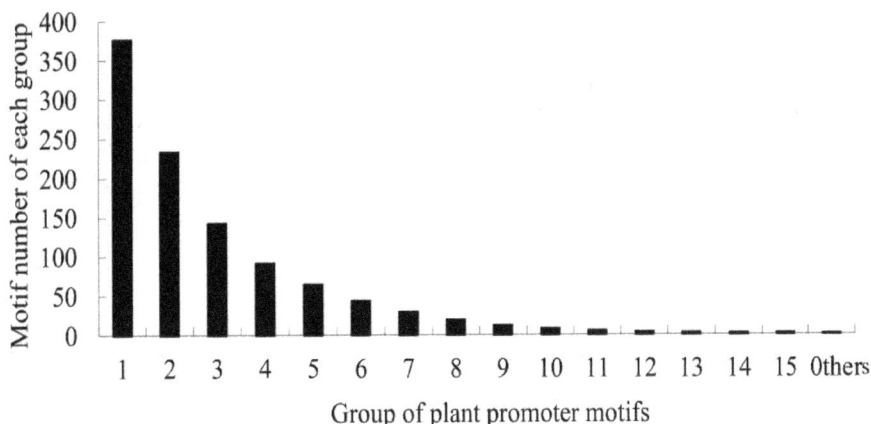

Figure 3. Mean number distribution of transcription factor binding site (TFBS) motifs in each plant promoter group. A total of 31259 motifs could be classified into 16 groups with identical motifs whole length. These groups are arranged by number of members for each motif, from greatest to least

Aside from the conserved motifs in the TFBS mentioned above, different cis-regulatory elements were also found in promoter sequences: 29201 cis-regulatory elements were identified in total. Although the frequency of most regulatory element types was low, some regulatory elements (G-box, GA-box, and ABRE motifs) were found at considerably higher frequencies (Figure 4). Among those three, G-box regulatory elements were the most common, accounting for 7.07% (2065 out of 29201) of the total regulatory elements. GA-box regulatory elements were the second most common at 5.00% (1460 out of 29201), and ABRE regulatory elements were the third most common at 4.81% (1405 out of 29201). Our results show that a small number of motifs with high affinities for binding proteins are widely distributed in promoter sequences.

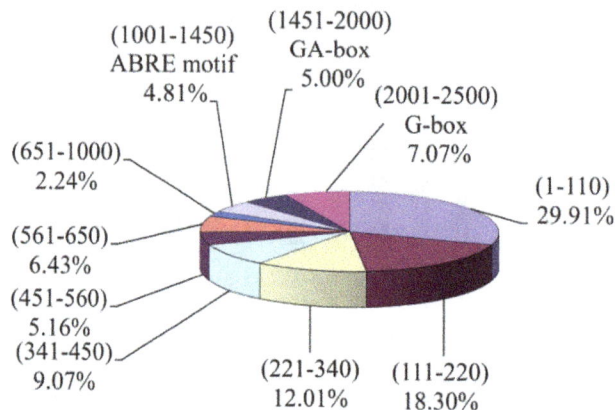

Figure 4. Number distribution of plant promoter regulatory elements detected in promoters. The values in the brackets represent the number range of all kinds of regulatory elements detected, and the percentages denote the total percentage of promoters in each regulatory element group

3.3.5 Putative Functional Annotation of the Transcribed Gene Regions with Promoters Containing the Corresponding TFBSs

NSITE (Version 2.2004; Softberry Inc.) was used to recognize TFBSs and provide information for the transcribed gene regions with promoters containing the corresponding TFBSs. Blast2Go was used to predict the functional annotation of the transcribed gene regions with promoters containing the corresponding TFBSs for biological process, molecular function and cellular component. For biological process annotation, the most common involvement was in metabolic processes (27.81%), followed by biological regulation (27.54%), and response to stimulus (17.77%) (Figure 5a). With respect to molecular functionality, the transcribed gene regions with the promoters containing the corresponding TFBSs mainly played a role in binding function (45.57%), followed by catalytic activity (23.86%) and other unknown molecular functions (17.57%) (Figure 5b). The transcribed gene regions with promoters containing the corresponding TFBSs most commonly functioned in the organelles (42.75%), followed by the intracellular (22.46%), and cellular components (20.53%) (Figure 5c). Hence, for biological process annotation, the transcribed gene regions with promoters containing the corresponding TFBSs were mainly involved in metabolic processes; with respect to molecular functionality, the most common function was binding; and with regard to cellular component annotation, transcribed gene regions most commonly functioned in the organelles.

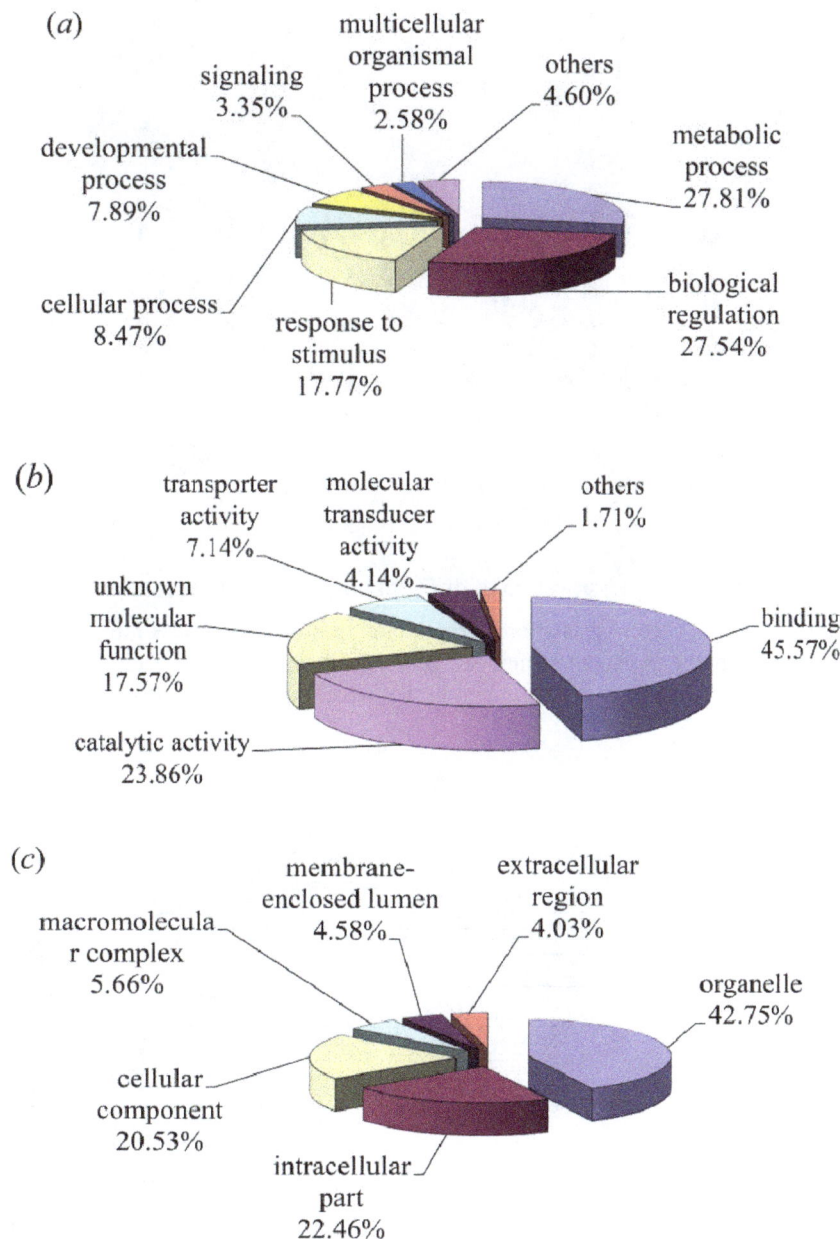

Figure 5. Functional annotation of the transcribed gene regions with promoters containing the corresponding transcription factor binding sites (TFBSs)

(a) Biological process; (b) Molecular function; (c) Cellular components.

3.3.6 Analysis of Alignment and Phylogenetic Dendrogram of Plant Promoter Sequences

All-by-all BlastN analysis of the plant promoters did not allow clear classification into different subclasses (Figure 6), indicating that the homology of these plant promoter sequences was relatively low. Nevertheless, according to the structure of the phylogenetic dendrogram, the ancestral lineages produced in MEGA 4 (Figure 7) and the species taxonomy, the plant promoter sequences could be classified into 8 groups containing 1172, 791, 60, 24, 136, 59, 287, and 1393 sequences, respectively (Figure 8). The genetic distance between the 8 groups was 0.19 on average, indicating greater divergence within the plant promoter sequence groups than between groups.

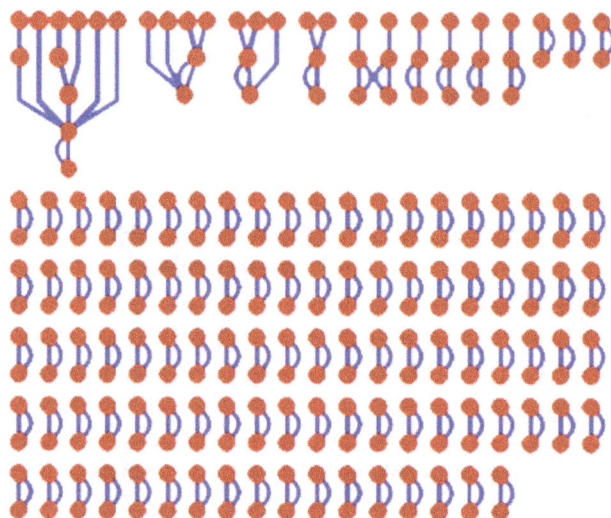

Figure 6. Classification of different plant promoter sequences

All-by-all BlastN analysis was used to classify different plant promoter sequences into different subclasses. The circles represent different plant promoter sequences, and the lines between the circles denote the homology between the two plant promoter sequences distributed in the two circles.

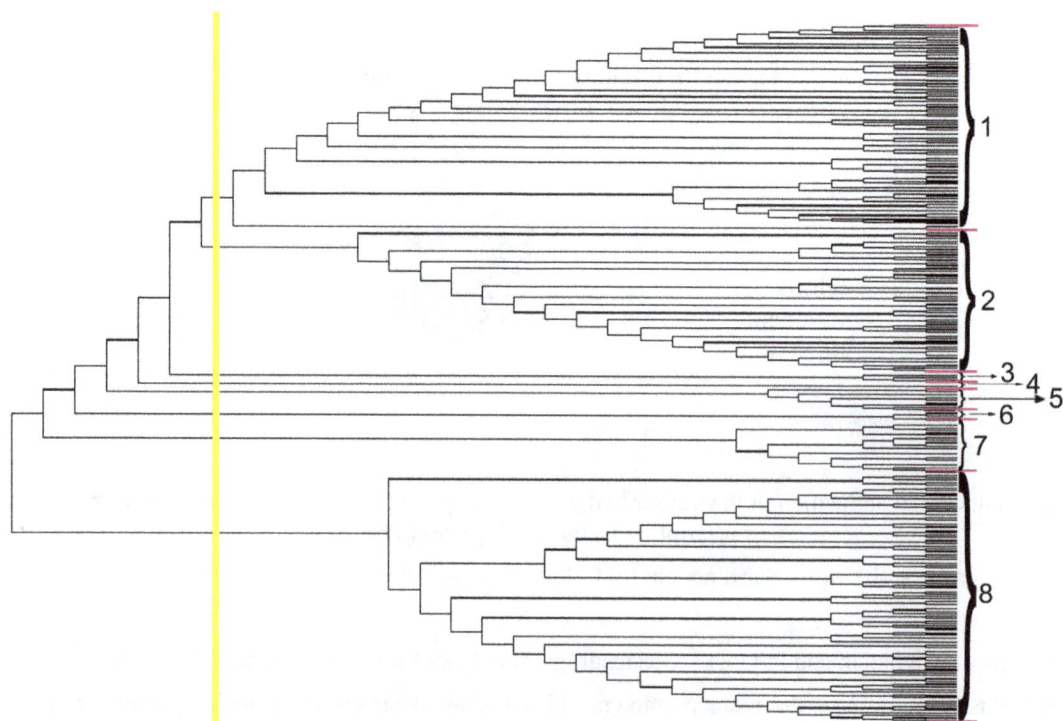

Figure 7. Phylogenetic dendrogram of the plant promoter sequences of 288 species

All plant promoter sequences were classified into 8 classes. The yellow line represents the demarcation of different classes. The numbers on the right represent the plant promoter sequence groups, with groups marked out using two pink lines.

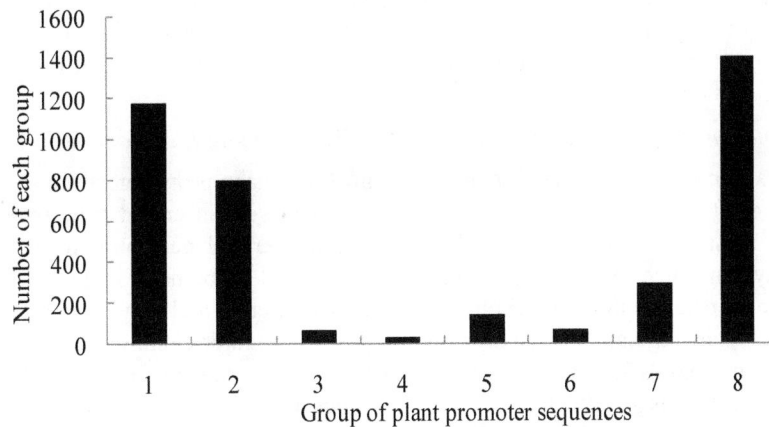

Figure 8. Distribution of the number of plant promoter sequences in each group

We divided the whole plant promoter sequences into eight groups according to the standard of uniform ancestral lineage in their phylogenetic dendrogram, and then the sequence number per group was counted and labeled in Y-axis.

4. Discussion

4.1 GC Content and Mutability of Plant Promoter Sequences

In the current study, the GC content of plant promoters was between 30% and 40% in most dicotyledon species, but was between 50% and 60% in most monocotyledon species, indicating that the GC content of plant promoters in monocotyledons is generally higher than that in dicotyledons. AT-rich regions are prone to mutate to generate diversity more often than GC-rich regions, and are inserted by exogenous gene fragments such as transposons (Gupta et al., 2005). Hence, more complex gene regulation may be required in dicotyledons compared to monocotyledons. AT-rich microsatellite sequences were also very common in the plant promoter sequences, suggesting that the mutability of plant promoters may have an important evolutionary adaptive role in diversification of gene expression. Nevertheless, some transcribed gene regions with GC-rich promoters are expressed more efficiently (Singh et al., 2012) suggesting that balancing selective pressure may exist for retention of GC-rich promoter sequences for genome stability.

4.2 Frequency and Possible Functionality of Key Promoter Motifs

According to the results, the length of most plant promoter TFBSs ranged from 6 to 17 bp, with AGAGAGAGA (1.6%; 495 out of 31259), TTAGGGTTT (1.3%; 392 out of 31259), and GCCGCC (1.1%; 336 out of 31259), being the most common. These high-frequency motifs may represent cis-regulatory elements which enhance the expression of sets of related genes. These common motifs detected may exist in the promoters of genes which have been highly conserved in species evolution, such as genes that play basic roles in plant growth and development. For example, the motif AGAGAGAGA is a known regulatory element participating in light-responsive regulation of phototransduction in plants (Parida et al., 2009). This motif is also present in the promoter of the *WRKY* gene which encodes the WRKY protein, one of the largest families of TFs, regulating processes such as response to biotic and abiotic stresses in plants (Zhang & Wang, 2005; Rushton et al., 2010). In rice, the WRKY gene family contains over 100 members (Pandey & Somssich, 2009). Likewise, the second most common TFBS motif (TTAGGGTTT) can directly interact with MYB2-box-like elements in the promoters of osmotic, drought, and ABA-induced genes (Yun et al., 2010). In contrast, different organisms may also have organism-specific but genome-wide TFBS motifs. For example, in Actinobacteria, the most significant TFBS motif is TCGAACA (Janky & van Helden, 2008). Similarly, the octamer AAAATTGA motif exists in the predicted core promoters of almost half the Mimivirus genes (Suhre et al., 2005). Therefore, high-frequency TFBS motifs may play multiple and comprehensive roles in many processes occurring in different organisms.

In addition, plant promoter motifs play important roles in accurate initiation of transcription. TFs can combine with DNA to orchestrate transcription of specific cis-regulatory elements (Rombauts et al., 2003). Only small numbers of TFs also combine with special promoter motifs to regulate expression of large numbers of genes (Smith et al., 2011a). Identification of such broad promoters may be useful for transgenic breeding, because the combination between these critical motifs and just a few TFs may allow for more effectively controlled expression of a batch of downstream transcribed gene regions. Critical promoter motifs with important roles can also be used to construct regulatory sequences which contribute to the spatio-temporal expression of transgenic plants. Thus, recombined regulatory sequences could not only accelerate the speed of breeding but also help in obtaining special gene products.

4.3 Functional Annotation of the Transcribed Gene Regions With Promoters Containing TFBSs

In this study, 31259 motifs of TFBS were detected from 3922 plant promoter sequences. On average, one promoter contained eight TFBS motifs. What are functions of these transcribed gene regions with promoters containing TFBSs? Blast2GO annotation revealed that the transcribed gene regions with TFBS-containing promoters commonly controlled metabolic processes during plant development, mainly had molecular binding functionality, and were operative in the organelles. We may characterize and mine critical TFBS and promoters from these transcribed gene regions to serve breeding purposes. Promoter cloning and subsequent manipulation of spatio-temporal gene expression offers significant promise as a developing research field in transgenic breeding. Promoter-based transgenic technologies have already been applied to great effect in wheat, where a heat-inducible promoter in transgenic wheat effectively controlled the spatio-temporal expression of a transgene (Freeman et al., 2011).

4.4 Some Microsatellites are Universally Distributed in Plant Promoters

Different species share common, prevalent motifs in promoters. The current study observed that $(A)_n$, $(T)_n$, $(AG)_n$, $(GA)_n$, $(CT)_n$, and $(TC)_n$ were the predominant mononucleotide and dinucleotide microsatellite motifs, respectively. This result suggests that microsatellites with specific motifs survived during natural selection due to positive selective advantages. The monomer microsatellites (almost all A and T motifs) accounted for the highest proportion of the microsatellite-containing promoter sequences. As the A/T-motif microsatellites are easily mutated (Gao et al., 2011), this may indicate a positive selection pressure due to the advantage provided by the extra diversity of gene expression in adapting to the environment and evolving into more complex higher organisms.

In summary, the GC content of plant promoters in monocotyledons appeared to be higher than that in dicotyledons. Most microsatellites and TEs were quite rare in promoter sequences, whereas microsatellites with A and T monomers were very commonly observed and may provide adaptive mutability potential in plant promoter sequences. Motifs of particular lengths occurred mainly on the TFBSs, and regulatory elements occurring with high frequency were mostly G-box, GA-box, and ABRE motifs. For biological process annotation, the transcribed gene regions with promoters containing the corresponding TFBSs were mainly involved in metabolic processes; with respect to molecular functionality, the most common function was binding; and with regards to cellular component annotation, the most common functional location was the organelles

The characteristics of higher A/T content, more microsatellites and a small quantity of TEs in plant promoters may play a role in evolution of plant promoters. The different TFBS motifs in plant promoters are a critical element of spatio-temporal expression of genes. These results are beneficial not only for elucidating the mechanisms of spatio-temporal gene expression and for cloning key plant promoters (or their main motifs), but also for investigating the basic structure of plant promoters and clarifying the evolutionary forces at work in plant promoter diversification.

Acknowledgments

This work was supported financially by National Natural Science Foundation of China (code: 31260335), and Research Fund for the Doctoral Program of Higher Education of China (code: 20123603120002). ASM is supported by an Australian Research Council Discovery Early Career Researcher Award (DE120100668).

References

Abdullah, S. N. A., Omidvar, V., Izadfard, A., Ho, C. L., & Mahmood, M. (2010). The oil palm metallothionein promoter contains a novel AGTTAGG motif conferring its fruit-specific expression and is inducible by abiotic factors. *Planta, 232*(4), 925-936. http://dx.doi.org/10.1007/s00425-010-1220-z

Anish, R., Hossain, M. B., Jacobson, R. H., & Takada, S. (2009). Characterization of transcription from TATA-less promoters: Identification of a new core promoter element XCPE2 and analysis of factor requirements. *PLOS One, 4*(4), e5103. http://dx.doi.org/10.1371/journal.pone.0005103

Bansal, M., & Kanhere, A. (2005). Structural properties of promoters: similarities and differences between prokaryotes and eukaryotes. *Nucleic Acids Research, 33*(10), 3165-3175. http://dx.doi.org/10.1093/nar/gki627

Calistri, E., Livi, R., & Buiatti, M. (2011). Evolutionary trends of GC/AT distribution patterns in promoters. *Molecular Phylogenetics and Evolution, 60*(2), 228-235. http://dx.doi.org/10.1016/j.ympev.2011.04.015

Cameron, M., & Williams, H. E. (2007). Comparing compressed sequences for faster nucleotide BLAST searches. *LEEE-ACM Transactions on Computational Biology and Bioinformatics, 4*(3), 349-364. http://dx.doi.org/10.1109/TCBB.2007.1029

Camp, E., Badhwar, P., Mann, G. J., & Lardelli, M. (2003). Expression analysis of a tyrosinase promoter sequence in zebrafish. *Pigment Cell Research, 16*(2), 117-126. http://dx.doi.org/10.1034/j.1600-0749.2003.00002.x

Carninci, P. (2006). Tagging mammalian transcription complexity. *Trends in Genetics, 22*(9), 501-510. http://dx.doi.org/10.1016/j.tig.2006.07.003

Carninci, P., Sandelin, A., Lenhard, B., Katayama, S., Shimokawa, K., Ponjavic, J., ... Hayashizaki, Y. (2006). Genome-wide analysis of mammalian promoter architecture and evolution. *Nature Genetics, 38*(6), 626-635. http://dx.doi.org/10.1038/ng1789

Chowdhury, S., Basundra, R., Kumar, A., Amrane, S., Verma, A., & Phan, A. T. (2010). A novel G-quadruplex motif modulates promoter activity of human thymidine kinase 1. *FEBS Journal, 277*(20), 4254-4264. http://dx.doi.org/10.1111/j.1742-4658.2010.07814.x

Conesa, A., Gotz, S., Garcia-Gomez, J. M., Terol, J., Talon, M., & Robles, M. (2005). Blast2GO: a universal tool for annotation, visualization and analysis in functional genomics research. *Bioinformatics, 21*(18), 3674-3676. http://dx.doi.org/10.1093/bioinformatics/bti610

Cooper, T. G., Georis, I., Tate, J. J., Feller, A., & Dubois, E. (2011). Intranuclear function forprotein phosphatase 2A: Pph21 and Pph22 are required for rapamycin-induced GATA factor binding to the DAL5 promoter in yeast. *Molecular and Cellular Biology, 31*(1), 92-104. http://dx.doi.org/10.1128/MCB.00482-10

Da Maia, L. C., Palmieri, D. A., de Souza, V. Q., Kopp, M. M., de Carvalho, F. I., & Costa de Oliveira, A. (2008). SSR Locator: Tool for simple sequence repeat discovery integrated with primer design and PCR simulation. *International Journal Plant Genomics, 2008*(2008), 412-426. http://dx.doi.org/10.1155/2008/412696

Deyneko, I. V., Kalybaeva, Y. M., Kel, A. E., & Blocker, H. (2010). Human-chimpanzee promoter comparisons: Property-conserved evolution? *Genomics, 96*(3), 129-133. http://dx.doi.org/10.1016/j.ygeno.2010.06.003

Freeman, J., Sparks, C. A., West, J., Shewry, P. R., & Jones, H. D. (2011). Temporal and spatial control of transgene expression using a heat-inducible promoter in transgenic wheat. *Plant Biotechnology Journal, 9*(7), 788-796. http://dx.doi.org/10.1111/j.1467-7652.2011.00588.x

Gao, C. H., Xiao, M. L., Jiang, L. Y., Li, J. N., Yin, J. M., Ren, X. D., ... Tang, Z. L. (2012). Characterization of transcriptional activation and inserted-into-gene preference of various transposable elements in the Brassica species. *Molecular Biology Reports, 39*(7), 7513-7523. http://dx.doi.org/10.1007/s11033-012-1585-0

Gao, C., Tang, Z., Yin, J., An, Z., Fu, D., & Li, J. (2011). Characterization and comparison of gene-based simple sequence repeats across Brassica species. *Molecular Genetics and Genomics, 286*(2), 161-170. http://dx.doi.org/10.1007/s00438-011-0636-x

Gupta, S., Gallavotti, A., Stryker, G. A., Schmidt, R. J., & Lal, S. K. (2005). A novel class of Helitron-related transposable elements in maize contain portions of multiple pseudogenes. *Plant Molecular Biology, 57*(1), 115-127. http://dx.doi.org/10.1007/s11103-004-6636-z

Halfon, M. S., & Zhu, Q. Q. (2009). Complex organizational structure of the genome revealed by genome-wide analysis of single and alternative promoters in Drosophila melanogaster. *BMC Genomics, 10*(9), 216-228. http://dx.doi.org/10.1186/1471-2164-10-9

Hernandez-Garcia, C. M., Bouchard, R. A., Rushton, P. J., Jones, M. L., Chen, X., Timko, M. P., & Finer, J. J. (2010). High level transgenic expression of soybean (Glycine max) GmERF and Gmubi gene promoters isolated by a novel promoter analysis pipeline. *BMC Plant Biology, 10*(10), 237-252. http://dx.doi.org/10.1186/1471-2229-10-237

Hoskins, R. A., Landolin, J. M., Brown, J. B., Sandler, J. E., Takahashi, H., Lassmann, T., ... Celniker, S. E. (2011). Genome-wide analysis of promoter architecture in drosophila melanogaster. *Genome Research, 21*(2), 182-192. http://dx.doi.org/10.1101/gr.112466.110

Hu, F., Wang, D., Zhao, X., Zhang, T., Sun, H., Zhu, L., ... Li, Z. (2011). Identification of rhizome-specific genes by genome-wide differential expression analysis in Oryza longistaminata. *BMC Plant Biology, 11*(18),

86-101. http://dx.doi.org/10.1186/1471-2229-11-18

Hwang, B. K., An, S. H., Choi, H. W., & Hong, J. K. (2009). Regulation and function of the pepper pectin methylesterase inhibitor (CaPMEI1) gene promoter in defense and ethylene and methyl jasmonate signaling in plants. *Planta, 230*(6), 1223-1237. http://dx.doi.org/10.1007/s00425-009-1021-4

Janky, R., & van Helden, J. (2008). Evaluation of phylogenetic footprint discovery for predicting bacterial cis-regulatory elements and revealing their evolution. *BMC Bioinformatics, 9*(37), 326-338. http://dx.doi.org/10.1186/1471-2105-9-37

Kumar, S., Tamura, K., Dudley, J., & Nei M. (2007). MEGA4: Molecular evolutionary genetics analysis (MEGA) software version 4.0. *Molecular Biology and Evolution, 24*(8), 1596-1599. http://dx.doi.org/10.1093/molbev/msm092

Martin, D. E. K. (2006). The exact joint distribution of the sum of heads and apparent size statistics of a "tandem repeats finder" algorithm. *Bulletin of Mathematical Biology, 68*(8), 2353-2364. http://dx.doi.org/10.1007/s11538-006-9146-0

Mastroeni, P., Janis, C., Grant, A. J., McKinley, T. J., Morgan, F. J. E., John, V. F., … Dougan, G. (2011). In vivo regulation of the vi antigen in salmonella and induction of immune responses with an in vivo-inducible promoter. *Infection and Immunity, 79*(6), 2481-2488. http://dx.doi.org/10.1128/IAI.01265-10

Molina, C., & Grotewold, E. (2005). Genome wide analysis of Arabidopsis core promoters. *BMC Genomics, 6*(25), 147-159. http://dx.doi.org/10.1186/1471-2164-6-25

Morgante, M., Hanafey, M., & Powell, W. (2002). Microsatellites are preferentially associated with nonrepetitive DNA in plant genomes. *Nature Genetics, 30*(2), 194-200. http://dx.doi.org/10.1038/ng822

Nozaki, T., Yachie, N., Ogawa, R., Kratz, A., Saito, R., & Tomita, M. (2011). Tight associations between transcription promoter type and epigenetic variation in histone positioning and modification. *BMC Genomics, 12*(1), 416-429. http://dx.doi.org/10.1186/1471-2164-12-416

Obara, N., Suzuki, N., Ki-Bom, K., Imagawa, S., Nagasawa, T., & Yamamoto, M. (2005). GATA motif on the erythropoietin gene promoter is essential for repression of ectopic constitutive erythropoietin production. *Blood, 106*(11), 878A-878A.

Pandey, S. P., & Somssich, I. E. (2009). The role of WRKY transcription factors in plant immunity. *Plant Physiology, 150*(4), 1648-1655. http://dx.doi.org/10.1104/pp.109.138990

Parida, S. K., Dalal, V., Singh, A. K., Singh, N. K., & Mohapatra, T. (2009). Genic non-coding microsatellites in the rice genome: characterization, marker design and use in assessing genetic and evolutionary relationships among domesticated groups. *BMC Genomics, 10*(6), 140-152. http://dx.doi.org/10.1186/1471-2164-10-140

Romania, M., Brigati, C., Banelli, B., Casciano, I., Di Vinci, A., Matis, S., … Allemanni, G. (2011). Epigenetic mechanisms regulate Delta NP73 promoter function in human tonsil B cells. *Molecular Immunology, 48*(4), 408-414. http://dx.doi.org/10.1016/j.molimm.2010.09.001

Rombauts, S., Florquin, K., Lescot, M., Marchal, K., Rouze, P., & van de Peer, Y. (2003). Computational approaches to identify promoters and cis-regulatory elements in plant genomes. *Plant Physiology, 132*(3), 1162-1176. http://dx.doi.org/10.1104/pp.102.017715

Rushton, P. J., Somssich, I. E., Ringler, P., & Shen, Q. X. J. (2010). WRKY transcription factors. *Trends in Plant Science, 15*(5), 247-258. http://dx.doi.org/10.1016/j.tplants.2010.02.006

Seliverstov, A. V., Lysenko, E. A., & Lyubetsky, V. A. (2009). Rapid evolution of promoters for the plastome gene ndhF in flowering plants. *Russian Journal of Plant Physiology, 56*(6), 838-845. http://dx.doi.org/10.1134/S1021443709060144

Shahmuradov, I. A., Gammerman, A. J., Hancock, J. M., Bramley, P. M., & Solovyev, V. V. (2003). PlantProm: a database of plant promoter sequences. *Nucleic Acids Research, 31*(1), 114-117. http://dx.doi.org/10.1093/nar/gkg112

Singh, D. P., Bhargavan, B., Chhunchha, B., Kubo, E., Kumar, A., & Fatma, N. (2012). Transcriptional protein Sp1 regulates LEDGF transcription by directly interacting with its cis-elements in GC-rich region of TATA-less gene promoter. *PLOS One, 7*(5), 3701-3712. http://dx.doi.org/10.1371/journal.pone.0037012

Smith, A. J., Chudnovsky, L., Simoes-Barbosa, A., Delgadillo-Correa, M. G., Jonsson, Z. O., Wohlschlegel, J. A., & Johnson, P. J. (2011a). Novel core promoter elements and a cognate transcription factor in the divergent unicellular eukaryote Trichomonas vaginalis. *Molecular and Cellular Biology, 31*(7), 1444-1458. http://dx.doi.org/10.1128/MCB.00745-10

Smith, C. L., Lee, S. C., & Magklara, A. (2011b). HDAC activity is required for efficient core promoter function at the mouse mammary tumor virus promoter. *Journal of Biomedicine and Biotechnology, 2011*(2011), 169-185. http://dx.doi.org/10.1155/2011/416905

Solovyev, V. V., Shahmuradov, I. A., Gammerman, A. J., Hancock, J. M., & Bramley, P. M. (2003). PlantProm: a database of plant promoter sequences. *Nucleic Acids Research, 31*(1), 114-117. http://dx.doi.org/10.1093/nar/gkg041

Suhre, K., Audic, S., & Claverie, J. M. (2005). Mimivirus gene promoters exhibit an unprecedented conservation among all eukaryotes. *Proceedings of the National Academy of Sciences of the United States of America, 102*(41), 14689-14693. http://dx.doi.org/10.1073/pnas.0506465102

Van Deursen, D., Botma, G. J., Jansen, H., & Verhoeven, A. J. (2007). Comparative genomics and experimental promoter analysis reveal functional liver-specific elements in mammalian hepatic lipase genes. *BMC Genomics, 8*(8), 99-112. http://dx.doi.org/10.1186/1471-2164-8-99

Van Oers, M. M., Nalcacioglu, R., Ince, I. A., Vlak, J. M., & Demirbag, Z. (2007). The Chilo iridescent virus DNA polymerase promoter contains an essential AAAAT motif. *Journal of General Virology, 2007*(88), 2488-2494. http://dx.doi.org/10.1099/vir.0.82947-0

Wang, Y. J., Xu, W. R., Yu, Y. H., Zhou, Q., Ding, J. H., Dai, L. M., ... Zhang, C. H. (2011). Expression pattern, genomic structure, and promoter analysis of the gene encoding stilbene synthase from Chinese wild vitis pseudoreticulata. *Journal of Experimental Botany, 62*(8), 2745-2761. http://dx.doi.org/10.1093/jxb/erq447

Wolf, E., Aigner, B., & Klymiuk, N. (2010). Transgenic pigs for xenotransplantation: selection of promoter sequences for reliable transgene expression. *Current Opinion in Organ Transplantation, 15*(2), 201-206. http://dx.doi.org/10.1097/MOT.0b013e328336ba4a

Wu, Q. Y., & Huang, W. (2004). The ManR specifically binds to the promoter of a Nramp transporter gene in Anabaena sp PCC 7120: a novel regulatory DNA motif in cyanobacteria. *Biochemical and Biophysical Research Communications, 317*(2), 578-585. http://dx.doi.org/10.1016/j.bbrc.2004.03.089

Yun, K. Y., Park, M. R., Mohanty, B., Herath, V., Xu, F., Mauleon, R., ... de Los Reyes, B. G. (2010). Transcriptional regulatory network triggered by oxidative signals configures the early response mechanisms of japonica rice to chilling stress. *BMC Plant Biology, 10*(16), 163-182. http://dx.doi.org/10.1186/1471-2229-10-16

Zhang, P., Li, W. L., Fellers, J., Friebe, B., & Gill, B. S. (2004). BAC-FISH in wheat identifies chromosome landmarks consisting of different types of transposable elements. *Chromosoma, 112*(6), 288-299. http://dx.doi.org/10.1007/s00412-004-0273-9

Zhang, Y. J., & Wang, L. J. (2005). The WRKY transcription factor superfamily: its origin in eukaryotes and expansion in plants. *BMC Evolutionary Biology, 5*(1), 1236-1248. http://dx.doi.org/10.1186/1471-2148-5-1

12

Effect of TiO₂ Nanoparticles on Antioxidant Enzymes Activity and Biochemical Biomarkers in Pinto Bean (*Phaseolus vulgaris* L.)

Ahmad Ebrahimi[1,2], Mohammad Galavi[3], Mahmood Ramroudi[3] & Payam Moaveni[4]

[1] Faculty of Agriculture, Zabol University, Iran

[2] Department of Agriculture, Islamic Azad University, Iranshahr Branch, Iranshahr, Iran

[3] Faculty of Agriculture, Zabol University, Zabol, Iran

[4] Department of Agriculture, Islamic Azad University, Shahr-e-Qods Branch, Tehran, Iran

Correspondence: Ahmad Ebrahimi, Department of Agronomy, Faculty of Agriculture, Zabol University, Iran. E-mail: Ae.iranshahr@gmail.com

Abstract

Tests were done on the effects of titanium dioxide spray on Pinto bean (Phaseolus vulgaris L. c.v 'c.o.s.16'). The study was conducted as a factorial experiment in a randomized complete block design with four replications for two years (2014 - 2015). Treatments consisted of two factors; the first factor was stage of plant growth that spraying was applied (rapid vegetative growth, flowering and pod filling); and the second factor was that of different concentrations of titanium dioxide nanoparticles (TiO₂) that consisted of spray with water (control), nano titanium dioxide at concentrations of 0.01%, 0.02%, 0.03% and 0.05%. Activity of guaiacol peroxidase (GPX), activity of superoxide dismutase (SOD), activity of catalase (CAT), activity of peroxidase (POD), malonyldialdehyde (MDA) Content and 8-deoxy-2-hydroxyguanosine (8-OHDG) content were assayed. Results showed that effect of nano TiO₂ was significant on activity of superoxide dismutase (SOD), activity of catalase (CAT), activity of peroxidase (POD), malonyldialdehyde (MDA) Content and 8-deoxy-2-hydroxyguanosine (8-OHDG) content. Results of combined analysis of variance showed that the effect year significantly affected on SOD and 8-OH-2-DG ($P \leq 0.05$). The effect of different amounts of titanium dioxide nanoparticles (TiO₂) significantly affected ($P \leq 0.05$) on MDA and 8-OH-2-DG. The effects of different amounts of titanium dioxide nanoparticles and year were significant on SOD, POD, MDA and the amount of 8-deoxy-2-hydroxyguanosine in $P \leq 0.05$. None of the physiological traits were affected by spraying of nano titanium dioxide. The effects of TiO₂ nanoparticles times of spraying and year were significant on SOD, CAT and 8-deoxy-2-hydroxyguanosine ($P \leq 0.05$). Interaction effects of nano TiO₂ concentrations × nano TiO₂ spraying times did not have a significant impact on SOD, CAT, POD, GPX, MDA and 8-OH-2-DG. Although, all trait were affected by interaction effects of year × nano TiO₂ concentrations × nano TiO₂ spraying times with the exception of GPX ($P \leq 0.05$).

Keywords: Tio₂ nanoparticles, various growth stages, Antioxidant Enzymes, MDA, 8-OH-2-DG and pinto bean (*Phaseolus vulgaris* L.)

1. Introduction

Common bean (*Phaseolus vulgaris* L.) is the most important food legume worldwide, providing the primary source of protein in human diets, supplying about 20% of the protein intake per person (Broughton et al., 2003). As half the grain legumes consumed worldwide are common beans. Unfortunately, the yields of common beans are low, and the quality of their seed proteins is sub-optimal. Most probably, common bean can be redressed by modern techniques (Broughton et al., 2003).

The use of nanoparticles for the growth of plants and control of plant diseases is a recent practice (Rico et al., 2011). During the last decade, an array of exploratory experiments has been conducted to gauge the potential impact of nanotechnology on crop improvement (Nair et al., 2010). Metal oxide nanoparticles are already being manufactured on a large scale for both industrial and household use. Titanium dioxide (TiO₂) nanoparticles are found in everything from cosmetics to sunscreen, paint, vitamins and water pollution treatment; therefore, TiO₂ Nanoparticles will inevitably reach bodies of water through wastewater and urban runoff (Yang et al., 2007). But, physiological effects, depending on the nanomaterial type, particle size, concentration, and plant species (Rico et al., 2011) For example, it is reported that TiO$_2$ nanoparticles in higher concentration had pronounced effects on photosynthetic pigments while lower concentration of Titanium dioxide (TiO$_2$) nanoparticle had significantly increased root length (Samadi et al., 2014). Nanoparticle-induced toxicity is mainly mediated through the generation of reactive oxygen species (ROS) in cells (Melegari et al., 2013). ROS are partly generated as byproducts of metabolic pathways in chloroplasts and are responsible for chlorophyll deterioration (Rico et al., 2015). Thus, disturbance in plant photosynthetic activity by MONPs can generate ROS and activate the plants' defense mechanisms to combat oxidative stress damage. Enzymes such as (superoxide dismutase (SOD), catalase (CAT), peroxidase (POD), guaiacol peroxidase (GPX), malondialdehyde (MDA) content and 8-deoxy-2-hydroxyguanosine (8-OHDG) are generally altered as a response to the alternation in ROS concentration (Du et al., 2016). In this regard, it is reported that metal oxide nanoparticles exposure can induce the generation of ROS, consequently causing oxidative stress and activating plant responses for detoxification such as enzymatic activity (Ma et al., 2015). Similarly, Servin et al. (2013), Hong et al. (2005) and Song et al. (2013) reported an increase in the activities of SOD, CAT, POD and decreased accumulation of ROS when plants were exposed to TiO$_2$ nanoparticle.

The ultraviolet-B (280–315 nm) light has long been recognized a detrimental factor for plants (Brosche & Strid, 2003; Jenkins, 2009). In addition to being a potential source of oxidative stress, solar UV-B is recognized as a key environmental signal, affecting development and metabolism (Hideg et al., 2013). Responses involve both UV-B-specific signaling and non-specific pathways. Photomorphogenic signaling in response to low intensity UV-B regulates the expression of genes involved in protection against UV (Jenkins, 2009), such as the synthesis of UV-absorbing phenylpropanoids (Brown, et al., 2005). The non-specific pathway involves reactive oxygen species (ROS) such as superoxide radicals (O$_2$·$^-$), singlet oxygen (·O$_2$), hydrogen peroxide (H$_2$O$_2$), and hydroxyl radicals (·OH) to accumulate in chloroplast during photosynthesis, which caused oxidative damage and photosynthesis reduction of plants (Zheng et al., 2008). One of the consequences of uncontrolled oxidative stress (imbalance between the prooxidant and antioxidant levels in favor of prooxidants) is cells, tissues, and organs injury caused by oxidative damage. It has long been recognized that high levels of free radicals or reactive oxygen species (ROS) can inflict direct damage to lipids. The primary sources of endogenous ROS production are the mitochondria, plasma membrane, endoplasmic reticulum, and peroxisomes (Moldovan and Moldovan, 2004). Lipid peroxidation or reaction of oxygen with unsaturated lipids produces a wide variety of oxidation products. The main primary product of lipid peroxidation is lipid hydroperoxides. Among the many different aldehydes which can be formed as secondary products during lipid peroxidation, including malondialdehyde (MDA), propanal, hexanal, and 4-hydroxynonenal (4-HNE) (Esterbauer et al., 1991), MDA appears to be the most mutagenic product of lipid peroxidation (Esterbauer et al., 1990).

8-deoxy-2-hydroxyguanosine (8-OH-2-DG) is a principal stable marker of hydroxyl radical damage to DNA. It has been related to a wide variety of disorders and environmental insults, and has been proposed as a useful systematic marker of oxidative stress (Bogdanov et al., 1999).

ROS can be activated by higher UV-B (Brosche and Strid, 2003). UV-B induced photomorphological changes in leaves including reduced leaf size, increased leaf thickness and the synthesis of phenolic compounds (Jansen et al., 1998). These changes also affect optical propertiesof leaves and thus may alter the amount of quanta reaching thephotosynthetic apparatus. The main influence of UV-B on photo-synthesis is believed to be more direct. Protein complexes engagedin the light reactions, as well as specific enzymes of the dark reac-tion are functionally impaired by UV-B (Jordan et al., 2016).

It has been reported that UV-B radiation induced oxidative damage of photosystem II and decreased electron transfer rate and thylakoid membrane stability (Renger, 1989; Eva, 1999). Nano-anatase is capable of undergoing electron transfer reactions under ultraviolet light, e.g., the electron was excited and transferred, then photogenerated electron holes in nano-anatase; the electron holes were reduced when the electron was captured by other molecule, while it was oxidized when self-captured (Crabtree, 1998). It is reported that nano-anatase treatments could markedly promote chlorophyll biosynthesis and the Rubisco activity and the photosynthesis efficiency of spinach (Zheng et al., 2005). Nano-anatase treatment could also activate superoxide dismutase (SOD), catalase (CAT), ascorbate peroxidase (APX), and guaiacol peroxidase (GPX), and remove ROS in the aged chloroplasts of spinach. Nano-anatase could absorb ultraviolet light and convert light energy to stable chemistry energy finally via electron transport in spinach chloroplasts. Therefore UV-B radiation on choloplasts could be reduced or could avoid the oxidative damage (Zheng et al., 2008).

According to the results obtained, we decided to investigate the effects of TiO$_2$ nanoparticles on the activities of antioxidant enzymes in pinto bean (*Phaseolus vulgaris* L.) in a two years study.

2. Materials and Methods

This survey was done as a factorial experiment in a complete randomized block design. Treatments consisted of two factors; 1- stage of plant growth that spraying was applied (rapid vegetative growth, flowering and pod filling); and 2- titanium dioxide nanoparticles (TiO_2) concentrations that sprayed; including water (control), 0.01%, 0.02%, 0.03% and 0.05% nano titanium dioxide. Pinto bean seeds were planted in May 2015 and 2016. Fertilization and plant feeding was done according to recommendations from results of a soil test. Spraying treatment was based on growth stages and concentrations of nanoTiO_2. Plants were treated with 240 ml titanium solution per square meter. Control plants were treated with distilled water. Evaluations were made for the parameters of (superoxide dismutase (SOD), catalase (CAT), peroxidase (POD), guaiacol peroxidase (GPX), malondialdehyde (MDA) content and 8-deoxy-2-hydroxyguanosine (8-OH-2-DG).

2.1 Enzyme Extraction

A quantity of 0.5 g of fresh foliar tissue from fresh seedlings (uppermost leaves) was harvested, weighed, washed with distilled water and homogenized in ice cold 0.1 M phosphate buffer (pH=7.5) containing 0.5 mM EDTA with prechilled pestle and mortar. Each homogenate was transferred to centrifuge tubes and was centrifuged at 4°C in Beckman refrigerated centrifuge for 15 min at 15000×g. The supernatant was used for enzyme activity assay. (Esfandiari et al., 2007).

2.2 Antioxidant Enzyme Activity Assays

SOD: The activity of superoxide dismutase (SOD) was assayed according to Misra and Fridovich (1972). About 500 mg of leaves were homogenized in 5 mL of 100 mmol L-1 K-phosphate buffer (pH 7.8) containing 0.1 mmol L-1 ethylenediaminetetracetic acid (EDTA), 0.1% (v/v) Triton X-100 and 2% polyvinylpyrrolidone (PVP) (w/v). The extract was filtered and centrifuged at 22,000 g for 10 min at 4°C, and the supernatant was utilized for assays. The assay mixture consisted of a total volume of 1 mL, containing glycine buffer (pH 10.5), 1 mmol L-1 epinephrine and enzyme material. Epinephrine was the last added component. Adrenochrome formation over the next 4 min was spectrophotometrically recorded at 480 nm. One unit of SOD activity is expressed as the amount of enzyme required to cause 50% inhibition of epinephrine oxidation under the experimental conditions used. This method is based on the ability of SOD to inhibit the autoxidation of epinephrine at an alkaline pH. Since the oxidation of epinephrine leads to the production of a pink adrenochrome, the rate of increase of absorbance at 480 nm, which represents the rate of autoxidation of epinephrine, can be conveniently followed. The enzyme has been found to inhibit this radical-mediated process.

CAT: Catalase activity was measured according to Aebi (1984). About 3 ml reaction mixture containing 1.5 ml of 100 mM potassium phosphate buffer (pH=7), 0.5 ml of 75 mM H2O2, 0.05 ml enzyme extraction and distilled water to make up the volume to 3 ml. Reaction started by adding H2O2 and a decrease in absorbance recorded at 240 nm for 1 min. Enzyme activity was computed by calculating the amount of H2O2 decomposed.

POD: Peroxidase activity was measured using modification of the procedure of McAdam et al. (1992). Guaiacol was used as the substrate. POD activity was measured in a reaction mixture (3 ml) that contained 0.1 ml enzyme extract, 12 mM H_2O_2, and 7.2 mM guaiacol in 50 mM phosphate buffer, pH 5.8. The kinetics of the reaction were followed at 470 nm. Activity was calculated using extinction coefficient (26.6 mM_1 cm_1 at 470 nm) for tetraguaiacol and expressed as units per gram of fresh weight (FW). One unit of POD activity was defined as 1 mmol tetraguaiacol produced per minute. Protein content in enzyme extracts was determined by the method of Bradford.

GPX: Glutathione peroxidase (GPX) activity was measured according to Paglia and Valentine (1987) in which 0.56 M (pH 7) phosphate buffer, 0.5 M EDTA, 1mM NaN3, 0.2 mM NADPH was added to the extracted solution. GPX catalyses the oxidation of glutathione by cumene hydroperoxide in the presence of glutathione reductase and NADPH, the oxidized glutathione is immediately converted to the reduced form with the concomitant oxidation of NADPH to NADP. The decrease in absorbance at 340 nm was measured with a spectrophotometer.

MDA: Malondialdehyde (MDA) was measured by colorimetric method (Stewart and Bewley, 1980). 0.5 g of leaf samples were homogenized in 5ml of distilled water. An equal volume of 0.5% thiobarbituric acid (TBA) in 20% trichloroacetic acid solution was added and the sample incubated at 95°C for 30 min. The reaction stopped by putting the reaction tubes in the ice bath. The samples then centrifuged at 10000×g for 30 min. The supernatant removed, absorption read at 532 nm, and the amount of nonspecific absorption at 600 nm read and subtracted from this value. The amount of MDA present was calculated through the extinction coefficient of 155 mM-1cm-1. Enzyme activity and MDA content of samples were recorded with duplication. The following formula was applied to calculate malondialdehyde (MDA) content using its absorption coefficient (ε) and expressed as nmol malondialdehyde g^{-1} fresh mass, following the formula:

$$MDA (nmol\ g\text{-}1\ FM) = [(A532\text{-}A600) \times V \times 1000/\varepsilon] \times W,$$

where ε is the specific extinction coefficient (=155 mM-1 cm-1), V is the volume of crushing medium, W is the fresh weight of leaf, A600 is the absorbance at 600 nm wavelength and A532 is the absorbance at 532 nm wavelength.

8-OH-2-DG: 8-deoxy-2-hydroxyguanosine (8-OH-2-DG) was measured according to Bogdanov et al. (1999). In this case, extract from the carbon column 8 (C8) is passed, this column is for the absorption of purines. After reaching equilibrium, passing all the solvent in the extraction, then column by passing the new mobile phase containing Tris Hcl with pH 8-deoxy-2-hydroxyguanosine of this column is removed and transferred to C8 new column. After equilibration column was washed with the mobile phase containing adenosine concentrations of 0.65 M. This has led to proprietary separation of the desired material as a private peak, so that the peak detector devices of the type Colometric were identified and transferred. The amount of 8-deoxy-2-hydroxyguanosine for a certain proportion of total purine peak will be assessed. To separate the extract, the leaves of beans are weighed and then determine the total protein ratio, a part of the leaf tissue in buffer solution of phosphate bicarbonate (mono sodium) 1.6 M (PH 7.4), crushed and then quickly the presence of ice and cold conditions were homogenized. Dimethyl sulfoxide at a concentration of 0.4 M was added to the solution. After adding the new buffer (pH 5.6) acetate monosodic, it is in the dialysis against distilled water for 6 hours and then the remaining content was centrifuged at 1500×g for 10 minutes. The supernatant in a spectrophotometer at wavelengths of 280 and 260 nm, respectively, were absorbed. Then, by adding trichloroacetic acid was free 35/0 mole of protein. This solution is then centrifuged at 300×g for 15 minutes was centrifuged and the supernatant is used for sensing of guanosine hydrate.

2.3 Characterization Analysis of TiO₂ NPs

The anatase TiO₂ NPs was purchased from Nano Pars Lima Company. The TiO₂ NPs had a purity of greater than 99.5%, average of particle diameter of 21 nm, and a surface area of 60 m2/g. Then, in order to prepare concentrations of nano TiO₂, 20 g nano TiO₂ was dissolved into water and then 0.01 ml of solution was filled up to 1000 ml. Thus, different concentrations of titanium dioxide (0.01%, 0.02%, 0.03% and 0.05%) were prepared. An ultrasound instrument was used to homogenize the solution. Titanium dioxide nanoparticles were sprayed on plants using a calibrated pressurized backpack sprayer (capacity 20 l).

2.4 Statistical Calculations

Data were subjected to analysis of variance (ANOVA) using software Statistical Analysis System 9.0 (SAS Institute 1988) and followed by Duncan's multiple range tests. Terms were considered significant at P ≤ 0.05.

3. Results

Results of combined analysis of variance showed that the effect year significantly affected on SOD and 8-OH-2-DG (P ≤ 0.05, Table 1). The effect of different amounts of titanium dioxide nanoparticles (TiO₂) significantly affected (P ≤ 0.05, Table 1) on MDA and 8-OH-2-DG. The effects of different amounts of titanium dioxide nanoparticles and year were significant on SOD, POD, MDA and the amount of 8-deoxy-2-hydroxyguanosine in P ≤ 0.05 (Table 1). None of the physiological traits were not affected by spraying of nano titanium dioxide. In the event that, the effects of TiO₂ nanoparticles times of spraying and year were significant on SOD, CAT and 8-deoxy-2-hydroxyguanosine (P ≤ 0.05, Table 1). Interaction effects of nano TiO₂ concentrations × nano TiO₂ spraying times did not have a significant impact on SOD, CAT, POD, GPX, MDA and 8-OH-2-DG (Table 1). Although, all traits were affected by interaction effects of year × nano TiO₂ concentrations × nano TiO₂ spraying times with the exception of GPX (P ≤ 0.05, Table 1).

Comparison of means demonstrated that nano TiO₂ at concentrations of 0.01% and 0.03% increased activity rate of SOD in leaves compared with control, although there was no significant difference between the nano TiO₂ concentrations regarding catalase enzyme and nano TiO₂ at concentrations of 0.02%, 0.03% and 0.05% boosted activity rate of CAT in leaf tissue (Table 2). In the case of nano titanium application, the highest amount of POD was obtained at concentrations of 0.01%, 0.03% and 0.05%, whereas the lowest amount of POD was found at concentrations of 0.02%. In the case of TiO₂ usage, the highest amount of GPX enzyme has been showed at concentrations of 0.03% and 0.05%, whereas the highest content of MDA was found in those plots which were treated with titanium dioxide 0.02% (Table 2). Finnaly, the highest content of 8-OH-2-DG (12.821 nm/ mg Protein) was found at concentration of 0.03% (Table 2).

According to Table 2, the highest amount of SOD, CAT, GPX and MDA were found in rapid vegetative growth stage in pinto bean leaves, whereas the highest content of POD was seen in pod filling stage and for 8-OH-2-DG, the highest content was found in flowering stage.

When it comes to the combined effects, results show that the maximum of the activity measure of the SOD enzyme (208.13 U mg $^{-1}$ Protein) in leaves was observed by the application of 0.02% nano titanium dioxide at the rapid vegetative growth stage and the minimum of the activity mount of the SOD enzyme (95.50 U mg $^{-1}$ Protein) in leaf tissue were found by the foliar of 0.05% nano titanium dioxide at the rapid vegetative growth stage (Table 3). Interaction effects of nano TiO$_2$ concentrations × nano TiO$_2$ spraying times on CAT showed the maximum of the activity measure of the catalase enzyme (232.75 U mg $^{-1}$ Protein) in leaves were observed by the application of 0.03% nano titanium dioxide at the rapid vegetative growth stage. By contrast, when nano titanium dioxide (0.03%) was applied on plants during the flowering stage, the lowest of CAT enzyme (111.50 U mg $^{-1}$ Protein) was produced (Table 3). The maximum and minimum POD enzyme were obtained when plants were treated with titanium dioxide nanoparticles (0.05%) at the pod filling stage and titanium dioxide nano-particles (0.02%) at the flowering stage, respectively (Table 3). The combined effects among growing stages and nano TiO$_2$ concentrations indicate that the most GPX enzym were produced when titanium dioxide nanoparticles (0.05%) were applied at the rapid vegetative growth stage, whereas the least GPX enzyme was obtained from nano titanium application at the flowering stage in pinto bean (Table 3). When the effects of tow experimental factors were combined, the highest MDA amount was obtained when titanium dioxide nanoparticles (0.02%) at the pod filling and the lowest MDA content was obtained when titanium dioxide nanoparticles (0.05%) at the pod filling stage (Table 3). The highest 8-OH-2-DG amount was observed when plants were treated with titanium dioxide nanoparticles (0.05%) at the flowering stage; on the other hand, the lowest 8-OH-2-DG amount was obtained from those plants which were not treated with nano TiO$_2$ at the flowering stage (Table 3). A significant positive correlation was found between GPX enzyme amount and CAT enzyme amount under experimental treatments of pinto bean (r0.01 = 0.266) (Table 4). Also, there was a significant positive correlation between GPX enzyme amount and 8-OH-2-DG content in leaves of pinto bean under experimental treatments (r0.05 = 0.202) (Table 4).

4. Discussion

In this two-year study, we found that foliar of TiO$_2$ nanoparticles at different growth stages induce some physiological and biochemical responses that can promote tolerance levels in pinto bean.

The nano TiO$_2$, attributed to their photocatalytic property and thermal conductivity, enhanced water absorption, improved light absorption in chlorophyll a, and induced oxygen evolution rate, consequently showed beneficial effects on photosynthesis (Rezaei et al., 2015; Rico et al., 2015b).

Our results showed that nano TiO$_2$ sprayed on pinto bean leaves inhibited ROS accumulation and increased antioxidant enzymes activity and also, showed a significant increase in the contents of each biomarkers MDA and 8-OH-2-DG.

Similarly, nano CuO caused significant oxidative stress with higher ROS, MDA content, but increased activities of antioxdative enzymes in rice (Da Costa & Sharma, 2016; Shaw & Hossain, 2013; Wang et al., 2015).

In the event of oxidative stress in plant, the leaves damaged by the collapse of chlorophyll, is reduced the rate of photosynthesis and eventually decline yields crops.

In this case, by increasing the amount and activity of antioxidant enzymes in plant leaf tissue, can be reduced the amount of damage and prevent yield losses. In this regard, it was reported that internal O$_2$ concentration is high during photosynthesis, and chloroplasts are especially prone to generate activated oxygen species; therefore, these cytotoxic active oxygen species can seriously disrupt normal metabolism through oxidative damage of lipids, nucleic acids, and proteins. Deleterious effects of ROS and lipid peroxidation products are counteracted by an antioxidant defense system (Pejic et al., 2009). From one aspect, it can be expressed that probably with the application of nano TiO$_2$ in the plant and then with the entry of it into the leaf tissue, causing phytotoxicity in plants, at the same time increases the amount of oxidants in plants, finally, increase the amount and activity of antioxidants (including antioxidant enzymes and biomarkers MDA and 8-OH-2-DG) to counteract the disadvantageous effects of ROS.

Malondialdehyde is the decomposition product of polyunsaturated fatty acids of bio membranes and its increase is the result of greater accumulation under high antioxidant stress. Malondialdehyde content serves as an indicator of the extent of lipid peroxidation and an indirect reflection of the extent of cell damage (Wang et al., 2011).

Servin et al. (2013) reported an increase in CAT activity (250-750 mg/kg) but a decrease in APX (500 mg/kg) when cucumber plants were exposed to nano TiO_2. Also, Laware and Raskar (2014) reported that CAT and GPX activities can be enhanced in the presence of 10-30 μgml^{-1} TiO_2 nanoparticles, but their activities decrease by higher concentrations of TiO_2 nanoparticles. Also, Lei et al. (2008) showed that nano-anatase treatment could activate SOD, CAT, APX, and GPX of spinach chloroplasts. SOD can convert O_2^{-} into H2O2 and O2; moreover, CAT, APX, and GPX can reduce H2O2 into H2O and O_2. Therefore, SOD, CAT, APX, and GPX can maintain a low level of ROS and prevent ROS toxicity and protect cells.

From other aspect, it can be declared that with the foliar application of titanium dioxide nanoparticles in the plant, nano TiO_2 can reduce the effects of photo oxidative stress and prevent the chloroplasts destruction and chlorophyll degradation of leaf tissue. About this issue, Lei et al. (2008) announced that nano-anatase could absorb ultraviolet (UV) light and convert light energy to steady chemistry energy finally via electron transport in spinach chloroplasts, so that UV-B radiation on choloplasts could be reduced or could avoid the oxidative damage. In other words, Zheng et al. (2007) reported that enriched energy electron from nano-anatase, which entered chloroplast under ultraviolet light, was transfered in photosynthetic electron transport chain and made $NADP^+$ reduced into NADPH, coupled to photophosphorylation, and made electron energy be transformed to ATP, and nano-anatase h^+, photogenerating electron holes, captured an electron from water that accelerated water photolysis and O_2 evolution.

Table 1. Combined analysis of variance on SOD, CAT, POD, GPX, MDA and 8-OH-2-DG of pinto bean affected by nano TiO_2 concentrations and different growth stages (2014 - 2015)

				Means square			
8-OH-2-DG	MDA	GPX	POD	CAT	SOD	df	Sources of variation
1.79*	0.02 n.s	0.67 n.s	0.92 n.s	1.41 n.s	1.89*	1	Y (Year)
0.90	0.40	0.16	0.77	0.70	0.37	6	Error
0.85*	0.70*	0.31 n.s	0.44 n.s	0.08 n.s	0.30 n.s	4	(A) Concentrations of nano TiO_2
0.79*	0.31*	0.25 n.s	0.72*	0.12 n.s	1.13*	4	Y × A
0.22 n.s	0.05 n.s	0.08 n.s	0.03 n.s	1.01 n.s	0.47 n.s	2	(B) Times of spraying
1.02*	0.32 n.s	0.06 n.s	0.15 n.s	0.00*	0.27*	2	Y × B
0.41 n.s	0.42 n.s	0.12 n.s	0.42 n.s	0.71 n.s	0.58 n.s	8	A × B
0.41*	0.82*	0.52 n.s	0.46*	0.27*	0.25*	8	Y × A × B
0.41	0.39	0.62	0.64	0.66	0.49	84	Error
30.41	17.45	17.74	24.22	16.98	14.73		C.V (%)

Note: Ns, Non Significant, * and **, Significant at 5% and 1% levels respectively.

Table 2. Mean comparison of physiological traits of pinto bean in nano TiO_2 concentrations and nano TiO_2 spraying times (2014 - 2015)

8-OH-2-DG (nm/ mg Protein)	MDA (nm/ mg Protein)	GPX (U mg $^{-1}$ Protein)	POD (U mg $^{-1}$ Protein)	CAT (U mg $^{-1}$ Protein)	SOD (U mg $^{-1}$ Protein)	Treatments
						Concentrations of nano TiO_2
8.317b	48.967ab	105.25a	0.040958a	149.92a	135.42a	(Distilled water) Control
9.738ab	38.200b	97.38a	0.042500a	141.38a	151.88a	(0.01%) Nano TiO_2
9.388ab	51.300a	97.50a	0.040417a	150.83a	134.75a	(0.02%) Nano TiO_2
12.821a	39.054ab	115.92a	0.051500a	157.22a	152.71a	(0.03%) Nano TiO_2
12.454ab	37.471ab	111.83a	0.064042a	160.71a	130.92a	(0.05%) Nano TiO_2
						Times of spraying of nano TiO_2
10.403a	44.360a	107.33a	0.045522a	171.35a	157.50a	Rapid vegetative growth
11.025a	42.113a	102.55a	0.045400a	140.35a	135.88a	Flowering
10.203a	42.523a	106.85a	0.053025a	144.33a	130.03a	Pod filling

Means in each column followed by similar letter(s) are not significantly different using Duncan's Multiple Range Test.

Table 3. Interaction effects of nano TiO$_2$ concentrations × nano TiO$_2$ spraying times on physiological traits of pinto bean (2014 - 2015)

8-OH-2-DG (nm/ mg Protein)	MDA (nm/ mg Protein)	GPX (U mg^{-1} Protein)	POD (U mg^{-1} Protein)	CAT (U mg^{-1} Protein)	SOD (U mg^{-1} Protein)	Times of spraying of nano TiO$_2$	Concentrations of nano TiO$_2$
8.688ab	50.605abc	104.13a	0.04150a	145.38ab	162.13ab	Rapid vegetative growth	Control
9.750ab	44.93abc	95.38a	0.04150a	132.00ab	153.13ab	Rapid vegetative growth	(0.01%)
9.888ab	45.28abc	90.38a	0.03425a	182.75ab	208.13a	Rapid vegetative growth	(0.02%)
11.250ab	40.55abc	114.00a	0.05713a	232.75a	168.63ab	Rapid vegetative growth	(0.03%)
12.438a	40.55abc	132.75a	0.05175a	163.88ab	95.50b	Rapid vegetative growth	(0.05%)
6.138b	51.665ab	95.00a	0.03938a	169.50ab	113.25ab	Flowering	Control
7.425ab	28.655c	92.13a	0.04113a	144.88ab	141.63ab	Flowering	(0.01%)
9.963ab	46.015abc	99.25a	0.02838a	144.25ab	98.13b	Flowering	(0.02%)
15.675a	38.365abc	119.88a	0.05200a	111.50b	154.25ab	Flowering	(0.03%)
15.925a	45.875abc	106.50a	0.06613a	131.63b	172.13ab	Flowering	(0.05%)
10.125ab	44.64abc	116.63a	0.04200a	134.88ab	130.88ab	Pod filling	Control
12.038ab	41.03abc	104.63a	0.04488a	147.25ab	160.88ab	Pod filling	(0.01%)
8.313ab	62.615a	102.88a	0.05863a	125.50b	98.00b	Pod filling	(0.02%)
11.538ab	38.25bc	113.88a	0.04538a	127.41b	135.25ab	Pod filling	(0.03%)
9.000ab	26.09c	96.25a	0.07425a	186.63ab	125.13ab	Pod filling	(0.05%)

Means in each column followed by similar letter(s) are not significantly different using Duncan's Multiple Range Test.

Table 4. Pearson's correlation coefficients among GPX, SOD, CAT, POD, MDA and 8-OH-2-DG of pinto bean affected by nano TiO$_2$ concentrations and different growth stages (2014 - 2015)

Trait	GPX	SOD	CAT	POD	MDA	DHG
GPX	1	-0.002	0.266**	0.155	-0.019	0.202*
SOD	-0.002	1	-0.002	-0.169	0.061	-0.137
CAT	0.266**	-0.002	1	-0.004	-0.059	0.042
POD	0.155	-0.169	-0.004	1	-0.068	0.057
MDA	-0.019	0.061	-0.059	-0.068	1	0.004
DHG	0.202*	-0.137	0.042	0.057	0.004	1

* and **, significant difference at 5 and 1%, respectively.

References

Aebi, H. (1984). In S. P. Colowick, & N. O. Kaplan (Eds.), *Catalase in Vitro. Methods in Enzymology* (Vol. 105, pp. 114–121). Florida: Acad. Press.

Bogdanov, M. B., Beal, M. F., Meccabe, D. R., Griffin, R. M., &Matson, W. R. (1999). A carbon column based LCEC approach to routine 8-hydroxy-2- deoxyguanosine measurements in urine and other biological matrices. *Free Rad Biol Med., 27*, 647-666.

Bradford, M. M. (1976). A rapid and sensitive method for the quantitation of microgram quantities of protein utilizing the principle of protein-dye binding. *Anal Biochem, May 7(72)*, 248–254.

Brosche, M., & Strid, A. (2003). Molecular events following perception of ultraviolet-Bradiation by plants. *Physiol. Plant, 117*, 1–10.

Broughton, W. J., Hernandez, G., Blair, M., Beebe, S., Gepts, P., & Vanderleyden, J. (2003). Beans (Phaseolus spp.)—model food legumes. *Plant Soil, 252*, 55–128.

Brown, B. A., Cloix, C., & Jiang, G. H. (2005). A UV-B-specific signaling componentorchestrates plant UV protection. *Proc. Natl. Acad. Sci. U. S. A., 102*, 18225–18230.

Crabtree, R. H. (1998). A new type of hydrogen bond. *Science, 282*, 2000 – 2001.

Da Costa, M. V. J., & Sharma, P. K. (2016). Effect of copper oxide nanoparticles on growth, morphology, photosynthesis, and antioxidant response in Oryza sativa. *Photosynthetica, 54*(1), 110-119.

Du, W., Tan, W., Peralta-Videa, J. R., Gardea-Torresdey, J. L., Ji, R., ... Guo, H. (2016). Interaction of metal oxide nanoparticles with higher terrestrial plants: Physiological and biochemical aspects. *Plant Physiology and Biochemistry,* 1-16.

Esfandiari, E., Shakiba, M. R., Mahboob, S., Alyari, H., & Toorchi, M. (2007). Water stress, antioxidant enzyme activity and lipid peroxidation in wheat seedling. *Journal of Food, Agriculture & Environment, 5,* 149-153.

Esterbauer, H., Eckl, P., & Ortner, A. (1990). Possible mutagens derived from lipids and lipid precursors. *Mutation Research, 238*(3), 223–233.

Esterbauer, H., Schaur, R. J., & Zollner, H. (1991). Chemistry and Biochemistry of 4 hydroxynonenal, malonaldehyde and related aldehydes. *Free Radical Biology and Medicine, 11*(1), 81–128.

Eva, H. (1999). Utilizing new adamantly spin traps in studying UV-B-induced oxidative damage of photosystem II. *J Photochem Photobiol B Biol, 48,* 174-179.

Hideg, E., Jansen, M. A. K., & Strid, A. (2013). UV-B radiation ROS and stress; inseparablecompanions or loosely linked associates? *Trends Plant Sci., 18,* 107–115.

Hong, F. S., Yang, F., Liu, C., Gao, Q., Wan, Z. G., Gu, F. G., ... Yang, P. (2005). Influences of nano-TiO₂ on the chloroplast aging of spinach under light. *Biol. Trace Elem. Res., 104*(3), 249-260.

Jansen, M. A. K., Gaba, V., & Greenberg, B. M. (1998). Higher plants and UV-B radiation:balancing damage, repair and acclimation. *Trends Plant Sci., 3,* 131–135.

Jenkins, G. I. (2009). Signal transduction in responses to UV-B radiation. *Annu. Rev. Plant Biol., 60,* 407–431.

Jordan, B. R., Strid, A., & Wargent, J. J. (2016). What role does UVB play in determiningphotosynthesis? In M. Pessarakli (Ed.), *Handbook of Photosynthesis* (pp. 275–286). CRCPress, Boca Raton.

Laware, S. L., & Raskar, S. (2014). Effect of titanium dioxide nanoparticles on hydrolytic and antioxidant enzymes during seed germination in onion. *Int J Curr Microbiol App Sci., 3*(7), 749-760.

Lei, Z., Mingyu, S., Xiao, W., Chao, L., Chunxiang, Q., Liang, C., ... & Fashui, H. (2008). Antioxidant stress is promoted by nano-anatase in spinach chloroplasts under UV-B radiation. *Biological Trace Element Research, 121*(1), 69-79.

Ma, C. X., White, J. C., Dhankher, O. P., & Xing, B. (2015. Metal-based nanotoxicity and detoxification pathways in higher plants. *Environ. Sci. Technol., 49*(12), 7109-7122.

MacAdam, J. W., Nelson, C. J., Sharp, R. E. (1992). Peroxidase activity in the leaf elongation zone of tall fescue. *Plant Physiol., 99,* 872-878.

Melegari, S. P., Perreault, F., Costa, R. H. R., Popovic, R., & Matias, W. G. (2013). Evaluation of toxicity and oxidative stress induced by copper oxide nanoparticles in the green alga Chlamydomonas reinhardtii. *Aquat. Toxicol., 142,* 431e440.

Misra, H. P., & Fridovich, I. (1972). The generation of superoxide radical during auto oxidation. *J. Biol. Chem., 247,* 6960-6966.

Moldovan, L., & Moldovan, N. I. (2004). Oxygen free radicals and redox biology of organelles. *Histochemistry and cell biology, 122*(4), 395-412.

Nair, R., Varghese, S. H., Nair, B. G., Maekawa, T., Yoshida, Y., & Kumar, D. S. (2010). Nanoparticulate material delivery to plants. *Plant Sci, 179,* 154–163.

Paglia, D. E., & Valentine, W. N. (1987). Studies on the quantitative and qualitative characterization of glutation proxidase. *J. Lab. Med., 70,* 158-165. Methods of soil analysis, 2. Chemical and Agronomy and Soil Science Society of America.

Pejic, S., Todorovic, A., Stojiljkovic, V., Kasapovic, J., & Pajovic, S. B. (2009). Antioxidant enzymes and lipid peroxidation in endometrium of patients with polyps, myoma, hyperplasia and adenocarcinoma. *Reprod Biol Endocrinol, 7,* 149.

Renger, G. (1989). On the mechanism of photosystem II deterioration by UV-Birradiation. *Photochem. Photobiol., 49,* 97–105,

Rezaei, F., Moaveni, P., & Mozafari, H. (2015). Effect of different concentrations and time of nano TiO₂ spraying on quantitative and qualitative yield of soybean (Glycine max L.) at Shahr-e-Qods. *Iran. Biol. Forum, 7*(1), 957-964.

Rico, C. M., Hong, J., Morales, M. I., Zhao, L. J., Barrios, A. C., Zhang, J. Y., ... Gardea-Torresdey, J. L. (2013b). Effect of cerium oxide nanoparticles on rice: a study involving the antioxidant defense system and in vivo fluorescence imaging. *Environ. Sci. Technol., 47*(11), 5635-5642.

Rico, C. M., Majumdar, S., & Duarte-Gardea, M. (2011). Interaction of nanoparticles with edible plants and their possible implications in the food chain. *J Agr Food Chem, 59*, 3485-98.

Rico, C. M., Peralta-Videa, J. R., & Gardea-Torresdey, J. L. (2015b). Chemistry, Biochemistry of Nanoparticles, and Their Role in Antioxidant Defense System in Plants. In *Nanotechnology and Plant Sciences* (pp. 1-17). Springer International Publishing.

Samadi, N., Yahyaabadi, S., & Rezayatmand, Z. (2014). Effect of TiO_2 and TiO_2 Nanoparticle on Germination, Root and Shoot Length and Photosynthetic Pigments of Mentha Piperita. *International Journal of Plant & Soil Science, 3*(4), 408-418.

Servin, A. D., Morales, M. I., Castillo-Michel, H., Hernandez-Viezcas, J. A., Munoz, B., Zhao, L. J., ... Gardea-Torresdey, J. L. (2013). Synchrotron verification of TiO_2 accumulation in cucumber fruit: a possible pathway of TiO_2 nanoparticle transfer from soil into the food chain. *Environ. Sci. Technol., 47*(20), 11592-11598

Shaw, A. K., & Hossain, Z. (2013). Impact of nano-CuO stress on rice (Oryza sativa L.) seedlings. *Chemosphere, 93*(6), 906-915.

Song, U., Jun, H., Waldman, B., Roh, J., Kim, Y., Yi, J., & Lee, E. J. (2013). Functional analyses of nanoparticle toxicity: a comparative study of the effects of TiO_2 and Ag on tomatoes (Lycopersicon esculentum). *Ecotoxicol. Environ. Safe, 93*, 60-67.

Stewart, R. R. C., & Bewley, J. D. (1980). Lipid peroxidation associated aging of soybean axes. *Plant Physiology, 65*, 245-248.

Wang, H. F., Zhong, X. H., Shi, W. Y., & Guo, B. (2011). Study of malondialdehyde (MDA) content, superoxide dismutase (SOD) and glutathione peroxidase (GSH-Px) activities in chickens infected with avian infectious bronchitis virus. *Afr J Biotechnol, 10*, 9213 – 9217.

Wang, S. L., Liu, H. Z., Zhang, Y. X., & Xin, H. (2015). The effect of CuO NPs on reactive oxygen species and cell cycle gene expression in roots of rice. *Environ. Toxicol. Chem., 34*(3), 554-561.

Yang, F., Liu, C., Gao, F., Su, M., Wu, X., Zheng, L., ... Yang, P. (2007). The importance of spinach growth by nano-anatase TiO_2 treatment is relate to nitrogen photoreduction. *Biol Trace Elem Res, 119*, 77 – 88.

Zheng, L., Su, M. Y., Liu, C., Chen, L., Huang, H., Wu, X., ... Hong, F. S. (2007). Effects of nano-anatase TiO_2 on photosynthesis of spinach chloroplasts under different light illumination. *Biol Trace Elem Res*.

13

Real Time PCR: The Use of Reference Genes and Essential Rules Required to Obtain Normalisation Data Reliable to Quantitative Gene Expression

Antônio J. Rocha[1], José E. Monteiro-Júnior[2], José E.C. Freire[1], Antônio J.S. Sousa[1] & Cristiane S.R. Fonteles[3]

[1] Departamento de Bioquímica e Biologia Molecular, Avenida Humberto Monte, s/n - Pici - CEP 60440-900, Fortaleza - CE, Brasil

[2] Departamento de Biologia, Avenida Humberto Monte, s/n - Pici - CEP 60440-900, Fortaleza - CE, Brasil

[3] Departmento de Clínica Odontológica, Universidade Federal do Ceará, Rua Monsenhor Furtado, s/n - Rodolfo Teófilo - CEP 60430-350, Fortaleza - CE, Brasil

Correspondence: Antônio J. Rocha, Departamento de Bioquímica e Biologia Molecular, Avenida Humberto Monte, s/n - Pici - CEP 60440-900, Fortaleza - CE, Brasil. E-mail: antonionubis@gmail.com

Abstract

Quantitative Real-time Polymerase Chain Reaction (qPCR) is an important tool for molecular biology and biotechnology research, widely used to determine the expression levels of mRNA. Two main methods to performing qPCR are largely used: The absolute quantification, in which the mRNA levels are determined by using a standard curve and the relative method, which is based on the use of reference genes. Reference genes are widely expressed in cells of animal and plant tissues and their expression pattern are theoretically unchanged within several situations, which makes them an excellent choice to normalize mRNA quantification data in relative qPCR studies. However, several reports are increasingly showing that the use of only one reference gene in relative qPCR studies should be avoided, because in the real world their expression levels can significantly change from tissue to tissue. Several softwares, such as geNorm, BestKeeper and NormFinder, have been developed to perform data normalisation, and these programs may assist in choosing the most stable reference genes. The aim of this review was to describe the current normalisation strategies used in qPCR assay, as well as to establish essential rules to perform reliable mRNA quantification. Finally, this review show some innovations in the advances on qPCR.

Keywords: primer design, DNA binding dyes, probes, normalisation

1. Introduction

The polymerase chain reaction (PCR) technique was first introduced by Kary Mullis (Saiki et al., 1985). PCR is historically used as a sensitive method for the detection and amplification of specific sequences of nucleic acids in a sample. Advances in the specificity and sensibility of PCR reactions gave birth to a more sensitive PCR technique, namely quantitative PCR (qPCR-quantitative real-time polymerase reaction), which utilizes mainly cDNA as template, a complementary DNA from RNA molecules through of the reverse transcriptase reaction. In these reactions, fluorescent reporters used include double-stranded DNA (dsDNA) binding dyes or probes that are incorporated into the product during amplification. The increase in fluorescent signal is directly proportional to the number of PCR product molecules generated in the reaction.

qPCR is amongst the best available methods to determining changes in gene expression, due their ability to rapidly and accurately quantify target genes, even in the presence of very low expression levels (Holland, 2002). Prior to the analysis of gene expression, the selection of an appropriate normalisation strategy is essential to control for non-specific variations between samples of cDNA. The most commonly used method to normalising qPCR data is relied on the use of one or more endogenous reference genes (Hamalainen et al., 2001; Rebouças et al., 2013).

Reference genes have uniform and stable expression in a wide variety of tissues and cell types, at different developmental stages, and comprise all genes that express protein products involved in basic cellular processes

(Reid et al., 2006), showing none or only minimal changes in the expression levels between individual samples and experimental conditions (Rebouças et al., 2013). These genes are largely used as internal controls for normalisation in gene expression studies in different tissues and/or condition as in plants and animals (Wong et al., 2005; Kumar et al., 2013; Sara et al., 2013; Rocha et al., 2013; Nakayama et al., 2004). Several reference genes, including those coding for biological products such as tubulins, actin, glyceraldehyde-3-phosphate dehydrogenase (GAPDH), phosphatases, albumin, cyclophilin, micro-globulin, ribosomal units (18S rRNA) or ubiquitin (UBQ) have been described in the literature (Foss et al., 2003; Rocha et al., 2013). The correct choice of reference genes is crucial to properly analyze the results of qPCR (Suzuki et al., 2000) and to measure and reduce the errors from variations among the samples (Barsalobres-Cavallari et al., 2009).

Several research groups have developed software tools to identify the most stable expressed genes across a set of samples in order to perform data normalisation. These tools include geNorm, NormFinder and BestKeeper (Vandesompele et al., 2002; Pfaffl et al., 2004; Andersen & Orntoft, 2004), which is freely available on the web and allows researchers to find the best reference gene for their experiments. A great number of studies describing the identification of multiple reference genes for normalisation of qPCR data using these algorithms have been performed on the animal and human health fields (Hong et al., 2008; De Boever et al., 2008), but similar reports are scarce in plant research (Jain et al., 2006; Ransbotyn et al., 2006; Exposito-Rodriguez et al., 2008).

The aim of this review was to evaluate the importance of the application of reference genes in normalisation strategies of qPCR assays, in different tissues or experimental conditions, as well as to describe essential rules necessary to conduct successful qPCR experiments. Besides, we pointed out several precautions required for a good qPCR. Finally, this review shows some innovations in the advances on qPCR in the last years.

2. DNA Binding Dyes Versus Hydrolysis Probes in qPCR

PCR is one of the most versatile technologies in molecular biology. The PCR reaction consists of 3 different stages which involve, (a) the DNA denaturation; (b) the primer annealing and (c) the extension phase (Mullis et al., 1987). In traditional (endpoint) PCR, the detection and quantification of amplified target sequences are performed at the end of the reaction, and it involves additional work, such as gel electrophoresis and image analysis. Nevertheless, in qPCR, the amount of PCR product is measured along each reaction cycle. The ability to monitor the reaction during its exponential phase enables users to determine the initial amount of target gene with great precision (Wong et al., 2005).

In qPCR, the amount of DNA is measured by the use of fluorescent markers that are incorporated into the PCR product. The increase in fluorescent signal is directly proportional to the number of PCR product molecules (amplicons) generated in the exponential phase of the reaction. Fluorescent reporters used include double-stranded DNA (dsDNA) binding dyes or probes that are incorporated into the product during amplification (Bustin et al., 2002). SYBR Green is an example of a fluorescent dye which binds to the double-stranded DNA and emits light upon excitation. Once the reaction proceeds and the PCR product is accumulated, the fluorescence levels increase proportionally to the amount of DNA present in the original sample (Livak et al., 1995; Pabla et al., 2008; Bustin et al., 2002). This dye is used to monitor the amplification of any DNA sequences and dispenses the use of a probe, thus reducing the cost of amplification and providing a great advantage in its application. On the other hand, since the dye binds not only to the target DNA, but to all dsDNA formed during qPCR, the use of SYBR Green, while simple lacks specificity (Figure 1a). The specificity of the reactions, however, can be easily accessed by the use of melting curve analysis (Dheda et al., 2004).

In addition to DNA binding dyes, there are probes, such as TaqMan®, which are designed to binds to specific DNA sequences. TaqMan® probes primarily consist in a oligonucleotide sequence complementary to some regions of the target DNA. The probe is complexed with a quencher and a reporter fluorophore dye at its 3' and 5' ends, respectively (Livak et al., 1995). During the amplification step the probe is associated to its complementary target DNA and then is cleaved by *Taq* DNA polymerase 5'-3' exonuclease activity (Figure 1b). This cleavage releases the reporter dye and generates a fluorescent signal that increases with each cycle (Bustin et al., 2002). TaqMan® provides higher specificity than DNA intercalating dyes, such as SYBR Green. In addition, these probes can also be labeled with distinct and distinguishable reporter dyes, which allows the amplification of two different sequences in the same reaction tube, eliminating the post-PCR processing, and reducing hand labor. The main drawback of this system is the requirement to synthesize specific probes to each target sequences, increasing the cost of the assay (La Cruz et al., 2013).

Another type of probe which is largely used in qPCR assay is molecular beacon. When free in solution molecular beacon probes assume a hairpin structure consisting of a quencher and a reporter dye (Tyagi et al., 1996). The reporter fluorescent dye and the quencher remain extremely close and therefore no fluorescence is detected when this structure is formed (Figure 1c). However, during the annealing step, Molecular Beacon hybridizes to the target sequence generating conformational changes leading to the separation of reporter and quencher dyes, which results in the emission of fluorescence (Tyagi et al., 1996; VanGuilder et al., 2008). The greater specificity for mismatch discrimination is due to structural constraints. However, the main disadvantage associated with Molecular Beacons is the accurate design of the hybridization probe. Optimal design of the Molecular Beacon stem annealing strength is crucial (Wong et al., 2005).

Scorpions consist of a single-stranded oligonucleotide probe of approximately 20 to 25 nt carrying a reporter fluorophore at its 5' end and a quencher at its 3' end. Their tridimensional conformation resembles a stem and loop structure, in which a PCR primer is attached (figure 1d). The stem-and-loop structure acts as a blocker to prevent DNA polymerase activity during the interaction of the probe with the target DNA (Bustin et al., 2002; Ng et al., 2005). The close proximity of the reporter to the quencher leads to a continuous suppression of the fluorescence emitted by the reporter. At the beginning of the PCR, *Taq*DNA polymerase extends the PCR primer and synthesizes the complementary strand of the target sequence (Whitcombe et al., 1999). During the next cycle, the stem-and-loop structure unfolds and the loop region of the probe hybridizes intra-molecularly to the newly synthesized target sequence. The reporter is excited by light from the qPCR instrument (Bustin et al., 2002). Once the reporter dye is no longer in close proximity to the quencher dye, fluorescence emission may take place. The significant increase of the fluorescent signal is detected by the qPCR instrument and it is directly proportional to the amount of target DNA (Holland, 2002; Wong et al., 2005; Kumar et al., 2013). Scorpions have the advantage to providing a stronger signal and lower level of background when in compared to other probes, such as molecular beacons (Bustin et al., 2002).

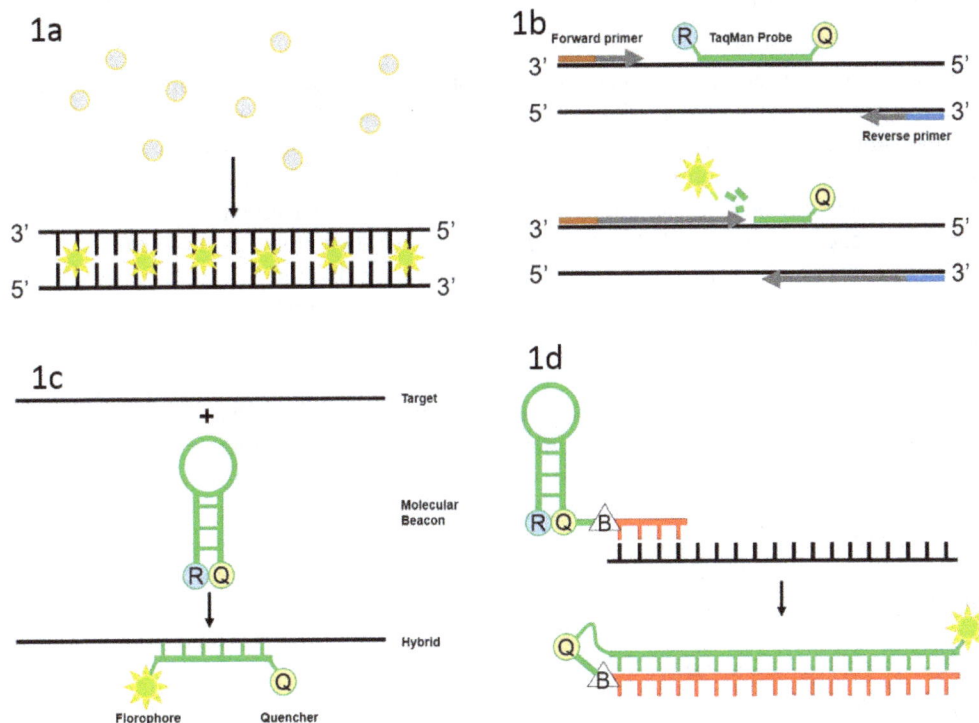

Figure 1. Probes and Dyes used in Real time PCR assay; a) SYBR Green; b) TaqMan; c) Molecular Beacon; d) Scorpions

3. The Use of Reference Genes to Normalize qPCR Data

Reference genes in qPCR are critical for normalisation of expression levels, thus, avoiding misinterpretation of results obtained by real time PCR data. In recent years, it has become clear that no single gene is constitutively expressed in all cell types and under all experimental conditions. This implies that the expression stability of a putative control gene (reference gene) must be verified before each qPCR assay and that the use of only one reference gene is generally not enough to normalize the expression data (Livak et al., 2001; Lee et al., 2010).

The choice of several reference genes to normalize and validate the final results may significantly influence the accuracy of gene expression. Consequently, the use of inappropriate reference genes for normalisation of expression data may lead to erroneous results and data misinterpretation (Suzuki & Higgins, 2000), because normalisation is a pivotal step that provides the Cq values-based differences between the reference and target genes, avoiding misinterpretation of the results and providing reliable Cqs, thus rendering a more accurate and reliable gene expression (Vandesompele et al., 2002). The Ct or threshold cycle value is the cycle number at which the fluorescence generated within a reaction crosses the fluorescence threshold, a fluorescent signal significantly above the background fluorescence. Therefore, the selection of appropriate reference genes is a critical step before evaluating gene expression in new species and/or tissues (Condori et al., 2001; Cordoba et al., 2001). The best candidate genes are those selected by programs used to establish reference genes, such as geNorm, BestKeeper and NormFinder. Therefore, the normalisation using appropriated reference genes are pivotal to acquire suitable data and avoid and misinterpretation of the experiments.

4. Algorithms Used to Normalize qPCR Data

In the last decade, relevant tools to select genes for normalisation have become available. Several research groups have developed softwares to identify the most stably expressed genes across a set of samples. Among these tools we will focus on the most cited articles as geNorm, NormFinder and Bestkeeper (Vandesompele et al., 2002; Andersen & Orntoft, 2004; Pfaffl et al., 2004), which are freely available on the web and allow researchers to find the best reference gene for their experiments. These programs allow the calculation of a normalisation factor over multiple reference genes, which improve the robustness of the normalisation even further (Dekkers et al., 2012). Different manners to access the stability of putative reference genes are available using the upon mentioned software. Hence, BestKeeper employs quantification cycle (C_q) values directly for stability calculations, whereas geNorm and NormFinder have these values transformed to relative quantities using normalisation factor (NF) (Mallona et al., 2004).

3.1 Genorm Analysis

The geNorm program has been recently reported to be one of the best statistical methods to identify stably expressed genes for qPCR analysis. The geNorm calculates a gene-stability measure M as the Average pairwise variation V of a particular gene reported to all other control genes. Genes with the lowest M values have the most stable expression. Stepwise exclusion of the gene with the highest M value allows ranking of the tested genes according to the stability (Condori et al., 2001; Cordoba et al., 2001; Vandesompele et al., 2002; Zhong et al., 2009). The analysis relies on the principle that the expression ratio of two proper control genes should be identical in all samples, regardless of the experimental conditions or cell type, and the M values below cutoff (< 1.5) are regarded the most stable genes among all candidate reference genes (Vandesompele et al., 2002).

The geNorm program estimates also the number of genes required to be used as appropriate controls for normalisation by evaluation of variation in pairs (V values), checking the variation of the expression of two by two possible genetic combinations . The optimal number of reference genes that should be used for accurate normalisation also depends on the specific experimental condition, which is determined by calculating V-values as a pairwise variation (Vn/Vn+1) between two consecutively ranked normalisation factors (NF) after the stepwise addition of the subsequent more stable reference gene (NFn and NFn+1) (Vandesompele et al., 2002). Actually, the geNorm is part of qBASEPlus (Biogazelle) program as tool important to provide the reference genes more stables (M value) and the number of genes suitable to normalisation (V value). Furthermore, the qBASEPlus (Biogazelle) also provides the relative expression on qPCR experiments based on the normalisation factor (NF). The use of the qBASEPlus (Biogazelle) is needed at least 8 reference genes and at least 2 samples (control and conditions) for to analyze the qPCR data.

3.2 Normfinder Analysis

The NormFinder is an algorithm used to identify the optimal normalisation gene among a set of candidates. It ranks the set of candidate normalisation genes according to their expression stability in a determined sample set and given experimental design. This algorithm is rooted in a mathematical model of gene expression and uses a solid statistical framework to estimate not only the overall expression variation of the candidate normalisation genes, but also the variation between sample subgroups of the sample set e.g. normal and cancer samples (Andersen & Orntoft, 2004). Notably, "NormFinder" provides a stability value for each gene, which is a direct measure for the estimated expression variation, enabling the user to evaluate the systematic error introduced when using the gene for normalisation (Dekkers et al., 2012; Selim et al., 2012).

Real Time PCR: The Use of Reference Genes and Essential Rule Required...

123

3.3 Bestkeeper Analysis

The BestKeeper calculates standard deviation (SD) and the coefficient of variation (CV) based on Cq values of all reference candidate genes. Genes with SD less than 1 are considered stable. Subsequently, the program calculates a pairwise correlation coefficient between each gene and the BestKeeper index–geometric mean between Ct values of stable genes grouped together. Genes with the highest coefficient of correlation with the BestKeeper Index indicates the highest stability (Pfaffl et al., 2004). The BestKeeper use raw Ct data and determines the most stably expressed genes based on a correlation coefficient (r) of the BestKeeper Index (BI) and standard deviation, whereas BI is the geometric mean of Ct values of best reference genes. Hence, this program relies on the "r" and "SD" values, and the higher the "r" value, the most stable is the gene; otherwise, the lower the standard deviation value, the most stable is the gene (Pfaffl et al., 2004; Demidenko et al., 2011; Niu et al., 2011; Petit et al., 2012).

These statistical algorithms have been developed for the evaluation of best suited reference gene(s) for normalisation of qPCR data in a set of biological samples. Recognizing the importance of reference genes in normalisation of qPCR data, various reference genes have been evaluated for their stable expression under specific conditions in various organisms. Many studies have been conducted in the animal and human health (De Boever et al., 2008; Hong et al., 2008) fields that describe the identification of multiple reference genes for normalisation of qPCR data, but similar reports are scarce in plant research (Ransbotyn et al., 2006; Exposito-Rodriguez et al., 2008).

The three algorithms are important for reference gene stability and normalisation data during qPCR experiments; however, geNorm is the best tool since in addition to providing the best reference genes in geNorm M, this software supplies the V value, which delivers the number of genes needed for use in normalisation data in a qPCR experiment. The algorithms NormFinder and BestKeeper will only identify the most stable genes. Generally, all three algorithms are used to render more reliable results for normalisation.

5. Essential Rules Required to Perform a Reliable qPCR

The efficiency and specificity of quantitative PCR depends on several parameters related to quantification of mRNA, which must be controlled to avoid errors of interpretation: purification of RNA, efficiency of primer specifics, normalisation of reference genes, tissue inhibitory factors, enzyme loading error (Rocha, Miranda, & Cunha, 2014), pipetting errors, among others (Thellin et al., 1999; Livak et al., 2001; Suzuki et al., 2000; Vandesompele et al., 2002) as described below.

5.1 Primers and Probes Design Considerations

The primers and probes design are essentials for amplification efficiency, specificity and fluorescence, respectively. The primers are specially needed in junction exon-exon to avoid an amplification of DNA genomic, ensuring the amplification of only a target gene specific cDNA sequence. In addition, it may be necessary to digest input DNA with an RNase free DNA in the following circumstances: (1) to avoid DNA amplification during qPCR; (2) to use primers that either flank an intron that is not present in the mRNA sequence or that span an exon-exon junction; (3) when the gene of interest has no introns; (4) if the intron positions are unknown; (5) when there are no suitable primers that span or flank introns (Udvardi et al., 2004). There are several programs used to design automatic primers, such as Perl Primer (Marshall, 2004) that will require previous annotation of genes, establishing the introns and exons of each sequence to input the program. Other programs are available in the web as primer BLAST, a tool available at http://www.ncbi.nlm.nih.gov/tools/primer-blast/index.cgi?LINK_LOC=BlastHome in the GenBank of NCBI, as well as Primer3 Plus available at http://www.bioinformatics.nl/ cgi-bin/primer3plus/primer3plus.cgi/. Moreover, an absence of primer-dimer and non-specific amplification is especially important to suitable data of qPCR. Therefore, the presence of homo-dimers, hetero-dimers, as well as self-dimers must be avoided, and the formation of harpin of a forward or reverse primer (Condori et al., 2001; Rocha et al., 2013).

The probes, such as TaqMan®, Molecular beacon and scorpions are primers marked with fluorophores to emit fluorescence. These probes are designed in different forms, but are used with the common purpose of emitting fluorescence to assess the increase on gene expression due to the number of probes that bind a double-stranded DNA (Bustin et al., 2002; Pabla et al., 2008; VanGuilder et al., 2008; Hwang et al., 2013). There are programs, such as primer BLAST, available in the GenBank of NCBI, as well as Primer3 Plus both available in the web. These are the same programs used to design primers and probes (Condori et al., 2001).

5.2 RNA Quality

The quality of RNA also is very important to provide accurate qPCR data. The quality of RNA depends on extraction and purification of RNA, for example, during the extraction of RNA contaminants such as proteins, carbohydrate, as well as phenolic compounds that will affect the PCR reaction by inhibiting the action of polymerases as reverse transcriptase and DNA polymerases, during qPCR must be avoided. Therefore, RNA of good quality is needed for further experiments. According to Sambrook et al. (1989), the best relations absorbance by spectrophotometer are as follows: RNA relation to $A_{260/280}$: 1.8-2.0, which is the acceptable limit of contamination with proteins, and $A_{260/230}$: > 2,0 for contamination with carbohydrates. These data have also been reported by Sambrook et al. (1989) and Romano (1998). Furthermore, to avoid contamination with genomic DNA, Digest purified RNA with DNase I is needed to remove contaminating genomic DNA, which can act as template during PCR and may lead to spurious results. It may also be necessary to perform PCR on the treated RNA, using gene-specific primers, to confirm absence of genomic DNA (Udvardi et al., 2008).

To complete a reliability of the extracted and purified RNA, the integrity of the RNA requires evaluation. The measure of RNA reliability is based on the integrity of 28S and 18S ribosomal RNA and the lack thereof shows a smear in the agarose gel, indicating that the total RNA is degraded. Thus, an electrophoresis in agarose gel at 0.8% to 1.0% is recommended to observe the integrity of the ribosomal RNA bands (Sambrook et al., 1989).

5.3 Optimization and Efficiency Curve of Primers

Other parameters such as the optimization of primer concentrations and efficiency primer curves that might be done in serial dilutions or standard curves are important to perform qPCR assays. A dilution series of known template concentrations can be used to establish a standard curve for determining the initial starting amount of the target template or for assessing the reaction efficiency. The log of each known concentration in the dilution series is plotted against the Cq value for that concentration. Information on the performance of the reaction as well as various reaction parameters (including slope, y-intercept, and correlation coefficient) can be derived from this standard curve. The slope is obtained by the linear equation of the graph constructed by plotting the Cq values on the y-axis and the log values of the dilutions on the x-axis. The concentrations chosen for the standard curve should encompass the expected concentration range of the target (Pfaffl et al., 2004). At the end of the qPCR assay, primer efficiency must be calculated, and the formula most frequently described in the literature for this purpose is as follows: Efficiency = $10^{(-1/slope)}$-1, in which the slope corresponds to the C_q value of the first dilution (concentration dilution) minus the C_q value of the last dilution divided by the number of dilutions. Hence, if the PCR is 100% efficient, the amount of PCR product will double with each cycle and the slope of the standard curve will be –3.33 (100 = 100% = $10^{(-1/-3.33)}$-1). The ideal slope is approximately -3.33 cycles; however, a slope between –3.9 and –3.0 (80-110% efficiency) is generally acceptable (Livak et al., 2001; Pfaffl et al., 2004). Calculated levels of target input may not be accurate if the reaction is not efficient. In order to improve efficiency, one must consider either (1) optimize primer concentrations or (2) design alternative primers.

Since SYBR Green binding dye is a non-specific dye that will detect any double-stranded DNA, it is important to verify if the qPCR is producing only the desired product. This can often be detected when PCR efficiencies are larger than 120% (Bustin et al., 2002; Bustin et al., 2009). Melting or dissociation curve is expressed during the last step of qPCR, following 40 cycles that only show one peak, revealing that a single multigene family isoform was amplified. These analyses can also be used to determine the approximate product size (Udvardi et al., 2004). If the melting curve has more than one major peak, the identities of the products should be determined by fractionating them on an ethidium DNA agarose gel electrophoresis to check for the presence of non-specific annealing. It must also be mentioned that lowering the primer concentrations will often reduce the amount of non-specific products. If the use of low primer levels still allow the detection of non-specific products in significant amounts, primer redesign may be a necessary measure. Once all cycles have been completed, the melting curve is added to evaluate the specificity of the primers. Melting curves with peaks lower than 78°C could indicate the presence of primer dimmers in the reaction or alternatively smaller non-specific amplicon products (Condori et al., 2001).

5.4 Normalisation and Analysis of qPCR Results

In gene-expression profile quantification, an assessment of the reliability of qPCR assay is required to normalize the target gene expression data. One of the most frequently used methods is the utilization of reference genes. Previous to the qPCR assay, it is necessary to design primers that amplify constitutive genes. The groups of reference genes are checked for stability to identify the most stable reference genes among all the selected genes that will be used to normalize the qPCR data, using programs such as geNorm, BestKeeper and NormFinder. Once the best reference genes are identified, data normalisation is required to ensure gene expression reliability (Zhong et al., 2011). Likewise, to confirm reliability of the results, biological and technical replicates must be obtained to provide data statistics, and evaluate the significance levels of gene expression analysis (Udvardi et al., 2004).

Relative quantification describes a real-time PCR experiment in which the gene of interest in one sample (i.e., treated) is compared to the same gene in another sample (i.e., untreated). The results are expressed as fold up- or down-regulation of the treated in relation to the untreated gene. Reference genes such as β-actin, GAPDH, elongation factor, among others are used as a control for experimental variability in this type of quantification (Tong et al., 2009). The most frequently used method for relative mRNA quantification by real time PCR has been described by Livak et al. (2001). This is a convenient method which presents the advantage of eliminating the need for standard curves. Thus, mathematical equations are used to calculate the relative expression levels of target relative to reference control or calibration, such as an untreated sample or RNA from normal tissue or a sample at time zero at qPCR experiments in time-course study. The amount of target gene in the sample normalized to a reference gene, relative to the normalized calibrator, is then given: $2^{-\Delta\Delta C_q}$, where $\Delta\Delta C_q = \Delta C_q$ (sample)- ΔCq (calibrator), and ΔC_q is the C_q of the target gene subtracted from the reference gene C_q, as describe by (Livak et al., 2001; Schmittgen et al., 2000; Schmittgen et al., 2008). In order to obtain reliable results, the target and reference gene must be approximately equal, and preferably at a percentage greater than 90%. This level of sequence equality is necessary to plot an efficiency curve based on the dilution serial method to given suitable results in experimental data, as described above. Finally, statistics methods, including student t-test, ANOVA, among others, must be applied to the concluding analysis. However, this method of Livak et al (2001) is limited due the use of only one reference gene. Actually, has been used more than one reference genes to normalisation data qPCR using algorithm that is based in the normalisation factor (NF) method, as geNorm, BestKeeper and NormFinder (Vandesompele et al., 2002; Andersen & Orntoft, 2004; Pfaffl et al., 2004) as described above.

6. Several Advances on the Real Time PCR

Several researchers are developing techniques to improve the quality of detection of DNA fluorescence. Recently, the manufacturer ELITE MGB™ has done a revolutionary advance in qPCR chemistry. The principle is based in the proprietary of the protein called Minor Groove Binder (MGB), Superbases™ and Eclipse®Dark Quencher technologies. These overlapping probes are much efficiently and accurately detect target DNA sequences, while offering greater sensitivity and specificity. According to manufacturer, the MGB protein is a synthetic molecule that binds to the minor groove of double stranded DNA molecules. In qPCR applications, MGB increases the stability of double stranded DNA complexes, specifically, the hybridization between the probe and the amplified DNA target. The increased DNA-DNA hybrid stability allows the design of shorter detection probes with higher specificity. Furthermore, The Eclipse®Dark Quencher is a proprietary fluorophore and dye quencher chemistry resulting in low background signals. Its key benefit is to ensure that every ELITe MGB™ assay will have the highest sensitivity by minimizing background signal interference. Together, show Real-Time PCR results of high accuracy.

Other works have shown important improving in the qPCR. Zheng et al. (2011) designed an aptamer-based sensing platform using a triple-helix molecular switch (THMS). The THMS consists of a central, target-specific aptamer sequence flanked by two arm segments and a dual-labeled oligonucleotide serving as a signal-transduction probe (STP). The STP is doubly labeled with pyrene at both ends and designed as a hairpin-shaped structure. Initially, the loop sequence of the STP binds with two arm segments of the aptamer, which forces the STP to form an ''open'' configuration and separate the two end labeled pyrene molecules, thus only emitting monomer fluorescence signal. The formation of the aptamer/target complex releases the STP, which switches to a ''closed'' hairpin configuration, bringing two pyrene molecules in close proximity and emitting excimer fluorescence signal. Hu et al. (2014) developed a modified Molecular Beacons–based multiplex qPCR Assay. In their work, all sets of primers and probes were combined, and the concentration of each reagent including primers, probes, magnesium, and Taq polymerase concentrations in the reaction mix were optimized. These modifications helped the sensitivity and specificity of the qPCR multiplex that were 100% and 99%, respectively.

Any need for fast and precise measurement of small amounts of nucleic acids represents a potential future niche for real-time PCR-based innovations. As machines become faster, cheaper, smaller, and easier to use through competition, standardized assay development, and advances in microfluidics (Mitchell et al., 2001), optics, and thermocycling, more in-field application needs are likely to be filled. In the commercial food industry and agriculture, real-time PCR will likely see expanded use for the detection and identification of microbes, parasites, or genetically modified organisms. Forensics will benefit from real-time PCR's sensitivity, specificity, and speed, especially because time is crucial to many criminal investigations and specimen size may be limited. Reduced cost and increased portability open the door for the diagnosis of diseases in remote areas along with on-site epidemiological studies and may facilitate the transfer of needed scientific technologies to developing countries, thereby contributing to their "scientific capacity".

7. Final Considerations

The polymerase chain reaction (PCR) is one of the most powerful technologies in molecular biology. qPCR is an efficient tool to measure the levels of mRNA expression in different types of samples; their use together with the reference genes are ideal for decreasing the possible errors in RNA extraction and contamination during sample manipulation, thus increasing the quality of cDNA. The qPCR has the sensible technical power to amplify target specific genes, but it is necessary to obtain reliable results in the gene expression profile. Several parameters must be considered, including good design of primers, evaluating their specificity and efficiency. In addition, the RNA extracted must be free of contaminants, such as carbohydrates, proteins and phenols, because these may interfere with the polymerases during PCR reaction. For the normalisation of qPCR data, the use of reference gene is needed to provide suitable results and reproducibility. Thus, the selection of a reference gene for each experimental condition is crucial. These precautions are pivotal to render reliable results during gene expression analysis.

Acknowledgements

The authors are grateful to CNPq and CAPES. This work was supported by the Universidade Federal do Ceará-UFC.

References

Andersen, C. L., Jensen, J. L., & Ørntoft, T. F. (2004). Normalization of real-time quantitative reverse transcription-PCR data: a model-based variance estimation approach to identify genes suited for normalization, applied to bladder and colon cancer data sets. *Cancer research*, *64*(15), 5245-5250. http://dx.doi.org/10.1158/0008-5472.CAN-04-0496

Barsalobres-Cavallari, C. F., Severino, F. E., Maluf, M. P., & Maia, I. G. (2009). Identification of suitable internal control genes for expression studies in Coffea arabica under different experimental conditions. *BMC molecular biology*, *10*(1), 1. http://dx.doi.org/10.1186/1471-2199-10-1

Bustin, S. A. (2002). Quantification of mRNA using real-time reverse transcription PCR (RT-PCR): trends and problems. *Journal of molecular endocrinology*, *29*(1), 23-39. http://dx.doi.org/10.1677/jme.0.0290023

Bustin, S. A., Benes, V., Garson, J. A., Hellemans, J., Huggett, J., Kubista, M., ... & Wittwer, C. T. (2009). The MIQE guidelines: minimum information for publication of quantitative real-time PCR experiments. *Clinical chemistry*, *55*(4), 611-622. http://dx.doi.org/10.1373/clinchem.2008.112797

Condori, J., Nopo-Olazabal, C., Medrano, G., & Medina-Bolivar, F. (2011). Selection of reference genes for qPCR in hairy root cultures of peanut. *BMC research notes*, *4*(1), 392. http://dx.doi.org/10.1186/1756-0500-4-392

Cordoba, E. M., Die, J. V., González-Verdejo, C. I., Nadal, S., & Román, B. (2011). Selection of reference genes in Hedysarum coronarium under various stresses and stages of development. *Analytical biochemistry*, *409*(2), 236-243. http://dx.doi.org/10.1016/j.ab.2010.10.031

De Boever, S., Vangestel, C., De Backer, P., Croubels, S., & Sys, S. U. (2008). Identification and validation of housekeeping genes as internal control for gene expression in an intravenous LPS inflammation model in chickens. *Veterinary immunology and immunopathology*, *122*(3), 312-317. http://dx.doi.org/10.1016/j.vetimm.2007.12.002

Dekkers, B. J., Willems, L., Bassel, G. W., van Bolderen-Veldkamp, R. M., Ligterink, W., Hilhorst, H. W., & Bentsink, L. (2012). Identification of reference genes for RT-qPCR expression analysis in Arabidopsis and tomato seeds. *Plant and Cell Physiology*, *53*(1), 28-37. http://dx.doi.org/10.1093/pcp/pcr113

Demidenko, N. V., Logacheva, M. D., & Penin, A. A. (2011). Selection and validation of reference genes for quantitative real-time PCR in buckwheat (Fagopyrum esculentum) based on transcriptome sequence data. *PLoS One*, *6*(5), e19434. http://dx.doi.org/10.1371/journal.pone.0019434

Dheda, K., Huggett, J. F., Bustin, S. A., Johnson, M. A., Rook, G., & Zumla, A. (2004). Validation of housekeeping genes for normalizing RNA expression in real-time PCR. *Biotechniques*, *37*, 112-119.

Expósito-Rodríguez, M., Borges, A. A., Borges-Pérez, A., & Pérez, J. A. (2008). Selection of internal control genes for quantitative real-time RT-PCR studies during tomato development process. *BMC Plant Biology*, *8*(1), 131. http://dx.doi.org/10.1186/1471-2229-8-131

Foss, D. L., Baarsch, M. J., & Murtaugh, M. P. (1998). Regulation of hypoxanthine phosphoribosyltransferase, glyceraldehyde‐3‐phosphate dehydrogenase and β‐actin mRNA expression in porcine immune cells and tissues. *Animal biotechnology*, *9*(1), 67-78. http://dx.doi.org/10.1080/10495399809525893

Hamalainen, H. K., Tubman, J. C., Vikman, S., Kyrölä, T., Ylikoski, E., Warrington, J. A., & Lahesmaa, R. (2001). Identification and validation of endogenous reference genes for expression profiling of T helper cell differentiation by quantitative real-time RT-PCR. *Analytical biochemistry*, *299*(1), 63-70. http://dx.doi.org/10.1006/abio.2001.5369

Holland, M. J. (2002). Transcript abundance in yeast varies over six orders of magnitude. *Journal of Biological Chemistry*, *277*(17), 14363-14366. http://dx.doi.org/10.1074/jbc.C200101200

Hong, S. Y., Seo, P. J., Yang, M. S., Xiang, F., & Park, C. M. (2008). Exploring valid reference genes for gene expression studies in Brachypodium distachyon by real-time PCR. *BMC plant biology*, *8*(1), 112. http://dx.doi.org/10.1186/1471-2229-8-112

Hwang, S., Kang, B., Hong, J., Kim, A., Kim, H., Kim, K., & Cheon, D. S. (2013). Development of duplex real - time RT - PCR based on Taqman technology for detecting simultaneously the genome of pan - enterovirus and enterovirus 71. *Journal of medical virology*, *85*(7), 1274-1279. http://dx.doi.org/10.1002/jmv.23588

Jain, M., Nijhawan, A., Tyagi, A. K., & Khurana, J. P. (2006). Validation of housekeeping genes as internal control for studying gene expression in rice by quantitative real-time PCR. *Biochemical and biophysical research communications*, *345*(2), 646-651. http://dx.doi.org/10.1016/j.bbrc.2006.04.140

Kumar, K., Muthamilarasan, M., & Prasad, M. (2013). Reference genes for quantitative real-time PCR analysis in the model plant foxtail millet (Setaria italica L.) subjected to abiotic stress conditions. *Plant Cell, Tissue and Organ Culture (PCTOC)*, *115*(1), 13-22. http://dx.doi.org/10.1007/s11240-013-0335-x

La Cruz, S., Lopez-Calleja, M. I., Alcocer, M., González, I., Martín, R., & García, T. (2013). TaqMan real-time PCR assay for detection of traces of Brazil nut (*Bertholletia excelsa*) in food products. *Food control*, 140, 382-389. http://dx.doi.org/10.1016/j.foodcont.2013.01.053

Lee, J. M., Roche, J. R., Donaghy, D. J., Thrush, A., & Sathish, P. (2010). Validation of reference genes for quantitative RT-PCR studies of gene expression in perennial ryegrass (Lolium perenne L.). *BMC Molecular Biology*, *11*(1), 8. http://dx.doi.org/10.1186/1471-2199-11-8

Livak, K. J., & Schmittgen, T. D. (2001). Analysis of relative gene expression data using real-time quantitative PCR and the 2− ΔΔCT method. *methods*, *25*(4), 402-408. http://dx.doi.org/10.1006/meth.2001.1262

Mallona, I., Lischewski, S., Weiss, J., Hause, B., & Egea-Cortines, M. (2010). Validation of reference genes for quantitative real-time PCR during leaf and flower development in Petunia hybrida. *BMC Plant Biology*, *10*(1), 4. http://dx.doi.org/10.1186/1471-2229-10-4

Marshall, O. J. (2004). PerlPrimer: cross-platform, graphical primer design for standard, bisulphite and real-time PCR. *Bioinformatics 20*(15): 2471–2472. http://dx.doi.org/10.1093/bioinformatics/bth254

Mitchell, P. (2001). Microfluidics-downsizing large-scale biology. Nature biotechnology, 19(8), 717-721. http://dx.doi.org/10.1093/bioinformatics/bth254

Mullis, K. B., & Faloona, F. A. (1987). Specific synthesis of DNA in vitro via a polymerase-catalyzed chain reaction. *Methods in enzymology*, *155*, 335-350. http://dx.doi.org/10.1016/0076-6879(87)55023-6

Ng, C. T., Gilchrist, C. A., Lane, A., Roy, S., Haque, R., & Houpt, E. R. (2005). Multiplex real-time PCR assay using Scorpion probes and DNA capture for genotype-specific detection of Giardia lamblia on fecal samples. *Journal of clinical microbiology*, *43*(3), 1256-1260. http://dx.doi.org/10.1128/JCM.43.3.1256-1260.2005

Niu, J. Z., Dou, W., Ding, T. B., Yang, L. H., Shen, G. M., & Wang, J. J. (2012). Evaluation of suitable reference genes for quantitative RT-PCR during development and abiotic stress in Panonychus citri (McGregor)(Acari: Tetranychidae). *Molecular biology reports*, *39*(5), 5841-5849. http://dx.doi.org/10.1007/s11033-011-1394-x

Pabla, S. S., & Pabla, S. S. (2008). Real-time polymerase chain reaction. *Resonance*, *13*(4), 369-377. http://dx.doi.org/10.1007/s12045-008-0017-x

Petit, C., Pernin, F., Heydel, J. M., & Délye, C. (2012). Validation of a set of reference genes to study response to herbicide stress in grasses. *BMC research notes*, *5*(1), 18. http://dx.doi.org/10.1186/1756-0500-5-18

Pfaffl, M. W., Tichopad, A., Prgomet, C., & Neuvians, T. P. (2004). Determination of stable housekeeping genes, differentially regulated target genes and sample integrity: BestKeeper–Excel-based tool using pair-wise correlations. *Biotechnology letters*, *26*(6), 509-515. http://dx.doi.org/10.1023/B:BILE.0000019559.84305.47

Ransbotyn, V., & Reusch, T. B. (2006). Housekeeping gene selection for quantitative real - time PCR assays in the seagrass Zostera marina subjected to heat stress. *Limnology and Oceanography: Methods, 4*(10), 367-373. http://dx.doi.org/10.4319/lom.2006.4.367

Rebouças, E. D. L., Costa, J. J. D. N., Passos, M. J., Passos, J. R. D. S., Hurk, R. V. D., & Silva, J. R. V. (2013). Real time PCR and importance of housekeepings genes for normalization and quantification of mRNA expression in different tissues. *Brazilian Archives of Biology and Technology, 56*(1), 143-154. http://dx.doi.org/10.1590/S1516-89132013000100019

Reid, K. E., Olsson, N., Schlosser, J., Peng, F., & Lund, S. T. (2006). An optimized grapevine RNA isolation procedure and statistical determination of reference genes for real-time RT-PCR during berry development. *BMC plant biology, 6*(1), 27. http://dx.doi.org/10.1186/1471-2229-6-27

Rocha, A. J., Miranda, R., & Cunha, R. M. S. (2014). Avaliação de DNA polimerases em ensaios de amplificação de microssatélites através do PowerPlex® 16 BIO System. *BBR-Biochemistry and Biotechnology Reports, 3*(2), 1-8. http://dx.doi.org/10.5433/2316-5200.2014v3n2p1

Rocha, A. J., Soares, E. L., Costa, J. H., Costa, W. L., Soares, A. A., Nogueira, F. C., ... & Campos, F. A. (2013). Differential expression of cysteine peptidase genes in the inner integument and endosperm of developing seeds of Jatropha curcas L.(Euphorbiaceae). *Plant science, 213*, 30-37. http://dx.doi.org/10.5433/2316-5200. 2014v3n2p1

Romano, E., Brasileiro, A. C. M. (1998). Extração de DNA de Tecidos Vegetais. In A. C. M. Brasileiro, & V. T. C. (Eds.), Carneiro *Manual de transformações Genéticas De Plantas.* Editora Embrapa: Brasília v. 40-43, 1998.

Saha, G. C., & Vandemark, G. J. (2013). Stability of expression of reference genes among different Lentil (Lens culinaris) genotypes subjected to cold stress, white mold disease, and aphanomyces root rot. *Plant Molecular Biology Reporter, 31*(5), 1109-1115. http://dx.doi.org/10.1007/s11105-013-0579-y

Saiki, R. K., Scharf, S., Faloona, F., Mullis, K. B., Horn, G. T., Erlich, H. A., & Arnheim, N. (1985). Enzymatic amplification of beta-globin genomic sequences and restriction site analysis for diagnosis of sickle cell anemia. *Science, 230*(4732), 1350-1354. http://dx.doi.org/10.1126/science.2999980

Sambrook, J., Fritsch, E. F., & Maniatis, T. (1989). *Molecular cloning* (Vol. 2, pp. 14-9). New York: Cold spring harbor laboratory press.

Schmittgen, T. D., & Livak, K. J. (2008). Analyzing real-time PCR data by the comparative CT method. *Nature protocols, 3*(6), 1101-1108. http://dx.doi.org/10.1038/nprot.2008.73

Schmittgen, T. D., & Zakrajsek, B. A. (2000). Effect of experimental treatment on housekeeping gene expression: validation by real-time, quantitative RT-PCR. *Journal of biochemical and biophysical methods, 46*(1), 69-81. http://dx.doi.org/10.1016/S0165-022X(00)00129-9

Selim, M., Legay, S., Berkelmann-Löhnertz, B., Langen, G., Kogel, K. H., & Evers, D. (2012). Identification of suitable reference genes for real-time RT-PCR normalization in the grapevine-downy mildew pathosystem. *Plant cell reports, 31*(1), 205-216. http://dx.doi.org/10.1007/s00299-011-1156-1

Suzuki, T., Higgins, P. J., & Crawford, D. R. (2000). Control selection for RNA quantitation. *Biotechniques, 29*(2), 332-337.

Thellin, O., Zorzi, W., Lakaye, B., De Borman, B., Coumans, B., Hennen, G., ... & Heinen, E. (1999). Housekeeping genes as internal standards: use and limits. *Journal of biotechnology, 75*(2), 291-295. http://dx.doi.org/10.1016/S0168-1656(99)00163-7

Tong, Z., Gao, Z., Wang, F., Zhou, J., & Zhang, Z. (2009). Selection of reliable reference genes for gene expression studies in peach using real-time PCR. *BMC Molecular Biology, 10*(1), 71. http://dx.doi.org/10. 1186/1471-2199-10-71

Udvardi, M. K., Czechowski, T., & Scheible, W. R. (2008). Eleven golden rules of quantitative RT-PCR. *The Plant Cell Online, 20*(7), 1736-1737. http://dx.doi.org/10.1105/tpc.108.061143

Vandesompele, J., De Preter, K., Pattyn, F., Poppe, B., Van Roy, N., De Paepe, A., & Speleman, F. (2002). Accurate normalization of real-time quantitative RT-PCR data by geometric averaging of multiple internal control genes. *Genome biology, 3*(7), research0034. http://dx.doi.org/10.1186/gb-2002-3-7-research0034

VanGuilder, H. D., Vrana, K. E., & Freeman, W. M. (2008). Twenty-five years of quantitative PCR for gene expression analysis. *Biotechniques, 44*(5), 619-626. http://dx.doi.org/10.2144/000112776

Wong, M. L., & Medrano, J. F. (2005). Real-time PCR for mRNA quantitation. *Biotechniques*, *39*(1), 75-85. http://dx.doi.org/10.2144/05391RV01

Zheng, J., Li, J., Jiang, Y., Jin, J., Wang, K., Yang, R., & Tan, W. (2011). Design of aptamer-based sensing platform using triple-helix molecular switch. *Analytical chemistry*, *83*(17), 6586-6592. http://dx.doi.org/10.2144/05391RV01

Zhong, H. Y., Chen, J. W., Li, C. Q., Chen, L., Wu, J. Y., Chen, J. Y., ... & Li, J. G. (2011). Selection of reliable reference genes for expression studies by reverse transcription quantitative real-time PCR in litchi under different experimental conditions. *Plant cell reports*, *30*(4), 641-653. http://dx.doi.org/10.1007/s00299-010-0992-8

Expression of Fat and Cholesterol Biomarkers in Meat Goats

M. M. Corley[1] & J. Ward[1]

[1] Agriculture Research Station, Virginia State University, Virginia, USA

Correspondence: M. M. Corley, Agriculture Research Station, Virginia State University, P. O. Box 9061, Petersburg, Virginia 23806, USA. E-mail: mcorley@vsu.edu

Abstract

Consumption of high amounts of saturated fatty acids in meat has been implicated in the onset of cardiovascular disease. Chevon (goat meat) is higher in mono-unsaturated and poly-unsaturated fatty acids than beef and lamb. Limited data is available on the expression of fat and cholesterol biomarkers in meat goats. The objective of this experiment was to determine expression of Acetyl-CoA Carboxylase (ACC1), Apoplipoproteins, A (ApoA1), and B (ApoB) in different breeds of meat goats. Protein sequence alignments were generated to determine conservation for antibody selection. The motif (SMS79^{p}GL) was conserved in the goat, human, mouse, rat and bovine ACC1 proteins. The ApoA1 and ApoB protein alignments (human, bovine and rabbit) revealed high protein sequence homology. The Enzyme-Linked Immunosorbent Assay (ELISA) was used to determine serum ACC1, ApoA1 and ApoB in Spanish and Myotonic goats. Spanish goats had higher ($P<0.05$) ACC1 than Myotonic goats. There was a gender effect ($P<0.05$) where females expressed more ACC1 than males. Breed and gender differences were detected in Spanish and Myotonic goats, with Spanish goats showing 37% more ($P<0.05$) ApoA1 expression in the blood than Myotonic goats and female goats with 47% higher expression of ApoA1 than males. Inversely, Myotonic goats expressed 35% higher ($P<0.05$) levels of ApoB than Spanish goats and males had 46% higher ($P<0.05$) ApoBexpression than females. These data demonstrate that inherent differences exist in lipid metabolism of meat goats and can lead to lipid biomarker assisted breeding programs to produce a heart, healthy red meat for human consumption.

Keywords: Acetyl Co-A carboxylase, apolipoproteins, ApoA1, ApoB, meat goat

1. Introduction

1.1 Significance of the Problem

A molecular genetics approach to enhance global meat goat production is very much in the infancy stage. Goat production continues to gain attention because of the potentially beneficial impact of chevon (goat meat) consumption on human health (Bourre, 2005). Chevon is a source of high quality animal protein with relatively less fat and marginal cholesterol when compared to similarly prepared beef or lamb. Goats deposit fat primarily internally as opposed to intramuscularly which is why chevon is a lean red meat compared to traditional beef, lamb or pork (Banskalieva, Sahlu, & Goetsch, 2000; Liméa, Alexandre, & Berthelot, 2012). The saturated fat content in cooked goat meat has been reported to be up to 40% lower than that of skinless chicken and 50-65% lower than similarly prepared beef (James & Berry, 1997). Goat meat also contains a relatively high amount of polyunsaturated fatty acids (PUFA) consisting mostly of linoleic, linolenic, and arachidonic acids (Banskalieva et al., 2000). According to the National Health and Nutrition Examination Survey (NHANES) statistics published by NIH, 12.5 million children and adolescents aged 2-19 years are obese. Consumption of high amounts of saturated fatty acids can affect the serum lipid profile (Mateo-Gallego et al., 2011) and the tissue lipid profile primarily in the total and low-density lipoprotein (LDL) cholesterol fractions (Glew et al., 2010). These factors are associated with an increased risk for developing obesity and subsequently cardiovascular disease (CVD) (Cascio, Schiera & Di Liegro, 2012). Genes that control the fatty acid profile in humans have been extensively studied in both humans and animals (Zulet & Martinez, 1995; Chilliard et al., 2001; Vincent et al., 2002; Viturro et al., 2009), but not much in meat goats. Much attention has been focused on Acetyl-CoA carobxylase (ACC1), which is the rate-limiting enzyme that catalyzes the carboxylation of acetyl-CoA to form malonyl-CoA, the first step in the synthesis of long chain fatty acids (LCFA) de novo. After malonyl-CoA formation another complex of enzymes, the fatty acid synthetase system (FAS), takes over catalyzing the rest of the synthesis. The end product of long chained fatty acid (LCFA) biosynthesis catalyzed by ACC1 and FAS in humans and animals, (Cánovas, Estany, Tor, Pena, & Doran, 2009; Zhang et al., 2010) is palmitic acid, one of the saturated fatty acids (SFA) implicated in coronary heart disease (Mensink, Temme, & Hornstra, 1994; Mensink, 1993). The apolipoprotein B (ApoB) gene controls the low density lipoprotein (LDL) levels in the body. The LDL cholesterol has been identified as the primary target for therapy in terms of reducing the risk for coronary heart disease (CHD) (Couvert et al., 2008). This LDL cholesterol can build up on the walls of arteries and increase the chances of getting heart disease. Consequently LDL cholesterol is referred to as "bad" cholesterol. The apolipoprotein A1 (ApoA1) gene regulates the level of high density lipoprotein (HDL) ("good" cholesterol) in the body. Research efforts are ongoing to find mechanisms to increase HDL cholesterol and the ratio of HDL: LDL (Gilmore et al., 2011; Gonzalez-Requejo et al., 1995; Harris et al., 2004; Sanchez-Muniz, Bastida, Viejo, & Terpstra, 1999). The HDL cholesterol protects against heart disease by taking the LDL cholesterol out of the blood and preventing it from building up in the arteries. The higher the HDL number, the lower the risk for CVD. High-density lipoprotein cholesterol accounts for about one-fourth to one-third of the total blood cholesterol. The HDL carries cholesterol and cholesterol esters away from the peripheral arteries and back to the liver, where it is passed from the body through reverse cholesterol transport (Daniels et al., 2010) Goats do not deposit fat intramuscularly, and therefore the unsaturated fat content is lower in chevon (Dhanda, Taylor, Murray, & McCosker, 1999).

1.2 Justification

It is known that diet can influence intramuscular fat in goats. Studies have shown that Boer x Spanish goats grazed on pasture without any grain supplementation is more saturated than intramuscular fat from goats fed a grain diet (Banskalieva et al., 2000; Ding, Kou, Cao, & Wei, 2010; Rhee, Waldron, Ziprin, & Rhee, 2000; Tan et al., 2011). In addition to diet, genotype can influence the lipid profile of both humans (Guzmán, Hirata, Quintão, & Hirata, 2000; Minihane et al., 2000; Salazar, Hirata, Quintão, & Hirata, 2000) and animals (Dhanda, Taylor, McCosker, & Murray, 1999; Hoashi et al., 2008; Maharani et al., 2012). Very limited molecular work has been done on the genes that control the lipid profile in meat goats. This study evaluated serum protein expression of genes that control the lipid profile in different breeds of meat goats.

2. Method

2.1 Animals and Experimental Design

Animals used for the study were housed at Virginia State University Randolph farm in accordance with animal care and use guidelines. A total of 20 Spanish and Myotonic goats (5 males, 5 non-pregnant females of each breed), grazing pasture and supplemented with hay, cracked corn and ground soybean meal were selected from a screened pool of over 100 goats (Corley & Jarmon, 2012). A total of three trials were conducted.

2.2 Blood Collection and Serum Preparation

Goats were adequately restrained and blood collected via jugular venipuncture. In brief, the vein was identified by palpation and visual inspection. The area was clipped and swabbed with 70% alcohol. Gentle pressure was applied at the thoracic inlet to produce distension of the vein. Blood samples (3 ml) were collected in vials without anti-coagulant using 16-20G needles. Blood samples were placed in a swinging bucket centrifuge and spun at10, 000 rpm. Serum was removed and subsequently stored at -80 °C for later protein analysis.

2.3 Screening and Selection of Antibodies for ACC1, ApoA1 and ApoB Protein Analysis

Before antibody selection, a comprehensive screening of the GenBank protein databases was performed. The ACC1, ApoB and ApoA1 protein target information was obtained from the product information published by the antibody manufacturing company (Abcam, Cambridge, MA). For the ACC1 antibody, specific target information (serine p[79] phosphorylated residue) was available. For the ApoA1 and ApoB antibodies only GenBank Accession numbers were available. Using this information as a guide, protein sequences (goat, sheep, bovine, human, rabbit) were retrieved from the GenBank and sequence alignments generated (Figure 1, 2, 3) using CLC Main Workbench Bioinformatics software (clcbio.com). Observation of conserved sequence homology enabled a more specific selection of the antibodies for this study.

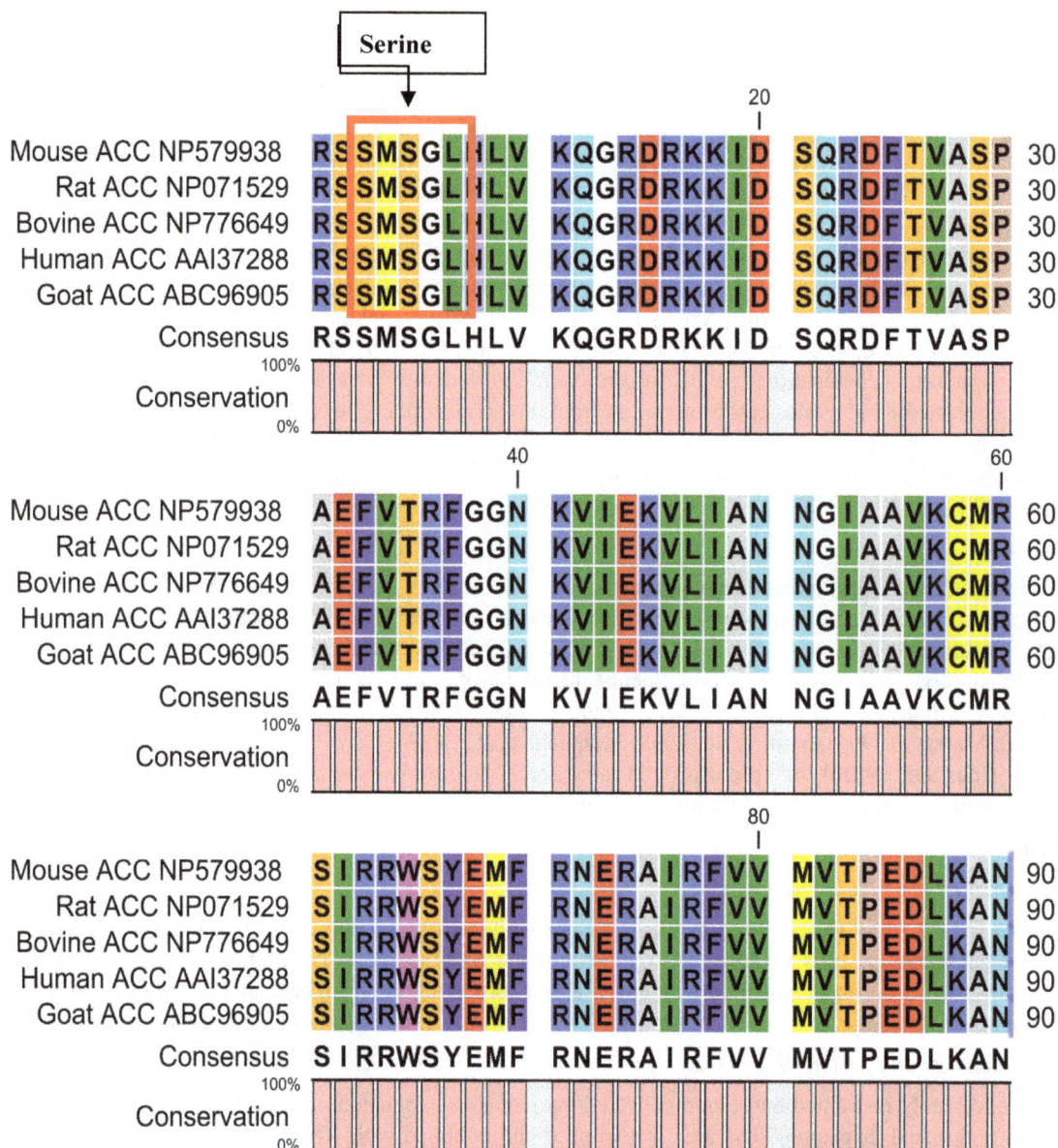

Protein sequence alignment of the Acetyl CoA Carboxylase1 gene, including the goat: The serine 79^p region of the protein from which the antibody was designed is conserved among all species and is shown in the outlined box. An alignment was done before antibody selection to determine sequence conservation among the ACC1 proteins and the possibility of the human ACC1 antibody's use to detect the ACC1 protein in goat serum.

Figure 1. Protein sequence alignment of Acetyl-CoA Carboxylase (ACC1) genes showing conserved ACC1 antibody target region

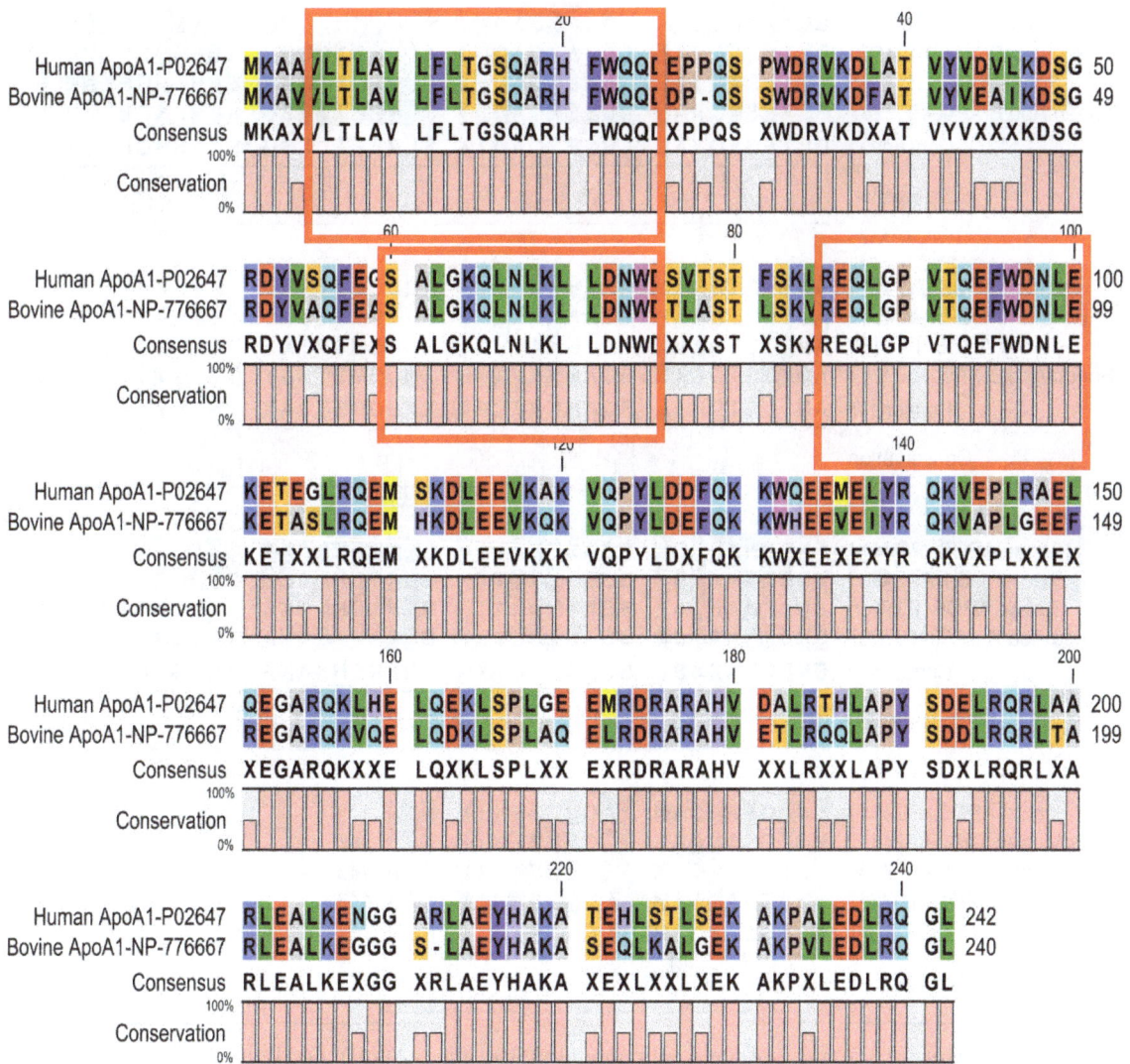

Protein sequence alignment of the Apolipoprotein A1 (ApoA1) gene: GenBank Accession numbers are shown on the labels. Stretches of protein sequence conservation are shown in boxes. An alignment was done before antibody selection to determine sequence conservation between the human and bovine ApoA1 proteins and the possibility of the human ApoA1 antibody's use to detect the ApoA1 protein in goat serum.

Figure 2. Protein sequence alignment of Apolipoprotein A1 (ApoA1) genes showing conserved ApoA1 potential antibody target regions

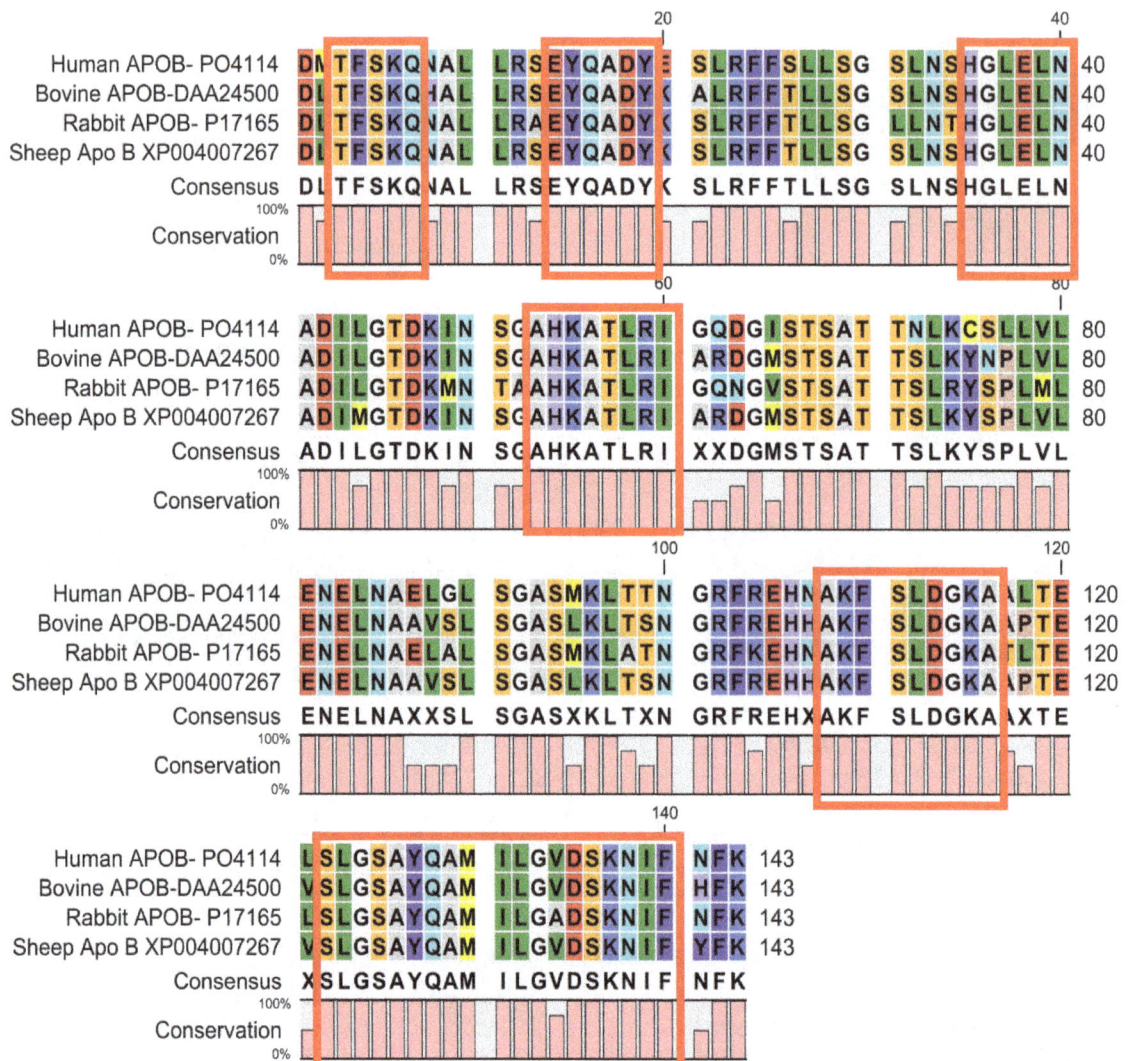

Protein sequence alignment of the Apolipoprotein B100 gene (ApoB): GenBank Accession numbers are shown on the labels. Stretches of protein sequence conservation are shown in boxes. An alignment was done before antibody selection to determine sequence conservation between the human, bovine, rabbit and sheep ApoA1 proteins, and the possibility of the human ApoB antibody's use to detect the ApoB protein in goat serum.

Figure 3. Protein sequence alignment of Apolipoprotein B (ApoB) genes showing conserved ApoB potential antibody target regions

2.4 Analysis of Acetyl-CoA Carboxylase (ACC1), ApoA1 and ApoB

To determine ACC1, APOA1 and APOB serum concentrations, an indirect ELISA was performed using Anti-Acetyl Coenzyme A Carboxylase, Rabbit polyclonal to Acetyl Coenzyme A Carboxylase, Goat polyclonal to Apolipoprotein A I and Goat polyclonal to Apolipoprotein B antibodies respectively (AbCam, Cambridge, MA). To the top wells of a PCV microtiter 96 well plate, 50 µl antigen standards (20 g/ml) and serum samples dissolved in coating buffer (100mM: 3.03g Na_2CO_3, 6.0g $NaHCO_3$) was added and serially diluted 10 fold. The plate was then covered with adhesive tape and incubated for two hours at room temperature. The coating solution was discarded in a decontamination pan. The plate was then washed a total of two times by filling all wells with 300 µl of wash solution PBS (pH 7.4) with 0.05% (v/v) Tween20, waiting 1 minute, and then discarding the liquid into a decontamination pan. The remaining drops were then removed by gently tapping the plate on a thick piece of paper towel. The remaining protein binding sites in the coated wells were blocked by adding 200 µl of blocking buffer (1% BSA). The plate was then covered with adhesive plastic and incubated at 4 °C for 1 hr. The plate was then washed twice as previously described. After tap drying, 100 µl of the primary antibody was added to the plate. The plate was covered with adhesive plastic and incubated for 1 hour at room temperature and washed three times as previously described. To each well 100 µl of the secondary antibody (HRP-anti-goatIgG) was added. Adhesive tape was placed over the plate and incubated at room temperature for thirty minutes. The plate was then washed three times after which 100 µl of TMB substrate (Bethyl Labs Inc.) solution was added to each well and incubated for 15 minutes in the dark. The chromogenic reaction was stopped by addition of 100 µl of stop solution to each well. Samples (triplicate) were read in an iMarkmicroplate reader (BioRad) at 450 nm. Standard curves were generated using the microplate manager 6 software and concentrations of ACC1, ApoA1 and ApoB proteins determined. Data were exported for statistical analysis.

2.5 Statistical Analysis

All data were analyzed using the General Linear Model procedure of SAS. To account for trial (n=3) differences, the data were analyzed in a Randomized Complete Block Design. Means were considered significant at the 5% level of probability.

3. Results and Discussion

Biomarkers of fat and cholesterol are of interest because chevon (goat meat) is naturally lower in fat and cholesterol than beef and lamb (Banskalieva et al., 2000; Rhee, Cho, & Pradahn, 1999), and can be the alternative heart healthy red meat in the hopes of controlling obesity and CVD. Goat producers would have the opportunity to engage in fat and cholesterol biomarker assisted breeding programs, and can therefore maximize production of a red meat that is low in fat and cholesterol. This study was conducted to test the hypothesis that different breeds of meat goats on the same diet would have different fatty acid and cholesterol profiles. Because the goat anti-ACC1, anti-ApoA1 and anti-ApoB were not readily available, a screening of the protein sequence databases (ncbi.nih.nlm.gov) of each biomarker was performed. Cross species protein sequence alignments of ACC1, ApoA1 and ApoB aided in antibody selection for the ELISA. As a result it was possible to quantify protein expression levels of ACC1, ApoA1 and ApoB in the serum of Spanish and Myotonic goats.

3.1 ACC1 Expression

The ACC1 antibody binds to mouse rat and human ACC1 and was derived from the human Acetyl CoA Carboxylase around the phosphorylation site of serine 79. In this study, the ACC1 antibody was successful in detecting ACC1 in goat serum. For selection of the anti-ACC1, assurance that the human anti-ACC1 could bind the goat ACC1 was evident in the presence of the SMS79PGL moiety in the goat, bovine, human, mouse, and rat protein sequence alignments (Figure 1). This same moiety was present in the ACC1 antibody from the manufacturer. Spanish goats had higher (P<0.05) ACC1 than Myotonic goats. More specifically, in Spanish goats, expression of ACC1 was 49% higher than in Myotonic goats (Figure 4).

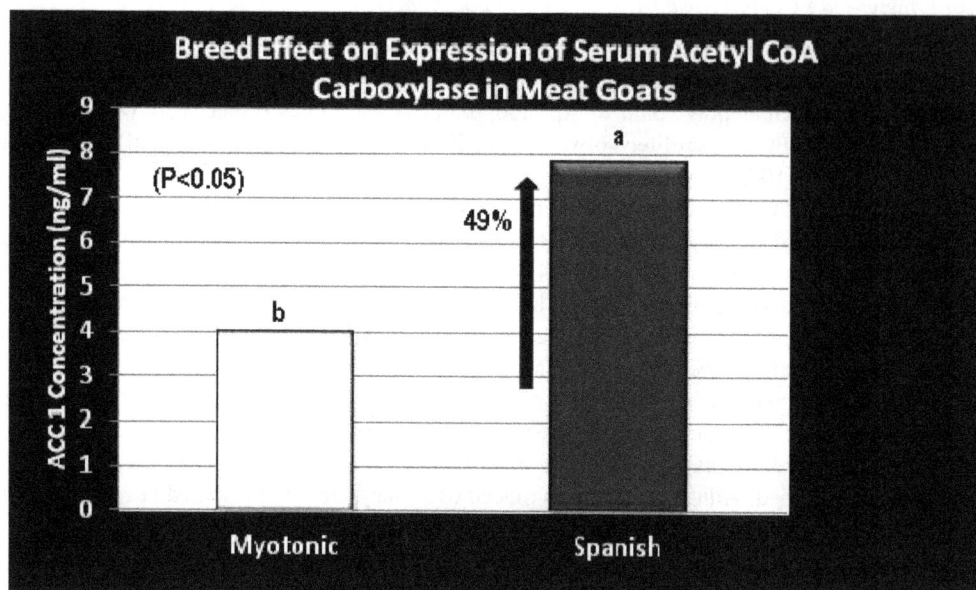

Figure 4. Breed effect on expression of Acetyl-CoA Carboxylase in meat goats

Breed Effect on Expression of AcetylCoA Carboxylase in Meat Goats as measured by ELISA: ab, Means with different letters differ (P<0.05). Arrows indicate the margin of increase in ACC1 expression. This increase was calculated as a percentage difference between mean serum ACC1 concentrations.

Results of the ELISA in this study showed that Spanish goats expressed higher ACC1 than Myotonic goats, indicating that different breeds of meat goats could inherently have different fatty acid metabolism. Studies have shown that genotype has an effect on the fatty acid profile and cholesterol levels of meat (Hoashi et al., 2008) (Daniels et al., 2010; Peña et al., 2009). In the case of gender, female goats expressed 45% more (P<0.05) ACC1 than males (Figure 5).

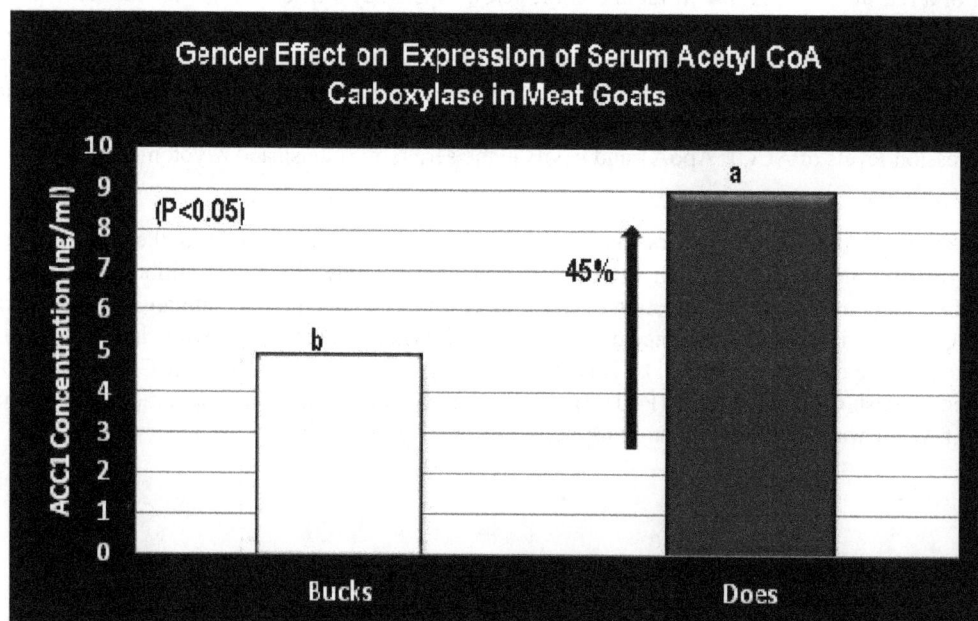

Figure 5. Gender effect on expression of Acetyl-CoA Carboxylase in meat goats

Expression of Acetyl CoA Carboxylase in male and female goats as measured by ELISA: ab, Means with different letters differ (P<0.05). Arrows indicate the margin of increase in ACC1 expression. This increase was calculated as a percentage difference between mean serum ACC1 concentrations.

It has been shown that gender can affect long chain fatty acid synthesis in humans. Significantly lower values of fatty acids were observed in men than in women (Knopp et al., 2005; Lohner, Fekete, Marosvölgyi, & Decsi, 2013). This is supporting evidence that gender needs to be taken into consideration when evaluating the lipid profile. To put this in the context of meat goats, if fat biomarker assisted breeding selection is used in production of chevon gender would have to be considered. Very few studies have addressed the expression of ACC1 in meat goats. Thus far expression of ACC1 has been conducted in other breeds of meat goats and only on the genomic level (Solaiman, Min, Gurung, Behrends, & McElhenney, 2012). Our study evaluated protein expression of ACC1 in Spanish and Myotonic goats focusing more on the translational level of ACC1 expression. These data demonstrate the first step in answering the question whether different breeds of meat goats on the same diet, have different fatty acid metabolism, therefore a different fatty acid profile.

3.2 Apo A1 and ApoB Expression

The selected anti-ApoA1 and anti-ApoB proteins successfully bound to the goat ApoA1 and ApoB as evidenced by the ELISA. In the case of ApoA1 and ApoB, the antibodies were designed from the whole native proteins and no specific target region information was available. However, the manufacturer did provide the UNIPROT database Accession numbers from which the antibodies were designed, as was observed from the protein sequence alignments (Figure 2 and 3). The ApoA1 antibody was the full length native ApoA1 and reacts with human ApoA1. Breed and gender differences were detected in Spanish and Myotonic goats, with Spanish goats showing 37% more ($P<0.05$) ApoA1 expression in the blood than Myotonic goats (Figure 6) and female goats with 47% higher expression of ApoA1 than males (Figure 7). In the case of HDL, Spanish goats expressed more of the HDL ("good cholesterol") biomarker (ApoA1) than Myotonic goats. This implies that genotype can play a role in expression of cholesterol biomarkers in the blood. Other studies have shown that breed does influence the level of cholesterol in the body (Anil, 2007; Ding et al., 2010; Wheeler, Davis, Stoecker, & Harmon, 1987).

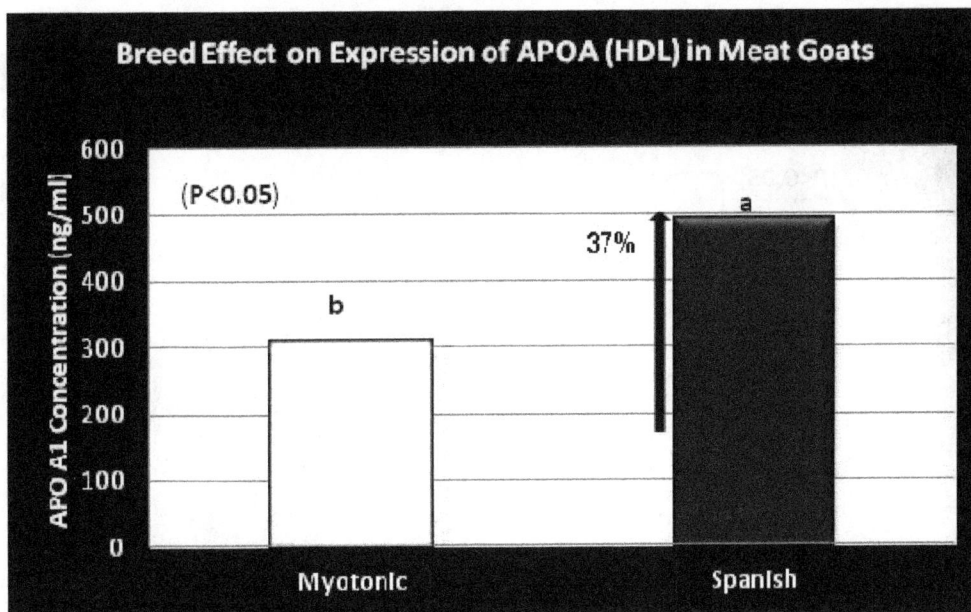

Figure 6. Breed effect on expression of apolipoprotein A1 (ApoA1) in meat goats

Expression of Apolipoprotein A 1 in Spanish and Myotonic goats as measured by ELISA: ab, Means with different letters differ ($P<0.05$). Arrows indicate the margin of increase in APOA1 expression. This increase was calculated as a percentage difference between mean serum APO A1 concentrations.

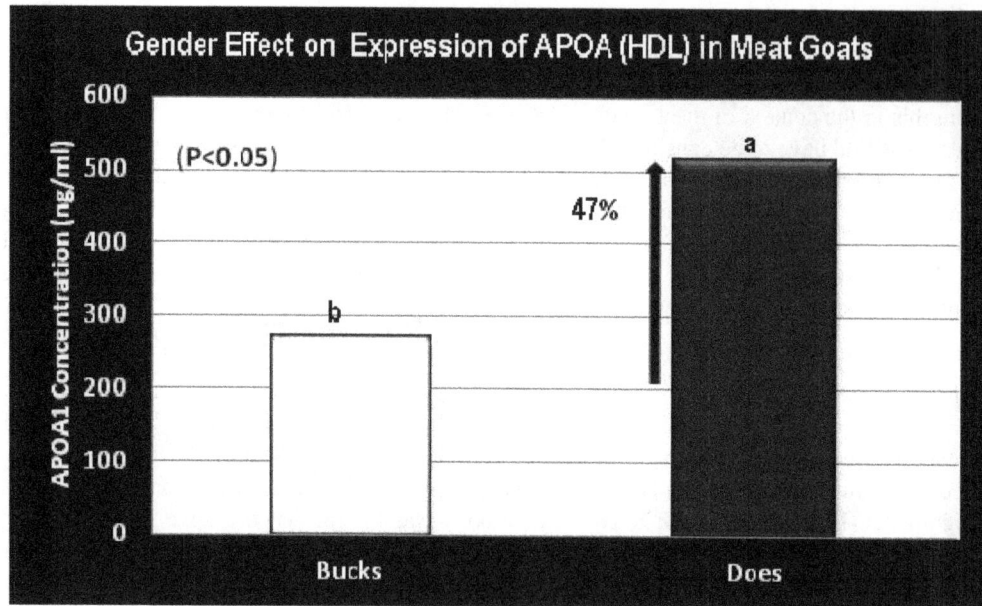

Figure 7. Gender effect on expression of apolipoprotein A1 (ApoA1) in meat goats

Expression of Apolipoprotein A1 in male and female goats as measured by ELISA: ab, Means with different letters differ (P<0.05). Arrows indicate the margin of increase in APOA1 expression. This increase was calculated as a percentage difference between mean serum APO A1 concentrations.

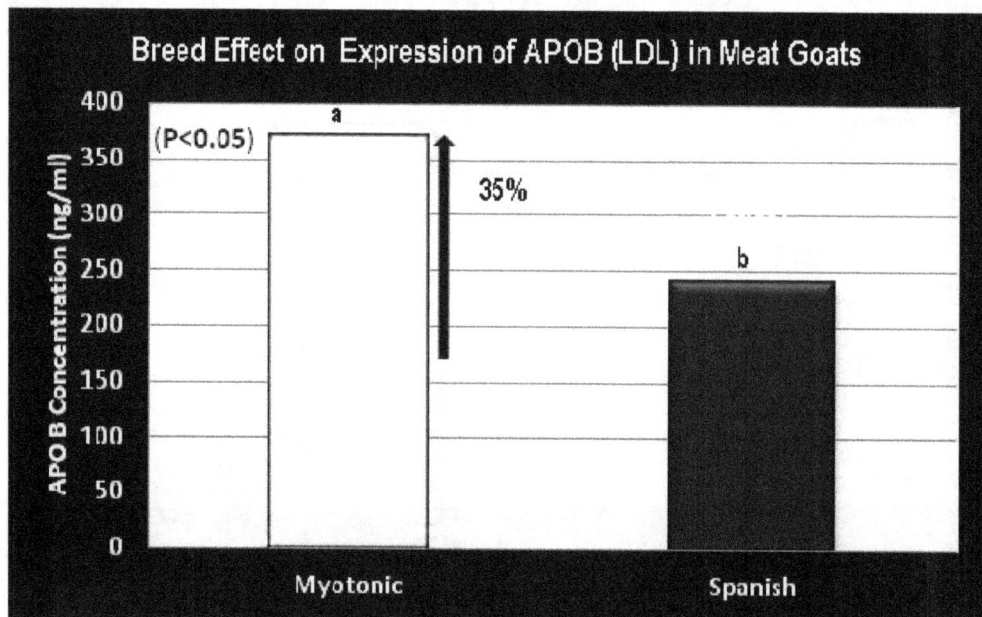

Figure 8. Breed effect on expression of apolipoprotein B (ApoB) in meat goats

Expression of Apolipoprotein B in Spanish and Myotonic goats as measured by ELISA: ab, Means with different letters differ (P<0.05). Arrows indicate the margin of increase in APOB expression. This increase was calculated as a percentage difference between mean serum APO B concentrations.

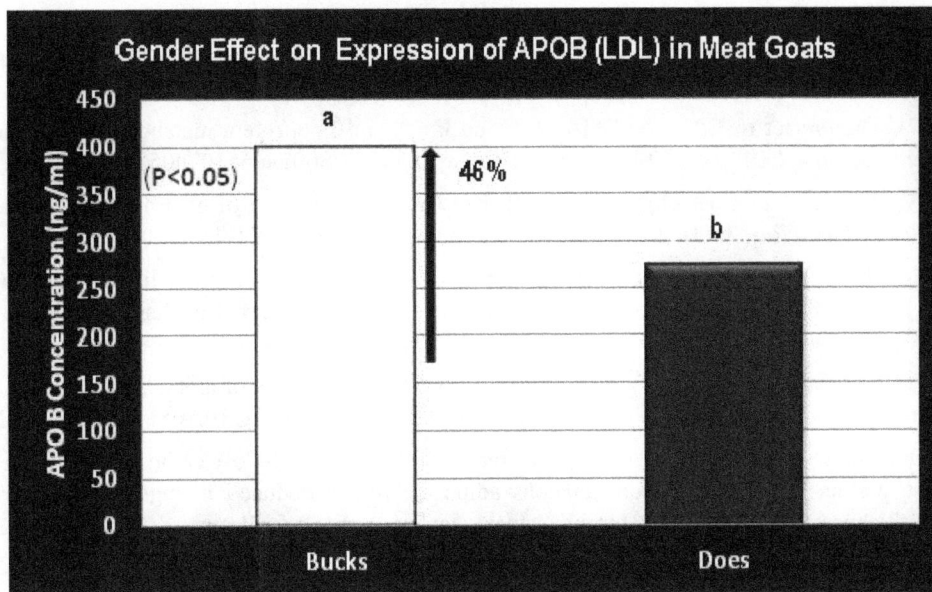

Figure 9. Gender Effect on Expression of Apolipoprotein B (ApoB) in Meat Goats

Expression of Apolipoprotein B in male and female goats as measured by ELISA: ab, Means with different letters differ (P<0.05). Arrows indicate the margin of increase in APOB expression. This increase was calculated as a percentage difference between mean serum APOB concentrations.

The ApoB (full length) antibody has binding affinity to both rabbit and human. Myotonic goats expressed 35% higher (P<0.05) levels of ApoB than Spanish goats (Figure 8) and males had 46% higher (P<0.05) ApoB expression than females (Figure 9). Myotonic goats expressed more of the LDL ("bad cholesterol") biomarker (ApoB) than Spanish goats. It has been demonstrated that defective apoB 100 gene results in increased plasma levels of total cholesterol and LDL cholesterol (Al-Khateeb, Al-Talib, Mohamed, Yusof, & Zilfalil, 2013) again demonstrating that genotype plays a role in cholesterol metabolism (Couvert et al., 2008; Daniels et al., 2010). Females had higher expression of the HDL biomarker than males. This finding supports that of other studies in which gender effect is linked more to protein expression of ApoA1. It was shown that in castrated mice testosterone and estrogen altered the protein synthesis of ApoA-I in castrated inbred strains of mice, but apoA-I mRNA levels remained unaltered, indicating post-transcriptional regulation of the ApoA-I production in liver. As with ApoA1 expression, gender was a contributing factor as males had higher ApoB expression than females. It is already established that gender can play a role in the development of hypercholesterolemia (Beauchesne-Rondeau, Gascon, Bergeron, & Jacques, 2003; Hoogerbrugge et al., 2001). Therefore it is evident from the baseline data given in this study that inherent differences can play an integral role in the outcome of the lipid profile in meat goats.

4. Conclusion

These data demonstrate that cross species binding affinity of antibodies can be useful in determining the level of protein expression of ACC1, ApoA1, and ApoB in meat goats. Therefore protein expression can be influenced by breed and gender, in meat goats and can therefore allow assessment of their lipid profile. Knowledge of the expression of biomarkers of fat and cholesterol in meat goats is essential to understanding the baseline from which an enhanced goat meat product can be produced. This would lead chevon producers to consider breed and gender selection in the production of an alternative healthy red meat for human consumption. Further studies are underway to test the effect of diets high in omega 3 and 6 fatty acids in the hopes of development of a heart healthy enhanced red meat (*omega*-chevon ©).

Acknowledgements

The authors would like to thank the Virginia State University animal care and laboratory staff. This research was funded by USDA-EVANS ALLEN grant at the Virginia State University Agricultural Research Station, Petersburg, Virginia. Journal Article Series No.: 307.

References

Al-Khateeb, A., Al-Talib, H., Mohamed, M. S., Yusof, Z., & Zilfalil, B. A. (2013). Phenotype-genotype analyses of clinically diagnosed Malaysian familial hypercholestrolemic patients. *Adv Clin Exp Med, 22*(1), 57-67.

Anil, E. (2007). The impact of EPA and DHA on blood lipids and lipoprotein metabolism: influence of apoE genotype. *Proc Nutr Soc, 66*(1), 60-68. http://dx.doi.org/10.1017/S0029665107005307

Banskalieva, V., Sahlu, T., & Goetsch, A. L. (2000). Fatty acid composition of goat muscles and fat depots: a review. *Small Rumin Res, 37*(3), 255-268. http://dx.doi.org/10.1016/S0921-4488(00)00128-0

Beauchesne-Rondeau, E., Gascon, A., Bergeron, J., & Jacques, H. (2003). Plasma lipids and lipoproteins in hypercholesterolemic men fed a lipid-lowering diet containing lean beef, lean fish, or poultry. *Am J Clin Nutr, 77*(3), 587-593.

Bourre, J. M. (2005). Effect of increasing the omega-3 fatty acid in the diets of animals on the animal products consumed by humans. *Med Sci (Paris), 21*(8-9), 773-779. http://dx.doi.org/10.1051/medsci/2005218-9773

Cánovas, A., Estany, J., Tor, M., Pena, R. N., & Doran, O. (2009). Acetyl-CoA carboxylase and stearoyl-CoA desaturase protein expression in subcutaneous adipose tissue is reduced in pigs selected for decreased backfat thickness at constant intramuscular fat content. *J Anim Sci, 87*(12), 3905-3914. http://dx.doi.org/10.2527/jas.2009-2091

Chilliard, Y., Bonnet, M., Delavaud, C., Faulconnier, Y., Leroux, C., Djiane, J., ... Bocquier, F. (2001). Leptin in ruminants. Gene expression in adipose tissue and mammary gland, and regulation of plasma concentration. *Domest Anim Endocrinol, 21*(4), 271-295. http://dx.doi.org/10.1016/S0739-7240(01)00124-2

Corley, M. M., & Jarmon, A. A. (2012). Interleukin 13 as a Biomarker for Parasite Resistance in Goats Naturally Exposed to Haemonchus contortus. *Journal of Agricultural Science, 4*(7), 31-40. http://dx.doi.org/10.5539/jas.v4n7p31

Couvert, P., Giral, P., Dejager, S., Gu, J., Huby, T., Chapman, M. J., & Carrié, A. (2008). Association between a frequent allele of the gene encoding OATP1B1 and enhanced LDL-lowering response to fluvastatin therapy. *Pharmacogenomics, 9*(9), 1217-1227. http://dx.doi.org/10.2217/14622416.9.9.1217

Daniels, T. F., Wu, X. L., Pan, Z., Michal, J. J., Wright, R. W., Killinger, K. M., ... Jiang, Z. (2010). The reverse cholesterol transport pathway improves understanding of genetic networks for fat deposition and muscle growth in beef cattle. *PLoS One, 5*(12), e15203. http://dx.doi.org/10.1371/journal.pone.0015203

Dhanda, J. S., Taylor, D. G., McCosker, J. E., & Murray, P. J. (1999). The influence of goat genotype on the production of Capretto and Chevon carcasses. 1. Growth and carcass characteristics. *Meat Sci, 52*(4), 355-361. http://dx.doi.org/10.1016/S0309-1740(99)00016-9

Dhanda, J. S., Taylor, D. G., Murray, P. J., & McCosker, J. E. (1999). The influence of goat genotype on the production of Capretto and Chevon carcasses. 4. Chemical composition of muscle and fatty acid profiles of adipose tissue. *Meat Sci, 52*(4), 375-379. http://dx.doi.org/10.1016/S0309-1740(99)00014-5

Ding, W., Kou, L., Cao, B., & Wei, Y. (2010). Meat quality parameters of descendants by grading hybridization of Boer goat and Guanzhong Dairy goat. *Meat Sci, 84*(3), 323-328. http://dx.doi.org/10.1016/j.meatsci.2009.04.015

Gilmore, L. A., Walzem, R. L., Crouse, S. F., Smith, D. R., Adams, T. H., Vaidyanathan, V., ... Smith, S. B. (2011). Consumption of high-oleic acid ground beef increases HDL-cholesterol concentration but both high- and low-oleic acid ground beef decrease HDL particle diameter in normocholesterolemic men. *J Nutr, 141*(6), 1188-1194. http://dx.doi.org/10.3945/jn.110.136085

Glew, R. H., Chuang, L. T., Berry, T., Okolie, H., Crossey, M. J., & VanderJagt, D. J. (2010). Lipid profiles and trans fatty acids in serum phospholipids of semi-nomadic Fulani in northern Nigeria. *J Health Popul Nutr, 28*(2), 159-166.

Gonzalez-Requejo, A., Sanchez-Bayle, M., Baeza, J., Arnaiz, P., Vila, S., Asensio, J., & Ruiz-Jarabo, C. (1995). Relations between nutrient intake and serum lipid and apolipoprotein levels. *J Pediatr, 127*(1), 53-57. http://dx.doi.org/10.1016/S0022-3476(95)70256-3

Guzmán, E. C., Hirata, M. H., Quintão, E. C., & Hirata, R. D. (2000). Association of the apolipoprotein B gene polymorphisms with cholesterol levels and response to fluvastatin in Brazilian individuals with high risk for coronary heart disease. *Clin Chem Lab Med, 38*(8), 731-736. http://dx.doi.org/10.1515/CCLM.2000.103

Harris, K. B., Pond, W. G., Mersmann, H. J., Smith, E. O., Cross, H. R., & Savell, J. W. (2004). Evaluation of fat sources on cholesterol and lipoproteins using pigs selected for high or low serum cholesterol. *Meat Sci, 66*(1), 55-61. http://dx.doi.org/10.1016/S0309-1740(03)00012-3

Hoashi, S., Hinenoya, T., Tanaka, A., Ohsaki, H., Sasazaki, S., Taniguchi, M., ... Mannen, H. (2008). Association between fatty acid compositions and genotypes of FABP4 and LXR-alpha in Japanese black cattle. *BMC Genet, 9*, 84. http://dx.doi.org/10.1186/1471-2156-9-84

Hoogerbrugge, N., van Domburg, R., van der Zwet, E., van Kemenade, M., Bootsma, A., & Simoons, M. L. (2001). High fat intake in hyperlipidaemic patients is related to male gender, smoking, alcohol intake and obesity. *Neth J Med, 59*(1), 16-22. http://dx.doi.org/10.1016/S0300-2977(01)00119-X

James, N. A., & Berry, B. W. (1997). Use of chevon in the development of low-fat meat products. *J Anim Sci, 75*(2), 571-577.

Knopp, R. H., Paramsothy, P., Retzlaff, B. M., Fish, B., Walden, C., Dowdy, A., ... Cheung, M. C. (2005). Gender differences in lipoprotein metabolism and dietary response: basis in hormonal differences and implications for cardiovascular disease. *Curr Atheroscler Rep, 7*(6), 472-479. http://dx.doi.org/10.1007/s11883-005-0065-6

Liméa, L., Alexandre, G., & Berthelot, V. (2012). Fatty acid composition of muscle and adipose tissues of indigenous Caribbean goats under varying nutritional densities. *J Anim Sci, 90*(2), 605-615. http://dx.doi.org/10.2527/jas.2010-3624

Lohner, S., Fekete, K., Marosvölgyi, T., & Decsi, T. (2013). Gender differences in the long-chain polyunsaturated fatty acid status: systematic review of 51 publications. *Ann Nutr Metab, 62*(2), 98-112. http://dx.doi.org/10.1159/000345599

Maharani, D., Jung, Y., Jung, W. Y., Jo, C., Ryoo, S. H., Lee, S. H., ... Lee, J. H. (2012). Association of five candidate genes with fatty acid composition in Korean cattle. *Mol Biol Rep, 39*(5), 6113-6121. http://dx.doi.org/10.1007/s11033-011-1426-6

Mateo-Gallego, R., Perez-Calahorra, S., Cenarro, A., Bea, A. M., Andres, E., Horno, J., ... Civeira, F. (2011). Effect of lean red meat from lamb v. lean white meat from chicken on the serum lipid profile: a randomised, cross-over study in women. *Br J Nutr*, 1-5.

Mensink, R. P. (1993). Effects of the individual saturated fatty acids on serum lipids and lipoprotein concentrations. *Am J Clin Nutr, 57*(5 Suppl), 711S-714S.

Mensink, R. P., Temme, E. H., & Hornstra, G. (1994). Dietary saturated and trans fatty acids and lipoprotein metabolism. *Ann Med, 26*(6), 461-464. http://dx.doi.org/10.3109/07853899409148369

Minihane, A. M., Khan, S., Leigh-Firbank, E. C., Talmud, P., Wright, J. W., Murphy, M. C., ... Williams, C. M. (2000). ApoE polymorphism and fish oil supplementation in subjects with an atherogenic lipoprotein phenotype. *Arterioscler Thromb Vasc Biol, 20*(8), 1990-1997. http://dx.doi.org/10.1161/01.ATV.20.8.1990

Peña, F., Bonvillani, A., Freire, B., Juárez, M., Perea, J., & Gómez, G. (2009). Effects of genotype and slaughter weight on the meat quality of Criollo Cordobes and Anglonubian kids produced under extensive feeding conditions. *Meat Sci.*

Rhee, K.S., Cho, S. H., & Pradahn, A. M. (1999). Composition, storage stability and sensory properties of expanded extrudates from blends of corn starch and goat meat, lamb, mutton, spent fowl meat, or beef. *Meat Sci, 52*(2), 135-141. http://dx.doi.org/10.1016/S0309-1740(98)00157-0

Rhee, K.S., Waldron, D. F., Ziprin, Y. A., & Rhee, K. C. (2000). Fatty acid composition of goat diets vs intramuscular fat. *Meat Sci, 54*(4), 313-318. http://dx.doi.org/10.1016/S0309-1740(99)00094-7

Salazar,L. A., Hirata, M. H., Quintão, E. C., & Hirata, R. D. (2000). Lipid-lowering response of the HMG-CoA reductase inhibitor fluvastatin is influenced by polymorphisms in the low-density lipoprotein receptor gene in Brazilian patients with primary hypercholesterolemia. *J Clin Lab Anal, 14*(3), 125-131. http://dx.doi.org/10.1002/(SICI)1098-2825(2000)14:3%3C125::AID-JCLA7%3E3.0.CO;2-S

Sanchez-Muniz, F. J., Bastida, S., Viejo, J. M., & Terpstra, A. H. (1999). Small supplements of N-3 fatty acids change serum low density lipoprotein composition by decreasing phospholid and apolipoprotein B concentrations in young adult women. *Eur J Nutr, 38*(1), 20-27. http://dx.doi.org/10.1007/s003940050042

Solaiman, S., Min, B. R., Gurung, N., Behrends, J., & McElhenney, W. (2012). Effects of breed and harvest age on feed intake, growth, carcass traits, blood metabolites, and lipogenic gene expression in Boer and Kiko goats. *J Anim Sci.* http://dx.doi.org/10.2527/jas.2011-3945

Tan, C. Y., Zhong, R. Z., Tan, Z. L., Han, X. F., Tang, S. X., Xiao, W. J., … Wang, M. (2011). Dietary inclusion of tea catechins changes fatty acid composition of muscle in goats. *Lipids, 46*(3), 239-247. http://dx.doi.org/10.1007/s11745-010-3477-1

Vincent, S., Planells, R., Defoort, C., Bernard, M. C., Gerber, M., Prudhomme, J., … Lairon, D. (2002). Genetic polymorphisms and lipoprotein responses to diets. *Proc Nutr Soc, 61*(4), 427-434. http://dx.doi.org/10.1079/PNS2002177

Viturro, E., Koenning, M., Kroemer, A., Schlamberger, G., Wiedemann, S., Kaske, M., & Meyer, H. H. (2009). Cholesterol synthesis in the lactating cow: Induced expression of candidate genes. *J Steroid Biochem Mol Biol, 115*(1-2), 62-67. http://dx.doi.org/10.1016/j.jsbmb.2009.02.011

Wheeler, T. L., Davis, G. W., Stoecker, B. J., & Harmon, C. J. (1987). Cholesterol concentration of longissimus muscle, subcutaneous fat and serum of two beef cattle breed types. *J Anim Sci, 65*(6), 1531-1537.

Zhang, S.,Knight, T. J., Reecy, J. M., Wheeler, T. L., Shackelford, S. D., Cundiff, L. V., & Beitz, D. C. (2010). Associations of polymorphisms in the promoter I of bovine acetyl-CoA carboxylase-alpha gene with beef fatty acid composition. *Anim Genet, 41*(4), 417-420.

Zulet, M. A., & Martinez, J. A. (1995). Corrective role of chickpea intake on a dietary-induced model of hypercholesterolemia. *Plant Foods Hum Nutr, 48*(3), 269-277. http://dx.doi.org/10.1007/BF01088448

15

Hydroxyl Radical (°OH) Scavenger Power of Tris (hydroxymethyl) Compared to Phosphate Buffer

Meysam Khosravifarsani[1], Ali Shabestani-Monfared[1], Mahdi Pouramir[1] & Ebrahim Zabihi[1]

[1] Cellular and Molecular Biology Research Center, Babol University of Medical Sciences, Babol, Iran

Correspondence: Ali Shabestani-Monfared, Cellular and Molecular Biology Research Center, Babol University of Medical Sciences, Babol, Iran. E-mail: monfared_ali@yahoo.com

Abstract

Tris and phosphate buffer are regularly used in experimental investigations. These buffers might have radical scavenger properties toward different kinds of Reactive Oxygen Species (ROS) produced in solutions during chemical reactions like Fenton reaction and gamma radiolysis. Hydroxyl radicals (°OH) are the most reactive and oxidizing agents having a great potential in oxidization of macromolecules like DNA and proteins. This *in vitro* study was aimed to evaluate radio-protective effects of Tris and phosphate buffer toward hydroxyl radicals generated by Fenton reaction. Hence, °OH radicals were produced using a mixture of Hydrogen Peroxide and Ferrous Sulfate, called Fenton system. Human serum albumin (500μM) was prepared in Tris (10mM) and phosphate buffer (10mM), separately. These two samples were incubated with Fenton reaction (Ferrous Sulfate + Hydrogen peroxide) (10 mM) for 30 minutes and carbonyl groups were quantified by spectrophotometric carbonylation assay. The results of this study revealed the values of 1.04 ± 0.02 and 1.73 ± 0.03 for Tris and phosphate buffer treated samples, respectively. In conclusion, these findings confirmed that Tris buffer is a stronger radical scavenger toward °OH radicals than phosphate buffer.

Keyswords: ROS, Fenton reaction, hydroxyl radicals, carbonylation assay

1. Introduction

Since several years ago, Tris and phosphate buffer have been used prevalently in most of experimental investigations due to their powerful buffering function in the range of physiological pH. These buffers have been frequentlyselected in protection of many chemical and biological systems, and also used in several examinations linked to free radicals and oxidative stress in vivo and in vitro studies (Cullis, Elsy, Fan, & Symons, 1993; Good et al., 1966; Greenwald & Moy, 1980). It is well documented that buffers have a profound impact on the tertiary and quaternary structure of proteins (Ugwu & Apte, 2004). Moreover, antioxidant properties of some buffers are explained due to their metal binding affinity (Porasuphatana, Weaver, Budzichowski, Tsai, & Rosen, 2001). During the study of effects of oxidative stress on different biological systems and reactions of free radicals, which are generated after gamma radiolysis, with various macromolecules like DNA and proteins, it is completely visible that different types of buffers give unequal results which complicate interpretation of our findings. In radiation biology and radiation protection field, also, these buffers are crucial in preparation of protein, DNA and other macromolecules during irradiation. The study of the reaction of these macromolecules with hydroxyl radicals (°OH) was therefore initiated in several experiments (Hicks & Gebicki, 1986). It was cleared that sodium phosphate is a good choice for radiation biology studies, as this compound has no effect on radiosensitivity of DNA (Achey, Duryea, & Michaels, 1974). Moreover, the values of constant rate for HEPES, Tricine and Tris were represented 5.1×10^9, 1.6×10^9 and 1.1×10^9 l.mol-1.S-1 toward °OH radicals, respectively (Hicks & Gebicki, 1986). Other investigations indicated that Tris buffer has radical scavenging capability toward °OH radicals (Finkelstein, Rosen, & Rauckman, 1980; Saprin & Piette, 1977). Hydroxyl radicals are highly reactive in direction of interaction with macromolecules in physiological conditions (Cheeseman & Slater, 1993; Du & Francisco, 2008; Harman, 1992; Loizos, 2004; Pryor et al., 2006; Winterbourn, 1995). These radicals can damage proteins by creating protein backbone cleavages, making inter and intra molecular cross-links, generating carbonyl groups, conversion of free thiol groups (-SH) to other forms and so on (Anraku, Yamasaki, Maruyama, Kragh-Hansen, & Otagiri, 2001; Leeuwenburgh, Hansen, Shaish, Holloszy, & Heinecke, 1998; Plowman, Deb-Choudhury, Grosvenor, & Dyer, 2013). Human serum albumin includes 582 amino-acids with a molecular weight of 66KD

(Meloun, Morávek, & Kostka, 1975). The normal concentration of this protein in human plasma is ranged between 35 and 50 g/l. Albumin has several important functions in physiological conditions like transferring metals, fatty acids (Curry, Brick, & Franks, 1999), cholesterol, bile pigments, and drugs (SUDLOW, BIRKETT, & WADE, 1976; Vallner, 1977). As this protein is continuously exposed to oxidative stress, a crucial part of antioxidant power of the human body is attributed to HSA (Bourdon & Blache, 2001; Friedrichs, 1997; Taverna, Marie, Mira, & Guidet, 2013). Considering above information, this study was planned to compare ºOH radical scavenger activity of Tris and phosphate buffer by measuring generated carbonyl groups on the most abundant protein in the humans blood plasma, called human serum albumin.

2. Materials and Methods

2.1 Materials

Human serum albumin (Sigma), Ferrous Sulfate (Merck), Hydrogen Peroxide (Merck), Tris (hydroxymethyl) (Merck), phosphate buffer (Potassium dihydrogen phosphate + Potassium hydrogen phosphate) (Merck), 2, 4 dinitrophenyl-hydrozine (DNPH), Hcl, Guanidine hydrochloride (GuHcl) (Merck), trichloroacetic acid (TCA) (Merck)

2.2 Fenton Reaction

Fenton reaction is a common method of generating highly reactive hydroxyl radicals as a final product in most chemical systems as follows (Thomas, Mackey, Diaz, & Cox, 2009):

$$Fe^{2+} + H_2O_2 \rightarrow Fe^{3+} + HO^{\bullet} + OH^-$$

In order to organize this reaction, a mixture of Ferrous Sulfate (10mM) and Hydrogen Peroxide (10mM) was incubated with human serum albumin (500µM) which have already been prepared in phosphate buffer (10mM) and Tris (10mM), pH 7, separately. An untreated sample was chosen as blank in each group. All experiments were carried out in triplicate.

2.3 Carbonylation Assay

This technique is a colorimetric method of measuring carbonyl groups which generate by oxidative stress (Luo & Wehr, 2009; Weber, Davies, & Grune, 2015). In this method, protein carbonyl groups are quantified as a result of interaction with 2,4-dinitrophenylhydrazine (DNPH) in order to create protein-bound 2,4-dinitrophenylhydrazones (Augustyniak et al., 2015; Luo & Wehr, 2009).

Here we describe the method of measuring DNPH content using a spectrophotometric assay proposed by Levine et al with some modifications (Levine et al., 1990). In the course of this technique, after oxidization of protein solutions, 200 µl of %0.2 DNPH (prepared in 2N HCL) was added to 200 µl of oxidized protein solution and incubated for 1hour in dark conditions, at room temperature. During that time, samples vortexed every 10 minutes. Protein pellets were sedimented by 20% TCA (ice cold) and spun at 10000rpm for 15 minutes. Washing step carried out three times using ethanol/ethyl acetate (1/1; v/v) in order to get rid of the extra of DNPH.

The acquired pellets were dissolved in 100 µl of 6M Guanidine hydrochloride and the absorbance was calculated by spectrophotometer at 370 nm against its blank (Hcl treated). The quantity of carbonyl groups in each sample and its control was measured by Beer-Lambert Law with the molar extinction coefficient for dinitrophenylhydrazine at 370nm (22000 M^{-1} cm^{-1}). The absorption values of DNPH treated samples were subtracted from blank and normalized to 15µM of protein. Finally, Carbonyl ratio was calculated by dividing carbonyl to protein concentration (mol/mol).

2.4 Data Analysis

All of statistical analyses were carried out using Microsoft office software (Excel 2007). Difference between two samples were analyzed by Student's t-test with the significance level at $P<0.05$. All results are shown as means ± SE.

3. Results

Solutions composed of different organic buffers can interact with ºOH radicals with unequal speeds or constant rates (Hicks & Gebicki, 1986). Here, in figure 1, Carbonyl/Protein (mol/mol) ratio is illustrated following ºOH radical oxidation of Human serum albumin.

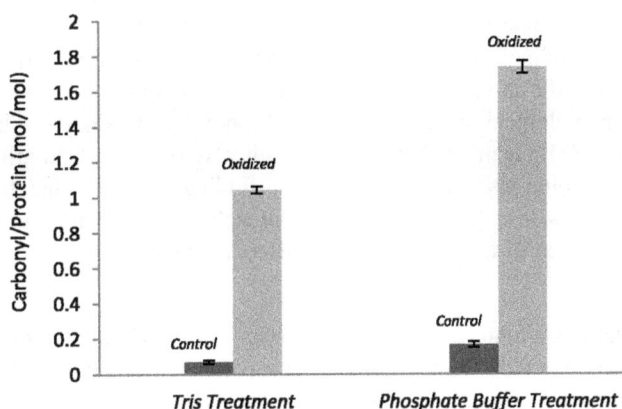

Figure 1. Carbonyl ratios of control and oxidized HSA treated with Tris and Phosphate Buffer

As can be seen in Figure 1, the ratio of carbonyl/protein is significantly increased in both Tris and phosphate buffer treated samples after treatment with Fenton reaction (p<0.001). Regarding the average values of carbonyl/protein in oxidized samples, these figures are 1.73 ± 0.03 and 1.04 ± 0.02 in phosphate buffer and Tris treated samples, respectively. Accordingly, phosphate buffer treated sample showed roughly 60% increase in carbonyl/protein ratio (mol/mol) compared to its counterpart after oxidization with °OH radicals (p< 0.001) (Figure1).

Table 1. Protein carbonyl concentrations (μM) in both Tris and phosphate buffer treated samples in control and oxidized samples

Type of Treatment	Protein Carbonyl Concentrations (μM)	
	Non-Oxidized	Oxidized
Tris Treatment	1.06 ± 0.02	15.90 ± 1.74
Phosphate Buffer Treatment	2.52 ± 0.18	26.46 ± 1.5

By comparing carbonyl concentrations in Tris and phosphate buffer treated samples after incubation with Fenton system it is clear that this figure is highly increased in phosphate buffer treated samples in comparison with that of tris samples(p≤0.001).

4. Discussion

Tris (hydroxymethyl) and phosphate buffer (Potassium dihydrogen phosphate + Potassium hydrogen phosphate) are the most extensively used buffers in biological and chemical studies, notably in preparation of protein solutions for radiation biology and chemistry experiments (Yukawa, Nagatsuka, & Nakazawa, 1983). In this fact, human serum albumin (HSA) is one of the most common proteins being investigated by scientists during recent decays (Kondakova, Ripa, & Sakharova, 1988; Maciazek-Jurczyk & Sulkowska, 2015; Sitar, Aydin, & Cakatay, 2013). A broad range of in vitro and in vivo methods have been launched to estimate radical scavenging power of buffers. Generation of formaldehydes was reported after °OH radical oxidation of Tris (hydroxymethyl) buffer using a Fenton system (Shiraishi, Kataoka, Morita, & Umemoto, 1993). Nonetheless, formaldehyde production was revealed as a result of °OH radical oxidation of dimethylsulfoxide (DMSO). It was suggested that the existence of formaldehyde depict the appearance of °OH radicals in biological solutions (Klein, Cohen, & Cederbaum, 1980). Fenton system serves as a °OH radical source in order to oxidize organic substances (Thomas et al., 2009; Zepp, Faust, & Hoigne, 1992). This reaction can generate advanced oxidation products (AOPP) and make conformational and structural changes on human serum albumin (Meucci, Mordente, & Martorana, 1991; Taverna et al., 2013). Carbonyl groups are one of the main generated end products during oxidative stress (Dalle-Donne, Rossi, Giustarini, Milzani, & Colombo, 2003). In this study, a spectrophotometric method for protein carbonyl assay was chosen to compare radical scavenger power of Tris (hydroxymethyl) versus phosphate buffer toward °OH radicals. It is clearly demonstrated in *figure 1* that Tris buffer is a stronger radical scavenger than phosphate buffer toward °OH radicals at pH=7 in vitro. These results are in agreement with former results reported by Mark Hicks et al. They revealed that 4-(2-hydroxyethyl)-1-piperazineethanesulfonic acid (Hepes), Tricine (N-[Tris(hydroxymethyl)methyl]glycine) and Tris buffers are efficient radical scavenger with the rate constants around 10^9 M^{-1}.s^{-1} (Hicks & Gebicki, 1986). According to carbonyl/protein ratio in our results, it is clearly illustrated that this value in phosphate buffer is much higher than Tris buffer by a factor of 1.66 at the same concentrations (10mM). These findings confirm that under these conditions Tris buffer can prohibit °OH radical damages more than phosphate buffer in vitro. However, there are numerous evidences that °OH radicals might be generated heterogeneously in biological systems (Samuni, Aronovitch, Godinger, Chevion, & Czapski, 1983; Shinar, Navok, & Chevion, 1983), meaning that hydroxyl radicals could produce at some inaccessible sites to buffer substances in aqueous solutions. In conclusion, our findings suggest that the type of buffer can be considered as an interfering item with radiation biology or chemistry experiments. Complimentary investigations are needed to unravel the interaction of free radicals with buffers in detail.

Acknowledgement

The authors would like to thank Cellular and Molecular Biology Research Center staffs at Babol University of Medical Sciences for their warmheartedly helps. This project was financially supported by deputy of Research and technology at Babol University of Medical Sciences.

References

Achey, P. M., Duryea, H. Z., & Michaels, G. S. (1974). Choice of Solvent for Studying the Role of Water in Ionizing Radiation Action on DNA. *Radiation Research, 58*(1), 83-90. http://dx.doi.org/ 10.2307/3573951

Anraku, M., Yamasaki, K., Maruyama, T., Kragh-Hansen, U., & Otagiri, M. (2001). Effect of Oxidative Stress on the Structure and Function of Human Serum Albumin. *Pharmaceutical Research, 18*(5), 632-639. http://dx.doi.org/10.1023/a:1011029226072

Augustyniak, E., Adam, A., Wojdyla, K., Rogowska-Wrzesinska, A., Willetts, R., et al. (2015). Validation of protein carbonyl measurement: A multi-centre study. *Redox Biology, 4*, 149-157. http://dx.doi.org/10.1016/j.redox.2014.12.014

Bourdon, E., & Blache, D. (2001). The importance of proteins in defense against oxidation. *Antioxid Redox Signal, 3*(2), 293-311. http://dx.doi.org/10.1089/152308601300185241

Cheeseman, K. H., & Slater, T. F. (1993). An introduction to free radical biochemistry. *Br Med Bull, 49*(3), 481-493.

Cullis, P. M., Elsy, D., Fan, S., & Symons, M. C. (1993). Marked effect of buffers on yield of single- and double-strand breaks in DNA irradiated at room temperature and at 77 K. *Int J Radiat Biol, 63*(2), 161-165. http://dx.doi.org/10.1080/09553009314550211

Curry, S., Brick, P., & Franks, N. P. (1999). Fatty acid binding to human serum albumin: new insights from crystallographic studies. *Biochim Biophys Acta, 1441*(2-3), 131-140. http://dx.doi.org/10.1016/S1388-1981(99)00148-1

Dalle-Donne, I., Rossi, R., Giustarini, D., Milzani, A., & Colombo, R. (2003). Protein carbonyl groups as biomarkers of oxidative stress. *Clinica Chimica Acta, 329*(1–2), 23-38. http://dx.doi.org/10.1016/S0009-8981(03)00003-2

Du, S., & Francisco, J. S. (2008). Interaction between OH Radical and the Water Interface. *The Journal of Physical Chemistry A, 112*(21), 4826-4835. http://dx.doi.org/10.1021/jp710509h

Finkelstein, E., Rosen, G. M., & Rauckman, E. J. (1980). Spin trapping of superoxide and hydroxyl radical: Practical aspects. *Arch Biochem Biophys, 200*(1), 1-16. http://dx.doi.org/10.1016/0003-9861(80)90323-9

Friedrichs, B. (1997). Th. Peters. Jr.: All about Albumin. Biochemistry, Genetics, and Medical Applications. XX and 432 pages, numerous figures and tables. Academic Press, Inc., San Diego, California, 1996. Price: 85.00 US $. *Food / Nahrung, 41*(6), 382-382. http://dx.doi.org/ 10.1002/food.19970410631

Good, N. E., Winget, G. D., Winter, W., Connolly, T. N., Izawa, S., et al. (1966). Hydrogen ion buffers for biological research. *Biochemistry, 5*(2), 467-477.

Greenwald, R. A., & Moy, W. W. (1980). Effect of oxygen-derived free radicals on hyaluronic acid. *Arthritis Rheum, 23*(4), 455-463.

Harman, D. (1992). Role of free radicals in aging and disease. *Ann N Y Acad Sci, 673*, 126-141.

Hicks, M., & Gebicki, J. M. (1986). Rate constants for reaction of hydroxyl radicals with Tris, Tricine and Hepes buffers. *FEBS Letters, 199*(1), 92-94. http://dx.doi.org/10.1016/0014-5793(86)81230-3

Klein, S. M., Cohen, G., & Cederbaum, A. I. (1980). The interaction of hydroxyl radicals with dimethylsulfoxide produces formaldehyde. *FEBS Letters, 116*(2), 220-222. http://dx.doi.org/10.1016/0014 -5793(80)80648-X

Kondakova, N. V., Ripa, N. V., & Sakharova, V. V. (1988). [Effect of ionizing radiation on the structure of human serum albumin. Effect of irradiation on the content of free amino groups and on the ability of HSA to be digested by proteinases]. *Radiobiologiia, 28*(6), 748-751.

Leeuwenburgh, C., Hansen, P., Shaish, A., Holloszy, J. O., & Heinecke, J. W. (1998). Markers of protein oxidation by hydroxyl radical and reactive nitrogen species in tissues of aging rats. *Am J Physiol, 274*(2 Pt 2), R453-461.

Levine, R. L., Garland, D., Oliver, C. N., Amici, A., Climent, I., et al. (1990). Determination of carbonyl content in oxidatively modified proteins. *Methods Enzymol, 186*, 464-478.

Loizos, N. (2004). Identifying protein interactions by hydroxyl-radical protein footprinting. *Curr Protoc Protein Sci, Chapter 19*, Unit 19.19. http://dx.doi.org/10.1002/0471140864.ps1909s38

Luo, S., & Wehr, N. B. (2009). Protein carbonylation: avoiding pitfalls in the 2,4-dinitrophenylhydrazine assay. *Redox Rep, 14*(4), 159-166. http://dx.doi.org/10.1179/135100009x392601

Maciazek-Jurczyk, M., & Sulkowska, A. (2015). Spectroscopic analysis of the impact of oxidative stress on the structure of human serum albumin (HSA) in terms of its binding properties. *Spectrochim Acta A Mol Biomol Spectrosc, 136 Pt B*, 265-282. http://dx.doi.org/ 10.1016/j.saa.2014.09.034

Meloun, B., Morávek, L., & Kostka, V. (1975). Complete amino acid sequence of human serum albumin. *FEBS Letters, 58*(1), 134-137. http://dx.doi.org/10.1016/0014-5793(75)80242-0

Meucci, E., Mordente, A., & Martorana, G. E. (1991). Metal-catalyzed oxidation of human serum albumin: conformational and functional changes. Implications in protein aging. *J Biol Chem, 266*(8), 4692-4699.

Plowman, J. E., Deb-Choudhury, S., Grosvenor, A. J., & Dyer, J. M. (2013). Protein oxidation: identification and utilisation of molecular markers to differentiate singlet oxygen and hydroxyl radical-mediated oxidative pathways. *Photochem Photobiol Sci, 12*(11), 1960-1967. http://dx.doi.org/10.1039/c3pp50182e

Porasuphatana, S., Weaver, J., Budzichowski, T. A., Tsai, P., & Rosen, G. M. (2001). Differential effect of buffer on the spin trapping of nitric oxide by iron chelates. *Anal Biochem, 298*(1), 50-56. http://dx.doi.org/10.1006/abio.2001.5389

Pryor, W. A., Houk, K. N., Foote, C. S., Fukuto, J. M., Ignarro, L. J., et al. (2006). Free radical biology and medicine: it's a gas, man! *Am J Physiol Regul Integr Comp Physiol, 291*(3), R491-511. http://dx.doi.org/10.1152/ajpregu.00614.2005

Samuni, A., Aronovitch, J., Godinger, D., Chevion, M., & Czapski, G. (1983). On the cytotoxicity of vitamin C and metal ions. A site-specific Fenton mechanism. *Eur J Biochem, 137*(1-2), 119-124. http://dx.doi.org/10.1111/j.1432-1033.1983.tb07804.x

Saprin, A. N., & Piette, L. H. (1977). Spin trapping and its application in the study of lipid peroxidation and free radical production with liver microsomes. *Arch Biochem Biophys, 180*(2), 480-492. http://dx.doi.org/10.1016/0014-5793(78)80216-6

Shinar, E., Navok, T., & Chevion, M. (1983). The analogous mechanisms of enzymatic inactivation induced by ascorbate and superoxide in the presence of copper. *J Biol Chem, 258*(24), 14778-14783.

Shiraishi, H., Kataoka, M., Morita, Y., & Umemoto, J. (1993). Interactions of hydroxyl radicals with Tris (hydroxymethyl) aminomethane and Good's buffers containing hydroxymethyl or hydroxyethyl residues produce formaldehyde. *Free Radic Res Commun, 19*(5), 315-321.

Sitar, M. E., Aydin, S., & Cakatay, U. (2013). Human serum albumin and its relation with oxidative stress. *Clin Lab, 59*(9-10), 945-952.

SUDLOW, G., BIRKETT, D. J., & WADE, D. N. (1976). Further Characterization of Specific Drug Binding Sites on Human Serum Albumin. *Molecular Pharmacology, 12*(6), 1052-1061. doi: Retrieved from http://molpharm.aspetjournals.org/content/12/6/1052.abstract

Taverna, M., Marie, A. L., Mira, J. P., & Guidet, B. (2013). Specific antioxidant properties of human serum albumin. *Ann Intensive Care, 3*(1), 4. http://dx.doi.org/10.1186/2110-5820-3-4

Thomas, C., Mackey, M. M., Diaz, A. A., & Cox, D. P. (2009). Hydroxyl radical is produced via the Fenton reaction in submitochondrial particles under oxidative stress: implications for diseases associated with iron accumulation. *Redox Rep, 14*(3), 102-108. http://dx.doi.org/10.1179/135100009x392566

Ugwu S. O., & Apte, S. P. (March 2004). The Effect of Buffers on Protein Conformational Stability. *Pharmaceutical Technology, 28*(3), 86.

Vallner, J. J. (1977). Binding of drugs by albumin and plasma protein. *J Pharm Sci, 66*(4), 447-465. http://dx.doi.org/10.1002/jps.2600660402

Weber, D., Davies, M. J., & Grune, T. (2015). Determination of protein carbonyls in plasma, cell extracts, tissue homogenates, isolated proteins: Focus on sample preparation and derivatization conditions. *Redox Biology, 5*, 367-380. http://dx.doi.org/10.1016/j.redox.2015.06.005

Winterbourn, C. C. (1995). Toxicity of iron and hydrogen peroxide: the Fenton reaction. *Toxicol Lett, 82-83*, 969-974.

Yukawa, O., Nagatsuka, S., & Nakazawa, T. (1983). Reconstitution studies on the involvement of radiation-induced lipid peroxidation in damage to membrane enzymes. *Int J Radiat Biol Relat Stud Phys Chem Med, 43*(4), 391-398. http://dx.doi.org/ 10.1080/09553008314550451

Zepp, R. G., Faust, B. C., & Hoigne, J. (1992). Hydroxyl radical formation in aqueous reactions (pH 3-8) of iron(II) with hydrogen peroxide: the photo-Fenton reaction. *Environmental Science & Technology, 26*(2), 313-319. http://dx.doi.org/10.1021/es00026a011

16

Combining Network Topological Characteristics With Sequence and Structure Based Features for Predicting Protein Stability Changes Upon Single Amino Acid Mutation

Lijun Yang[1], Qifan Kuang[1], Yanping Jiang[1], Ling Ye[1], Yiming Wu[1], Menglong Li[1] & Yizhou Li[1]

[1] College of Chemistry, Sichuan University, Chengdu, China

Correspondence: Yizhou Li, College of Chemistry, Sichuan University, Chengdu 610064, China.
E-mail: liyizhou_415@163.com

Abstract

It has been shown that the stability of protein structure could be significantly changed by single amino acid substitution. Accurate prediction of protein stability changes caused by single amino acid substitutions is valuable for understanding the relationship between protein structures and functions as well as designing new proteins. Currently, various computational methods have been developed to study the effect of single amino acid mutation on protein stability. In this study, by combining network topological characteristics extracted from Protein Structure Network (PSN) with other physicochemical features obtained from protein sequence or structure, a Support Vector Machine (SVM) model was developed to distinguish the stabilizing mutants from the destabilizing mutants. 20-fold cross-validation was implemented for performance evaluation. An accuracy of 0.88 and a Matthews Correlation Coefficient (MCC) of 0.71 were obtained for the dataset with 1925 variants.

Our method is superior to the existing machine learning approaches evaluated under the same datasets. It suggests that such a combining strategy should be valuable in predicting protein stability changes upon amino acid mutation. In our study, the topological parameters are informative for prediction upon substitutions. Moreover, it is indicated that the Protein Structure Network (PSN) could be effectively used for representing the three-dimensional structure of protein and such network parameters are associated with the changes of protein function and structure.

Keywords: amino acid mutation, protein stability prediction, Protein Structure Network (PSN), Support Vector Machine (SVM), topological features

1. Introduction

Single amino acid substitution shows great impact on protein. Such substitutions can cause a series of changes of proteins, including the physicochemical properties, structures as well as functions and etc. (Shirley, Stanssens, Hahn, & Pace, 1992), which may lead to protein destabilization and even diseases. Variations are mostly owing to non-synonymous single nucleotide polymorphisms (nsSNPs). Each person may have 24,000-40,000 nsSNPs, and there are a total of more than 67,000 common nsSNPs in the human population (Cargill et al., 1999). These nsSNPs may result in amino acid changes in proteins. The majority of nsSNPs are neutral for protein function, while the remains may affect protein function and lead to diseases. It is estimated that about 25% of nsSNPs may be deleterious to protein function (Yue & Moult, 2006). Moreover, the majority of disease-associated mutations are caused by protein destabilization (Wang & Moult, 2001). Therefore, if a mutation causes the protein destabilization, it is more likely to cause disease. So, accurate prediction of how single amino acid mutation affects the stability of proteins is invaluable for understanding the protein structure, function and designing new proteins. However, with the increasing number of mutant data, it is not enough to determine the effect of each mutation on protein through experiment alone for its time and labor-consuming. Thus, more and more machine learning based methods have been developed owing to its quick and effective prediction performance.

The selections of feature vectors and algorithms are really important in machine learning methods. The machine learning methods can be designed to predict either the sign of the stability free energy change ($\Delta\Delta G$, classification) upon mutation (stabilizing if $\Delta\Delta G > 0$, destabilizing if $\Delta\Delta G < 0$) or the actual value of the stability free energy change ($\Delta\Delta G$, regression). Until now, many machine learning methods have been applied in the prediction of protein stability changes upon single amino acid mutation. These methods took advantage of sequence or structural information of proteins for classification or regression or for both. However, for most biological applications, accurate prediction of the sign of $\Delta\Delta G$ (classification) is more relevant than estimating the actual value of $\Delta\Delta G$ (regression) (Capriotti, Fariselli, & Casadio, 2004). In this research, the classification models were constructed to predict the sign of $\Delta\Delta G$ instead of predicting the actual value of $\Delta\Delta G$.

Cheng et al. (Cheng, Randall, & Baldi, 2006) used Support Vector Machine (SVM) with sequence and structural information of proteins to predict protein stability changes upon single amino acid substitutions, and achieved 84% accuracy. Teng et al. (Teng, Srivastava, & Wang, 2010) developed a sequence feature-based model to predict protein stability changes upon amino acid substitutions and an overall accuracy of 84.59% was obtained. It was shown that the relevant sequence features can be used for the prediction of protein stability changes upon single amino acid substitutions. Folkman et al. (Folkman, Stantic, & Sattar, 2013) proposed a number of evolutionary features and predicted structural features acquired from protein sequence to predict the protein stability changes and achieved satisfying results. Yang et al. (Yang, Chen, Tan, Vihinen, & Shen, 2013) reported a structure-based method with the best accuracy of 87% which combined the contact energy with other physicochemical properties of amino acids. Based on the studies mentioned above, it is clear that either the protein sequence or structure implies useful information for predicting protein stability changes upon single amino acid substitutions.

From another perspective, a protein is truly an interacting network system because of the connections between nodes (amino acids). Both the location of residues and the interactions among them are important for the function and stability of protein. Bagler et al. and Greene et al. described the small-word property of protein structure network (Bagler & Sinha, 2005; Greene & Higman, 2003). Dokholyan et al. and Vendruscolo et al. demonstrated that a selected set of residues played an important role in protein folding from protein structure network (Dokholyan, Li, Ding, & Shakhnovich, 2002; Vendruscolo, Dokholyan, Paci, & Karplus, 2002). Also, another study showed that a set of hub residues in protein structure network primarily contributed to both the folding and stability of proteins (Brinda & Vishveshwara, 2005). Amitai et al. took advantage of such protein structure network for studying functional residues (Amitai et al., 2004). Li et al. predicted disease-associated substitution upon single site mutation by analyzing residue interactions in protein structure network (Li et al., 2011). Therefore, the protein structure network involves much potential information for biochemical studies.

Overall, the studies mentioned above suggest that it is feasible for predicting protein stability changes based on sequence or structure information of proteins and that the protein structure network is a valuable method for utilizing the structure information of proteins. In this study, the protein structure network was introduced which regards a protein structure as a network and then four topological parameters were calculated for each residue in a protein. Our results indicate that only using topological features can achieve a satisfying accuracy in prediction with SVM (accuracy of 0.83 for S1925 dataset). Better results can be obtained when combining the topological features with other physicochemical properties. It suggests that the topological features we introduced are useful for predicting protein stability changes upon single amino acid substitutions. It further indicats that the network parameters of amino acids are associated with the changes of protein function and structure. In addition, the calculated environmental features are also informative for the prediction from the research and eight sequence neighboring residues are enough for providing environmental information of the variant.

2. Materials and Methods

2.1 Data Sets

In this work, two datasets were used for training and testing the models. The datasets S1925 and S2760 were originally extracted from ProTherm database (Bava, Gromiha, Uedaira, Kitajima, & Sarai, 2004) (http://gibk26.bio.kyutech.ac.jp/jouho-u/Protherm/protherm.html). The dataset S2760 was collected from (Yang et al., 2013) and S1925 dataset was originally compiled by (Masso & Vaisman, 2008).

S1925 contained 1925 single mutations for 55 proteins with 582 positive and 1343 negative cases. While in S2760, there were 2758 single mutations for 75 different proteins with 872 positive and 1886 negative cases, in which two mutations were removed due to the mismatch (renamed as SR2760). The two data sets both contained six attributes for each record including the PDB ID, the mutation, pH, Temperature (T), $\Delta\Delta G$ and the solvent accessibility (ASA of the variant residue).

2.2 Protein Structure Network

In this study, a network was constructed from a protein PDB structure where the nodes were residues and the interactions between residues were edges. If the distance between any two atoms from two residues was smaller than the sum of their Vander Waals radius plus 0.5Å (Greene et al., 2003), the two residues were considered to interact with each other. Based on the constructed protein structure network, then, the igraph package in R (Ihaka & Gentleman, 1996) was used to extract the topological parameters of each residue from the network.

2.3 Features

Thirty-four attributes were used to encode each data instance, including network topological features, environmental features and other physicochemical properties. We divided these features into the following classes.

2.3.1 Network Topological Features

In this study, four network topological features were calculated, including Degree, Clustering Coefficient, Betweenness and Closeness.

Degree is the number of edges that directly connect to node *i*. It reflects the number of the nearest neighbors for vertex *i*. It is calculated as

$$D(i) = \Sigma_{j \in N} a_{i,j} \tag{1}$$

Where $a_{i,j}$ is the number of edges between nodes *i* and *j*, and N is the total number of vertices incident to vertex *i*.

The *Clustering Coefficient* reflects how well connected are the neighbors of node *i*. It can be calculated as

$$CC(i) = \frac{2e_i}{D_i(D_i - 1)} \tag{2}$$

Where e_i is the virtual number of edges among the neighborhoods of vertex *i*, and D_i is the degree of node *i*.

Betweenness reflects the probability of a vertex occurs on the shortest paths between other vertices. It is calculated as

$$B(i) = \sum_{j,k \in N, j \neq k} \frac{n_{j,k}(i)}{n_{j,k}} \tag{3}$$

Where $n_{j,k}$ is the number of shortest paths connecting vertices *j* and *k*. $n_{j,k}(i)$ is the number of shortest paths linking *j* and *k* which pass through vertex *i*.

Closeness describes the adjacent level of vertex *i* and all other vertices. It is calculated as

$$C(i) = \frac{N-1}{\sum_{i \neq j} d_{i,j}} \tag{4}$$

Where $d_{i,j}$ is the shortest path between vertices *i* and *j* and N is the total number of vertices.

For more detailed information of the four parameters, please refer to the references (Newman, 2003; Watts & Strogatz, 1998)

2.3.2 Amino Acid Properties of the Variant

Eleven amino acid properties were introduced in this work, which were obtained from AAindex (Kawashima & Kanehisa, 2000) (http://www.genome.jp/aaindex/), Protscal (Gasteiger et al., 2005) (http://web.expasy.org/protscale/) and the reference (Collantes & Dunn, 1995). These properties consist of Bulkiness (Bu), Average area buried on transfer from standard state to folded protein (Aa), Conformational parameter for alpha helix (Al), Beta-sheet (Be) and Coil (Co), Polarity (P), Hydropathicity (H), Transmembrane tendency (Tt), Flexibility (F), Electronic charge index (ECI) and Isotropic surface area (ISA).

Bulkiness (Bu) (Zimmerman, Eliezer, & Simha, 1968), reflecting the role of individual residues in the entire protein configuration, may influence the local conformation of protein. Average area buried on transfer from standard state to folded protein (Aa) (Rose, Geselowitz, Lesser, Lee, & Zehfus, 1985) estimates a residue's mean area buried upon folding. Protein secondary structures can be divided into three types, including alpha-helix, beta-sheet and coil conformations. Each type of amino acid has a different preference to form one of the three secondary structures. So, the Conformational parameters for alpha helix (Al) (Chou & Fasman, 1978), beta-sheet (Be) (Chou et al., 1978), and coil (Co) (Deleage & Roux, 1987) were introduced. Polarity (P) (Grantham, 1974) reflects the intermolecular interactions between the residues of positive and negative charge. Hydropathicity（H） (Kyte & Doolittle, 1982) takes the hydrophilic and hydrophobic properties of each 20 amino acid side-chains into consideration. It is important for keeping the protein structure as it is critical for amino acid side chain packing and protein folding. Transmembrane tendency (Tt) is related to biological hydrophobicity, which was reported by Zhao and London (Zhao & London, 2006). Flexibility (Vihinen, Torkkila, & Riikonen, 1994) of protein structure is important for catalysis, binding, and allostery, which shows correlation to protein stability. Electronic charge index (ECI) (Collantes et al., 1995) is a measure of the charge density for amino acid. Isotropic surface area (ISA) (Collantes et al., 1995) approximates the hydrophobic property of the side chain substituent, which directly reflects the effect of the variant on structure.

The difference between the variant and the original in each amino acid property was calculated as the variant characterization and the prefix d was used to characterize the information. For example, the attribute dBu was defined as the difference between the variant and the original in amino acid Bulkiness (Bu). So, the remaining ten attributes of the variant were dAa, dAl, dBe, dCo, dP, dH, dTt, dF, dECI and dISA.

2.3.3 Environmental Features

Additionally, the environmental features were also considered by using four network topological features and eleven amino acid properties of each residue in protein. The prefix Env was used to characterize the environmental features. There were fifteen environmental features defined as Env_CC, Env_D, Env_B, Env_C, Env_Aa, Env_Al, Env_Be, Env_Co, Env_P, Env_H, Env_Tt, Env_F, Env_ECI and Env_ISA.

To calculate the environmental features, we took a subsequence of w consecutive residues from a protein sequence into consideration, where w was called the window size. The mutation site was located in the middle of the subsequence, and the other (w-1) neighboring residues provided the environment information for the substitution site. Then, the average value obtained from neighboring residues under each attribute was reported to be the corresponding environmental features. Here, took the attribute Env_P for instance.

The Polarity (P) for each of the 20 amino acid can be obtained from AAindex. When the window size was set to 7, 6 neighboring residues' polarity were averaged to represent the environmental features of the variant. The Env_P was calculated as

$$Env\_P\,(i) = \frac{\sum P_j}{w-1} \tag{5}$$

Where P_j is the polarity of residue j and w is the window size.

In this study, we investigated nine kinds of window size to obtain different environmental features and analyzed their effect on classifier.

2.3.4 Evolutionary Feature

In a protein sequence, residues in different sites bear different evolutionary constraint. Specifically, functionally important sites tend to be more conserved. Here, we introduced an evolutionary feature, SIFT score(S).

SIFT (Ng & Henikoff, 2001) is a method for predicting whether an amino acid substitution affects the function of a protein and the prediction is based on the degree of conservation of amino acids in the alignment derived from PSI-BLAST. SIFT scores were scaled from 0 to 1. The amino acid substitutions are considered to be damaging if the scores below 0. 05. The SIFT Sequence tool was used to calculate the SIFT scores online (http://sift.jcvi.org/www/SIFT_seq_submit2.html). It was run on the Uniprot-TrEMBL database with sequences more than 90% identical to the query removed and the median conservation of sequences was set to 2.75.

2.3.5 Other Features

Additional three features were the solvent accessibility of the variant residue (ASA), the experimental pH and Temperature (T). All of the three attributes were extracted from ProTherm database.

2.4 Support Vector Machine

Support Vector Machine (SVM) has been successfully applied in a wide range of biological applications. It can learn from examples to assign labels to objects. Indeed, for understanding SVM classification, one only needs to comprehend the four basic concepts: (I) the separating hyperplane, (II) the maximum-margin hyperplane, (III) the soft margin and (IV) the kernel function (Noble, 2006). There are four basic kernels for SVM: linear, sigmoid, polynomial and radial basis function (RBF). In general, the RBF kernel is an optimal choice as demonstrated by Capriotti et al (Capriotti, Fariselli, Calabrese, & Casadio, 2005b). So, we chose the RBF kernel as SVM kernel function. The RBF kernel was calculated as

$$K\left(\vec{x}, \vec{y}\right) = e^{-\gamma \left\| \vec{x} - \vec{y} \right\|^2} , \gamma > 0 \tag{6}$$

Where $\vec{x}$ and $\vec{y}$ are two vectors, and γ is a training parameter.

In the research, the SVMlight software package (http://svmlight.joachims.org/) was used to construct the classifiers. Various values of C and γ parameters were examined to optimize classifier performance. The best C and γ for each dataset were then used to construct the final classifier models.

2.5 Classifier Evaluation

The 20-fold cross-validation method was employed to evaluate the performance of the classifier under variant wise as many other studies did. In the case, all positive and negative instances were randomly divided into 20 folds without considering proteins. 19 folds were used to train a model, and then the remaining one fold was used to test the model. The process was repeated 20 times until each fold was used as the test set, then the evaluation value was obtained from the average result of the 20-fold cross validation. We computed the following performance measures:

$$\text{Accuracy (ACC)} = \frac{TP+TN}{TP+TN+FP+FN}$$

$$\text{True positive rate (TPR)} = \frac{TP}{TP+FN}$$

$$\text{True negative rate (TNR)} = \frac{TN}{TN+FP}$$

$$\text{Positive predictive value (PPV)} = \frac{TP}{TP+FP}$$

$$\text{Negative predictive value (NPV)} = \frac{TN}{TN+FN}$$

$$\text{Matthews Correlation Coefficient (MCC)} = \frac{(TP \times TN)-(FP \times FN)}{\sqrt{(TP+FN)(TP+FP)(TN+FN)(TN+FP)}}$$

where TP, TN, FP, and FN are true positives, true negatives, false positives and false negatives, respectively. Matthews Correlation Coefficient (MCC) reflects the relationship between predictions and the real class labels.

In this paper, the area under the Receiver Operating Characteristic (ROC) curve (AUC) was also used as a measure of a classifier's performance (Bradley, 1997). Weak classifiers are with AUC values close to 0.5 and a perfect classifier has the maximum AUC value of 1.

3. Results and Discussion

3.1 Analysis of the Network Topological Features for Stabilization and Destabilization-Associated Variants

We constructed a classifier with four network topological features, including the Clustering Coefficient, Degree, Betweenness and Closeness of the variant. The results of S1925 and SR2760 dataset were listed in Table 1. For S1925, the obtained accuracy was 0.83 and MCC was 0.57. For SR2760 dataset, the accuracy of the model was about 0.81 and MCC was 0.54. Therefore, it is believed that the network topological features contained valuable information for predicting protein stability changes upon single amino acid mutation. Moreover, it can be demonstrated that the parameters reflecting the importance of nodes in protein structure network are related with the changes of protein function and structure.

Table 1. Predictive performance of classifiers constructed by topological features of the variants

Dataset	TPR	TNR	PPV	NPV	Accuracy	MCC	AUC
S1925	0.64	0.91	0.75	0.85	0.83	0.57	0.81
SR2760	0.62	0.90	0.74	0.83	0.81	0.54	0.80

Figure 1 shows that the two types of mutations differ in distributions of network topological features. The frequency was the ratio of mutations in the range of (x, x+d) to the total number of mutations. *Degree* describes the number of direct connections to a residue. Figure 1a shows that the destabilization-associated variants tend to have more neighbors than the stabilization-associated ones in the protein structure network. Residues with higher degree are more likely to be hub residues and therefore their mutations have more important influence on the protein structure and protein stability.

Similarly, in figure 1b, it can be observed that the destabilization-associated variants are with higher closeness compared with the stabilization-associated variants. It had been reported that residues associated with protein function have higher closeness than those neutral to protein function (Amitai et al., 2004). So, a centrally located residue in the protein structure network is more likely to associate with protein stability.

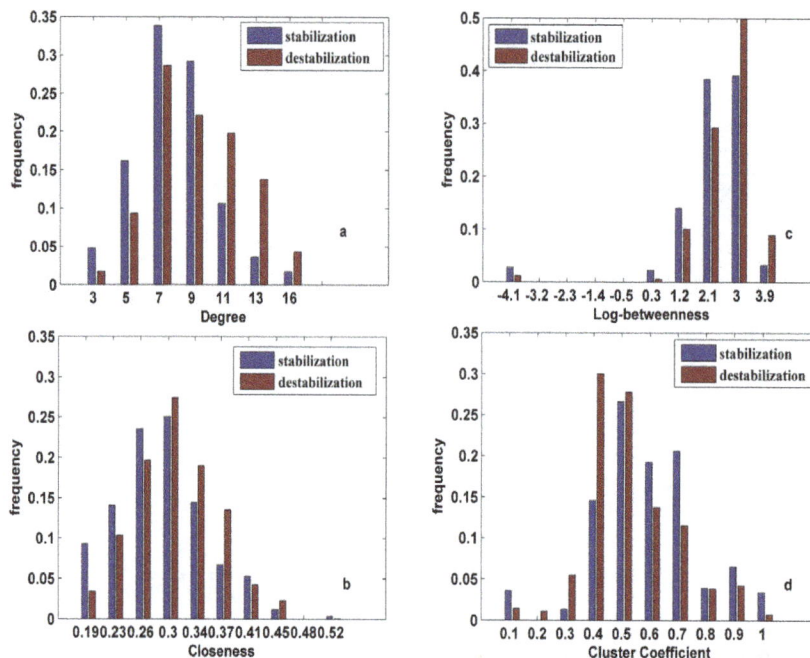

Figure 1. The frequency distributions of a) Degree; b) Closeness; c) Betweenness; d) Cluster Coefficient for stabilization and destabilization-associated mutants of S1925

For Betweenness and Clustering coefficient, the distributions of the two types of mutations are less distinct. However, as shown in figure 1c, higher frequencies were obtained for destabilization-related variants in the high-scoring region. Betweenness is an important centrality index of the interaction network. Residues with higher betweenness are more central residues, which make more important contribution for protein structure (Del Sol & O'Meara, 2005). Thus, the protein stability is more likely to be changed by mutations on the residues with higher betweenness. Figure 1d shows the distribution of *Clustering Coefficient*. As shown, higher frequencies were obtained for stabilization-associated variants in the high-scoring region. The clustering coefficient shows how well connected are the neighbors of a vertex in a network. Residues with high clustering coefficient reflect their compact environment. It can be found that the effect of substitution is related with the environment of the mutant site. More compact environment of the mutant site is of higher tolerance to substitution.

As mentioned above, the distributions of the four topological features between destabilization-associated and stabilization-associated mutations are different. The differences of the four parameters between the two types of substitutions are statistically significant (P value = 2.87E-19 for CC, P value = 5.86E-21 for D, P value = 4.43E-16 for B, P value = 1.15E-09 for C). Therefore, they can be identified as features with distinction power in discriminating destabilization-associated substitutions from other substitutions. The frequency distributions of the network topological features for SR2760 dataset can be obtained in Supplementary Figure 1.

3.2 The Optimal Choice of Window Size for Environmental Features Calculation

To calculate the environmental features, we took a subsequence of w consecutive residues into consideration, where w was the window size. The details about the calculation can be found in section 2.3.3. An optimal window size was chose for environmental features calculation by performing a conditional experiment. As the 11 amino acid properties and four topological parameters can be used to calculate the corresponding environmental features, we divided the environmental features into two types. One was the amino acid environmental features and the other was the topological environmental features. Then, two individual experiments were implemented. One took the 11 amino acid properties into consideration, and the other took the amino acid properties as well as topological parameters into consideration.

First, the amino acid properties were used for analyzing the optimal window size for environmental features calculation. For all the models, the amino acid properties of the variant were treated as baseline features. A model with the baseline features (i.e. no environmental features, $w = 1$) was constructed. Then, the classifiers with the baseline features and corresponding environmental features calculated from different window size were constructed for comparison. The results for S1925 dataset were listed in Table 2. As shown, the prediction was affected by environmental features. The model constructed without any environmental information ($w = 1$) gave a prediction accuracy of 0.75, but when the environmental information was considered ($w > 1$), the models' prediction strength was improved significantly with the accuracy of 83%-84%. The window sizes of 7 and 9 were both suitable for environmental features calculation,

Table 2. Effect of window sizes on classifiers constructed by amino acid properties for S1925 dataset

Window size	TPR	TNR	PPV	NPV	Accuracy	MCC	AUC
1	0.42	0.90	0.64	0.78	0.75	0.37	0.69
3	0.67	0.88	0.72	0.86	0.82	0.56	0.86
5	0.66	0.91	0.75	0.86	0.83	0.59	0.86
7	0.67	0.91	0.76	0.86	0.84	0.61	0.86
9	0.68	0.91	0.77	0.87	0.84	0.61	0.87
11	0.67	0.92	0.78	0.86	0.84	0.61	0.87
13	0.59	0.93	0.80	0.84	0.83	0.58	0.86
15	0.67	0.91	0.77	0.86	0.84	0.61	0.87
31	0.66	0.90	0.74	0.86	0.83	0.58	0.86

The effect of window sizes on SVM classifiers prediction strength was also illustrated by ROC curves. As shown in Figure 2, the ROC curves of the models for $w = 7$, 9 were better than that for $w = 1$. However, the classifier performance was not improved when using $w = 31$. It indicates that the environment of variants may also affect their influence on protein stability and .the window size is not the bigger the better.

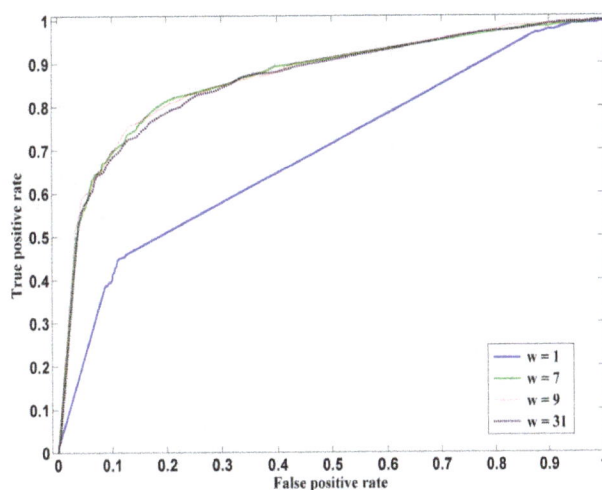

Figure 2. ROC curves to show the effect of window size on the performance of classifiers constructed with amino acid properties for S1925 dataset

To further study the best choice of window size, we combined the topological parameters and amino acid properties to detect their performance on classifiers using S1925 dataset. For these models, the baseline features were the amino acid properties and topological features of the variant ($w = 1$). A classifier with the baseline features (i.e. no environmental features, $w = 1$) was constructed. Then, the classifiers with the baseline features and corresponding environmental features calculated from different window size were constructed for comparison. Results were shown in Table 3. As shown, when no environmental features were considered ($w = 1$), the accuracy decreased, and other evaluation indexes also declined.

Table 3. Effect of window sizes on classifiers constructed with amino acid properties and topological parameters for S1925 dataset

Window size	TPR	TNR	PPV	NPV	Accuracy	MCC	AUC
1	0.69	0.87	0.71	0.86	0.82	0.57	0.82
3	0.64	0.94	0.82	0.86	0.85	0.63	0.87
5	0.65	0.94	0.84	0.86	0.85	0.64	0.88
7	0.73	0.92	0.80	0.88	0.86	0.66	0.89
9	0.74	0.92	0.80	0.89	0.86	0.67	0.89
11	0.64	0.94	0.83	0.86	0.85	0.64	0.88
13	0.64	0.94	0.83	0.86	0.85	0.63	0.87
15	0.64	0.94	0.83	0.86	0.85	0.63	0.87
31	0.64	0.94	0.82	0.86	0.85	0.63	0.87

From Table 2 and Table 3, it can be found that the use of w = 9 was marginally better than the use of other window sizes. As shown, the results of w = 7 and w = 9 were similar with each other. But the AUC of w = 9 was slightly better than w = 7 in Table 2. In addition, the difference between TPR and TNR of w = 9 was marginally smaller than that of w = 7 which represented that the classifier constructed with w = 9 was more balance in predicting positive and negative cases. Though the excellence is small, we also choose w = 9 as a better window size in order to improve the predictive strength in details. It is clear that the environmental features made contribution to the classifier. Moreover, it demonstrates that eight sequence neighbors are enough for providing environment information of the variant and that such environment may also have influence on protein structure and function.

3.3 The Classifier Performance Assessment Using the Dataset SR2760 and S1925

34 features were finally used to construct the model. Besides the network topological features and amino acid properties of the variants and their corresponding environmental features, the features ASA, SIFT score(S), T and pH were also included. The performance of the model was evaluated by 20-fold cross-validation with SVM. With $C = 16$, $\gamma = 0.0625$, the best result (accuracy of 0.86 and MCC of 0.68) was obtained when the window size was set to 9 (w = 9) for SR2760 dataset. We also compared the result with other methods, like M47 (Yang et al., 2013), FoldX (Schymkowitz et al., 2005), I-Mutant 2.0 (Capriotti, Fariselli, & Casadio, 2005a) and MUpro (Cheng et al., 2006). As the M47 model was also compared with the other three methods, we just got the comparison results from literature (Yang et al., 2013). The comparison was listed in Table 4. It reveals that our method outperforms in the comparison.

Table 4. Prediction performance for the SR2760 dataset

Method	TPR	TNR	PPV	NPV	Accuracy	MCC
Our method	0.75	0.91	0.80	0.89	0.86	0.68
M47	0.66	0.94	0.84	0.86	0.85	0.65
FoldX	0.64	0.72	0.52	0.81	0.70	0.35
I-Mutant 2.0	0.64	0.93	0.81	0.85	0.84	0.61
MUpro	0.65	0.92	0.79	0.85	0.84	0.61

Table 5. Prediction performance for the S1925 dataset

Method	TPR	TNR	PPV	NPV	Accuracy	MCC
Our method	0.76	0.93	0.83	0.90	0.88	0.71
M47	0.76	0.92	0.80	0.9	0.87	0.68
AUTO-MUTE	0.70	0.9	0.75	0.87	0.84	0.61
FoldX	0.55	0.69	0.43	0.78	0.66	0.22
I-Mutant 2.0	0.56	0.91	0.73	0.83	0.80	0.51
Mupro	0.68	0.92	0.79	0.87	0.85	0.63

To further compare our method with other methods, we used the S1925 dataset to test the model. With $C = 4$, $\gamma = 0.0625$ and window size of 9 ($w = 9$), the accuracy of our model was 0.88 for S1925 dataset with 0.71 MCC. The result was compared with other five methods, M47, AUTO-MUTE (Masso et al., 2008), FoldX, I-Mutant 2.0 and Mupro. Our approach obtained the best prediction accuracy (Table 5).

3.4 Feature Importance Measures

The analysis of feature importance was addressed by adopting the feature estimation module of random Forest package in R (Breiman, 2001). Permutation accuracy importance measure was adopted to evaluate the importance of each of the 34 features used in this research. According to the principle of permutation importance, the higher score a feature gets, the more important it will be.

The evaluation process was repeated 100 times. The importance score of each parameter was shown as box plots in Figure 3. In a box plot, the middle bar represents the mean value of scores. The top three features were dBu, dP and ASA. It had been reported that Bu, P and ASA were relevant for predicting protein stability changes on single site mutation by (Teng et al., 2010; Yang et al., 2013). The attributes Betweenness (B) and Clustering Coefficient (CC) ranked in top 10. Moreover, the scores between topological features and the top rank parameters were not in huge difference. It can be believed that the topological parameters are comparable with previously reported valuable attributes. In addition, the environmental features we calculated were not in high score as they were used for measuring individually. When combining the environmental features together, they make valuable contribution for predicting as demonstrated in section 3.2.

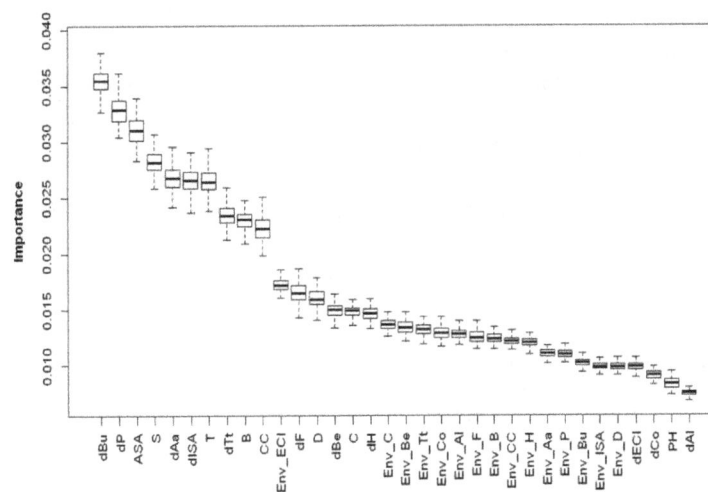

Figure 3. Importance score of each feature evaluated by the random Forest algorithm in the R package

3.5 The Analysis of the Mutant Rules

As SR2760 dataset almost covered the S1925 dataset, so we just used the SR2760 dataset for analyzing the mutant rules. Owing to the dataset contained more than one experimental ddG values for a variant under the same experimental conditions, the redundant data were removed to ensure that a unique ddG value for a variant under the same experimental conditions can be acquired. After a rigorous selection process, the final dataset contained 2218 mutations. We renamed the dataset as NSR2760 dataset. Based on NSR2760 dataset, we got the frequency of stabilizing and destabilizing mutations. The results were shown in Supplementary Table 1.

As shown in Supplementary Table 1, several rules can be observed. If the deleted residue is L while the introduced residue is A, the stability change will be negative (frequency = 88.6%). The mutants from A to S (frequency = 90.3%), Y to A (frequency = 100%), V to G (frequency = 90.5%), I to F (frequency = 100%), I to M (frequency = 93.8%) and so on also cause the protein destabilization. It may be believed that different mutations have their specific effect on protein stability changes. We hope such statistics will give valuable information for the protein engineering.

3.6 The Model Performance Assessment Using the New Cut Datasets

As the two datasets both contained more than one experimental ddG values for a variant under the same experimental conditions, in order to ensure that our method was not significantly affected by the redundant data, we made a selection for the datasets. After a rigorous selection process, the final dataset contained 1752 mutations with 531 positive and 1221 negative cases for S1925 dataset. We renamed the new dataset as NS1925 dataset. SR2760 which was renamed as NSR2760 dataset contained 2218 mutations with 696 positive and 1522 negative cases. The new datasets just involved a unique ddG value for a variant under the same experimental conditions.

We used the new datasets with 34 features to construct the classifier. The window size was set to 9 ($w = 9$). The model achieved accuracy of 0.86 with 0.67 MCC for NS1925 dataset. For NSR2760 dataset, the accuracy was 0.84 with MCC of 0.63. The results were shown in Table 6. For both datasets, the accuracy declined about 2% compared with the results obtained from original datasets, which may be caused by the change of dataset. However, it also achieved satisfying prediction performance. It could be believed that our method was effective for predicting protein stability changes upon single amino acid mutation.

Table 6. Results for NS1925 and NSR2760

Dataset	TPR	TNR	PPV	NPV	Accuracy	MCC	AUC
NS1925	0.74	0.91	0.79	0.89	0.86	0.67	0.90
NSR2760	0.73	0.89	0.75	0.88	0.84	0.63	0.88

4. Conclusions

In this study, the topological characteristics extracted from the protein structure network were introduced in predicting protein stability changes upon single-site mutations. These parameters exhibit satisfying predictive strength. It further indicates that the network parameters reflecting the importance of nodes in protein structure network are related with their importance in protein function and structure. Based on our study, eight sequence neighbors centered on the mutant site were enough for providing the mutant site's environment information in protein sequence. In addition, from the analysis of the mutant rules, some substitutions were observed to have the preference for protein stability changes. Such as the mutants of Y to A, I to F and etc are more likely to cause protein destabilization. Our result could be anticipated valuable for designing new proteins.

Acknowledgements

We would like to thank the anonymous reviewers for their patient review and constructive suggestions. This study was supported by the National Natural Science Foundation of China (21175095), Doctoral Fund of Ministry of Education of China (20120181110051).

References

Amitai, G., Shemesh, A., Sitbon, E., Shklar, M., Netanely, D., Venger, I., & Pietrokovski, S. (2004). Network analysis of protein structures identifies functional residues. *Journal of Molecular Biology, 344*(4), 1135-1146. http://dx.doi.org/ 10.1016/j.jmb.2004.10.055

Bagler, G., & Sinha, S. (2005). Network properties of protein structures. *Physica a-Statistical Mechanics and Its Applications, 346*(1-2), 27-33. http://dx.doi.org/ 10.1016/j.physa.2004.08.046

Bava, K. A., Gromiha, M. M., Uedaira, H., Kitajima, K., & Sarai, A. (2004). ProTherm, version 4.0: thermodynamic database for proteins and mutants. *Nucleic Acids Research, 32*, D120-D121.

http://dx.doi.org/ 10.1093/nar/gkh082

Bradley, A. P. (1997). The use of the area under the roc curve in the evaluation of machine learning algorithms. *Pattern Recognition, 30*(7), 1145-1159. http://dx.doi.org/ 10.1016/s0031-3203(96)00142-2

Breiman, L. (2001). Random forests. *Machine Learning, 45*(1), 5-32. http://dx.doi.org/ 10.1023/a:1010933404324

Brinda, K. V., & Vishveshwara, S. (2005). A network representation of protein structures: Implications for protein stability. *Biophysical Journal, 89*(6), 4159-4170. http://dx.doi.org/ 10.1529/biophysj.105.064485

Capriotti, E., Fariselli, P., & Casadio, R. (2004). A neural-network-based method for predicting protein stability changes upon single point mutations. *Bioinformatics (Oxford, England), 20 Suppl 1*, i63-68. http://dx.doi.org/ 10.1093/bioinformatics/bth928

Capriotti, E., Fariselli, P., & Casadio, R. (2005a). I-Mutant2.0: predicting stability changes upon mutation from the protein sequence or structure. *Nucleic Acids Research, 33*, W306-W310. http://dx.doi.org/ 10.1093/nar/gki375

Capriotti, E., Fariselli, P., Calabrese, R., & Casadio, R. (2005b). Predicting protein stability changes from sequences using support vector machines. *Bioinformatics (Oxford, England), 21 Suppl 2*, ii54-58. http://dx.doi.org/ 10.1093/bioinformatics/bti1109

Cargill, M., Altshuler, D., Ireland, J., Sklar, P., Ardlie, K., Patil, N., . . . Lander, E. S. (1999). Characterization of single-nucleotide polymorphisms in coding regions of human genes. *Nature Genetics, 22*(3), 231-238.

Cheng, J. L., Randall, A., & Baldi, P. (2006). Prediction of protein stability changes for single-site mutations using support vector machines. *Proteins-Structure Function and Bioinformatics, 62*(4), 1125-1132. http://dx.doi.org/ 10.1002/prot.20810

Chou, P. Y. & Fasman, G. D. (1978). *Adv. Enzym, 47*, 45-148

Collantes, E. R., & Dunn, W. J. (1995). Amino acid side chain descriptors for quantitative structure-activity relationship studies of peptide analogues. *Journal of Medicinal Chemistry, 38*(14), 2705-2713. http://dx.doi.org/ 10.1021/jm00014a022

Del Sol, A., & O'Meara, P. (2005). Small-world network approach to identify key residues in protein-protein interaction. *Proteins, 58*(3), 672-682. http://dx.doi.org/ 10.1002/prot.20348

Deleage, G., & Roux, B. (1987). An algorithm for protein secondary structure prediction based on class prediction. *Protein engineering, 1*(4), 289-294. http://dx.doi.org/ 10.1093/protein/1.4.289

Dokholyan, N. V., Li, L., Ding, F., & Shakhnovich, E. I. (2002). Topological determinants of protein folding. *Proceedings of the National Academy of Sciences of the United States of America, 99*(13), 8637-8641. http://dx.doi.org/ 10.1073/pnas.122076099

Folkman, Lukas, Stantic, Bela, & Sattar, Abdul. (2013). Sequence-only evolutionary and predicted structural features for the prediction of stability changes in protein mutants. *Bmc Bioinformatics, 14*. http://dx.doi.org/ 10.1186/1471-2105-14-s2-s6

Gasteiger, E., Hoogland, C., Gattiker, A., Duvaud, S., Wilkins, M. R., Appel, R. D., & Bairoch, A. (2005). In J. M. Walker (Ed.), *The Proteomics Protocols Hadbook* (pp. 571-607). Humana Press.

Grantham, R. (1974). Amino acid difference formula to help explain protein evolution. *Science (New York, N.Y.), 185*(4154), 862-864. http://dx.doi.org/ 10.1126/science.185.4154.862

Greene, L. H., & Higman, V. A. (2003). Uncovering network systems within protein structures. *Journal of Molecular Biology, 334*(4), 781-791. http://dx.doi.org/ 10.1016/j.jmb.2003.08.061

Ihaka, R., & Gentleman, R. (1996). R: A Language for Data Analysis and Graphics. *Journal of Computational and Graphical Statistics, 5*(3), 299-314.

Kawashima, S., & Kanehisa, M. (2000). AAindex: Amino acid index database. *Nucleic Acids Research, 28*(1), 374-374. http://dx.doi.org/ 10.1093/nar/28.1.374

Kyte, J., & Doolittle, R. F. (1982). A simple method for displaying the hydropathic character of a protein. *Journal of molecular biology, 157*(1), 105-132. http://dx.doi.org/ 10.1016/0022-2836(82)90515-0

Li, Y. Z., Wen, Z. N., Xiao, J. M., Yin, H., Yu, L. Z., Yang, L., & Li, M. L. (2011). Predicting disease-associated substitution of a single amino acid by analyzing residue interactions. *Bmc Bioinformatics, 12*.

http://dx.doi.org/ 10.1186/1471-2105-12-14

Masso, M., & Vaisman, I. I. (2008). Accurate prediction of stability changes in protein mutants by combining machine learning with structure based computational mutagenesis. *Bioinformatics, 24*(18), 2002-2009. http://dx.doi.org/ 10.1093/bioinformatics/btn353

Newman, M. E. J. (2003). The structure and function of complex networks. *Siam Review, 45*(2), 167-256. http://dx.doi.org/ 10.1137/s003614450342480

Ng, P. C., & Henikoff, S. (2001). Predicting deleterious amino acid substitutions. *Genome Research, 11*(5), 863-874. http://dx.doi.org/ 10.1101/gr.176601

Noble, W. S. (2006). What is a support vector machine? *Nature Biotechnology, 24*(12), 1565-1567. http://dx.doi.org/ 10.1038/nbt1206-1565

Rose, G. D., Geselowitz, A. R., Lesser, G. J., Lee, R. H., & Zehfus, M. H. (1985). Hydrophobicity of amino acid residues in globular proteins. *Science (New York, N.Y.), 229*(4716), 834-838. http://dx.doi.org/ 10.1126/science.4023714

Schymkowitz, J., Borg, J., Stricher, F., Nys, R., Rousseau, F., & Serrano, L. (2005). The FoldX web server: an online force field. *Nucleic Acids Research, 33*, W382-W388. http://dx.doi.org/ 10.1093/nar/gki387

Shirley, B. A., Stanssens, P., Hahn, U., & Pace, C. N. (1992). Contribution of hydrogen bonding to the conformational stability of ribonuclease T1. *Biochemistry, 31*(3), 725-732. http://dx.doi.org/ 10.1021/bi00118a013

Teng, S. L., Srivastava, A. K., & Wang, L. J. (2010). Sequence feature-based prediction of protein stability changes upon amino acid substitutions. *Bmc Genomics, 11*. http://dx.doi.org/ 10.1186/1471-2164-11-s2-s5

Vendruscolo, M., Dokholyan, N. V., Paci, E., & Karplus, M. (2002). Small-world view of the amino acids that play a key role in protein folding. *Physical Review E, 65*(6). http://dx.doi.org/ 10.1103/PhysRevE.65.061910

Vihinen, M., Torkkila, E., & Riikonen, P. (1994). Accuracy of protein flexibility predictions. *Proteins, 19*(2), 141-149. http://dx.doi.org/10.1002/prot.340190207

Wang, Z., & Moult, J. (2001). SNPs, protein structure, and disease. *Human Mutation, 17*(4), 263-270. http://dx.doi.org/ 10.1002/humu.22

Watts, D. J., & Strogatz, S. H. (1998). Collective dynamics of 'small-world' networks. *Nature, 393*(6684), 440-442. http://dx.doi.org/ 10.1038/30918

Yang, Y., Chen, B., Tan, G., Vihinen, M., & Shen, B. R. (2013). Structure-based prediction of the effects of a missense variant on protein stability. *Amino Acids, 44*(3), 847-855. http://dx.doi.org/ 10.1007/s00726-012-1407-7

Yue, P., & Moult, J. (2006). Identification and analysis of deleterious human SNPs. *Journal of Molecular Biology, 356*(5), 1263-1274. http://dx.doi.org/ 10.1016/j.jmb.2005.12.025

Zhao, Gang, & London, Erwin. (2006). An amino acid "transmembrane tendency" scale that approaches the theoretical limit to accuracy for prediction of transmembrane helices: Relationship to biological hydrophobicity. *Protein Science, 15*(8), 1987-2001. http://dx.doi.org/ 10.1110/ps.062286306

Zimmerman, J. M., Eliezer, N., & Simha, R. (1968). The characterization of amino acid sequences in proteins by statistical methods. *Journal of theoretical biology, 21*(2), 170-201. http://dx.doi.org/10.1016/0022-5193(68)90069-6

Supplementary Materials

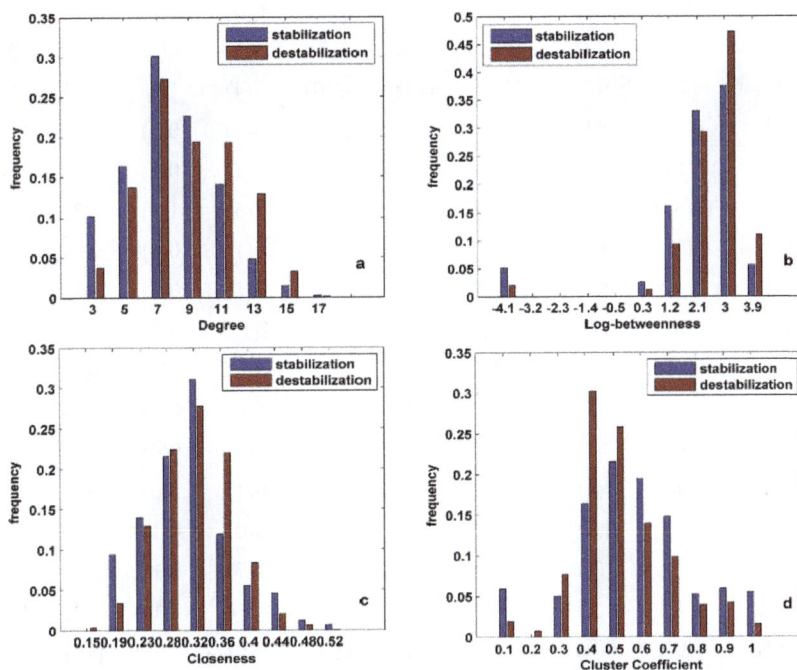

Supplementary Figure 1. The frequency distributions of a) Degree; b) Betweenness; c) Closeness; d) Cluster Coefficient for stabilization and destabilization-associated mutants of SR2760

Supplementary Table 1. Frequency of occurrence of stabilizing mutants obtained from NSR2760 dataset

	A	R	N	D	C	Q	E	G	H	I
A	-	0(2)	1(3)	2(5)	1(3)	0(4)	1(5)	6(19)	0(4)	3(5)
R	3(15)	-	0(0)	0(0)	2(3)	1(6)	3(7)	2(4)	1(7)	0(0)
N	8(27)	0(0)	-	12(21)	0(0)	0(1)	3(3)	1(5)	1(4)	7(7)
D	15(37)	4(4)	19(44)	-	1(2)	3(5)	10(13)	6(12)	9(12)	3(4)
C	2(8)	0(0)	0(0)	0(0)	-	0(0)	0(0)	0(1)	0(0)	0(1)
Q	7(18)	1(1)	2(3)	0(0)	1(1)	-	3(6)	2(8)	0(1)	1(1)
E	9(30)	2(3)	1(7)	1(7)	2(4)	9(25)	-	3(6)	3(4)	3(3)
G	17(44)	2(6)	0(2)	0(5)	0(1)	1(3)	2(2)	-	1(4)	0(0)
H	3(9)	1(4)	0(3)	1(4)	1(1)	3(6)	1(2)	4(8)	-	0(0)
I	1(51)	0(0)	2(2)	0(2)	0(2)	0(0)	0(5)	1(10)	0(1)	-
L	**8(70)**	2(4)	0(1)	0(3)	0(3)	0(1)	0(2)	0(7)	1(2)	1(10)
K	11(29)	5(9)	6(7)	0(2)	0(0)	1(8)	6(19)	3(10)	0(5)	1(2)
M	3(15)	1(1)	0(0)	3(3)	1(1)	0(0)	3(3)	0(5)	0(0)	3(10)
F	2(24)	0(0)	0(0)	0(0)	0(0)	0(0)	0(0)	0(2)	0(1)	0(2)
P	3(25)	0(1)	0(1)	0(0)	0(0)	0(0)	0(0)	0(9)	0(0)	0(0)
S	12(35)	2(2)	2(4)	5(8)	0(2)	1(2)	3(3)	2(6)	1(5)	2(2)
T	3(27)	3(6)	3(11)	3(9)	1(5)	1(4)	4(14)	0(12)	1(5)	7(16)
W	0(0)	0(0)	0(0)	0(0)	0(0)	0(0)	0(0)	0(0)	0(2)	0(0)
Y	**0(11)**	0(1)	0(2)	1(2)	2(3)	1(2)	0(0)	0(6)	1(1)	0(0)
V	12(86)	1(4)	1(6)	2(4)	0(9)	0(1)	1(5)	**2(21)**	0(7)	17(35)

	L	K	M	F	P	S	T	W	Y	V
A	2(8)	1(6)	2(6)	1(3)	13(24)	**3(31)**	4(11)	0(2)	0(3)	11(18)
R	0(1)	1(8)	1(4)	0(0)	0(0)	0(3)	0(0)	0(0)	0(0)	0(0)
N	1(1)	2(5)	3(3)	1(1)	0(0)	3(7)	3(4)	0(0)	0(0)	2(2)
D	3(4)	6(11)	1(2)	3(5)	0(5)	7(10)	3(4)	1(2)	2(2)	2(3)
C	0(1)	0(0)	0(1)	0(0)	0(0)	2(11)	4(5)	0(0)	0(0)	0(5)
Q	4(5)	5(5)	0(1)	0(0)	0(4)	0(2)	0(1)	0(0)	0(0)	0(1)
E	5(7)	12(33)	3(4)	2(4)	3(4)	3(6)	1(5)	1(3)	2(3)	7(8)
G	0(1)	0(1)	0(0)	0(1)	1(3)	1(9)	0(1)	0(1)	0(0)	4(8)
H	5(5)	0(1)	0(0)	0(0)	7(7)	1(2)	0(2)	0(1)	7(8)	0(0)
I	11(22)	0(0)	**1(16)**	**0(11)**	0(3)	0(3)	2(16)	0(2)	0(3)	**8(58)**
L	-	0(1)	2(13)	2(9)	1(2)	0(1)	0(6)	1(1)	1(1)	**1(16)**
K	0(0)	-	3(5)	2(2)	2(3)	0(1)	0(0)	0(1)	1(1)	3(3)
M	4(23)	3(8)	-	2(3)	0(0)	0(0)	0(2)	0(0)	0(1)	1(8)
F	3(10)	0(1)	0(4)	-	0(0)	0(0)	0(1)	5(6)	5(8)	0(5)
P	3(4)	0(0)	0(0)	0(0)	-	1(5)	0(0)	0(1)	0(1)	0(1)
S	1(2)	2(2)	1(1)	4(4)	1(2)	-	4(7)	1(1)	1(2)	3(5)
T	1(6)	1(1)	1(1)	3(4)	0(3)	8(19)	-	1(1)	3(3)	11(27)
W	0(4)	0(0)	0(0)	0(15)	0(0)	0(0)	0(0)	-	0(7)	0(0)
Y	1(2)	1(1)	0(0)	14(35)	0(1)	0(3)	0(0)	2(5)	-	1(1)
V	11(21)	0(4)	4(13)	**0(10)**	2(4)	0(6)	**0(33)**	0(0)	0(8)	-

(1) Mutations are from columns of residue to rows of residue.

(2) The number of the stabilizing mutants is given outside the parentheses with the total number of the corresponding mutants in parentheses.

17

Nystatin Modulates Genes in Immunity and Wingless Signaling Pathways in Cow Blood

Emmanuel K Asiamah[1], Sarah Adjei-Fremah[1], Kingsley Ekwemalor[1] & Mulumebet Worku[1]

[1] North Carolina A&T State University, USA

Correspondence: Mulumebet Worku, North Carolina A&T State University, USA. E-mail: worku@ncat.edu

Abstract

Nystatin is an antifungal agent isolated from bacteria found in the dairy cow environment. It disrupts small platforms in the cell membrane, composed of sphingolipids and cholesterol known as lipid rafts. Pathogen recognition receptors (PRR) may be embedded in these lipid rafts. This study was conducted to evaluate the *in vitro* effects of the lipid raft inhibitor Nystatin, on the expression of genes in the innate and adaptive immunity and wnt signaling pathway in cow peripheral blood. Blood collected from four adult female Holstein-Friesian cows (n=4) was treated with 100ng/mL of Nystatin *in vitro*. Samples treated with Phosphate Buffer Saline served as control. Total protein concentration and prostaglandin E2in plasma were determined. Total RNA was isolated from cells and was used for cDNA synthesis. The effect of Nystatin on the expression of 84 genes on the cow Wingless signaling pathway and human innate and adaptive immunity arrays were assessed in cow blood using real-time PCR. Fold change in transcript abundance was calculated using Livak's method. Nystatin was found to modulate transcription and translation of genes involved in homeostasis and immunity in cow blood. It also increased the concentration of total plasma protein and PGE2in cow blood and may thus have had a pro-inflammatory effect. This study provides evidence for the association between lipid raft inhibition and alterations in the wingless signaling pathway in ruminant blood. Furthermore, the results presented may inform antifungal drug design and use in cows.

Keywords: Nystatin, lipid raft, cow blood, wingless

1. Introduction

There are over 300 species of fungi that are recognized as real or potential pathogens responsible for mycoses in animals (Drouhet, 1998). In bovine dairy herds, inflammation of the mammary gland is one of the most important health problems. Bovine mycotic mastitis is reported to be responsible for 1—12% of all mastitis cases (Costa et al., 1998; Lagneau et al., 1996). Drugs such as Amphotericin B (Amp B) and Nystatin are used for the treatment of mycotic mastitis and other fungus related infections. However, Amp B has to be administered in higher doses and therefore is not favorable because of toxicity constraints (Cheng et al., 1982).

Nystatin was originally isolated from bacteria in the dairy cow environment and is now widely used as an prophylactic. It has been reported to be less toxic and superior to Amp B against experimental candidiasis in model animals (Massood et al., 2003). Nystatin has also been used widely to demonstrate the involvement of lipid rafts in biological processes (Smart et al., 2002). Lipid rafts are small platforms, composed of sphingolipids and cholesterol content of cell membranes. These assemblies are fluid but more ordered and tightly packed than the surrounding bilayer. They are evolutionarily conserved structures that play a role in a number of signaling processes involving receptors expressed by a variety of cell types (Brown & London, 2000). Nystatin disrupts lipid rafts by binding to cholesterol in the lipid raft thereby degrading its integrity through hole formations in cells.

Nystatin stimulates innate immunity by binding to Toll-like receptor 2 (TLR2) (Razonable et al., 2009). Toll-like receptors are pattern recognition receptors (PRRs) that recognize highly conserved structural motifs known as pathogen-associated molecular patterns (PAMPs). The PAMPs are involved in the mediation of innate immunity and the activation of adaptive immunity. TheTLR2-driven induction of anti-inflammatory cytokines is important in microbial recognition and control (Netea et al., 2004).

The wingless homolog gene family (Wnt), are a family of secreted glycoprotein signal transducers involved in embryonic development and cell polarity. They have also been identified as potential mediators of inflammation

(Sen, 2000). The Wnts affect diverse processes such as embryonic induction, generation of cell polarity and the specification of cell fate (Logan and Nusse, 2004). The Wnt signaling pathway is activated by the binding of Wnt to Frizzled and LRP5/6 co-receptors to affect transcription of target genes. A key role for Wnt signaling during adult homeostasis is the maintenance of stem cell pluripotency (Reya & Clevers, 2005). Both the Toll-like receptor (TLR) and Wnt signaling pathways play important roles in health and diseases (Logan,2004) and their interaction with receptors on the cell surface is the first step in transducing an extracellular signal into intracellular responses (Cong et al., 2004).

In this study, the objective was to evaluate the *in vitro* effects of the lipid raft inhibitor Nystatin, on the expression of genes in the innate and adaptive immunity and wnt signaling pathway in cow peripheral blood.

2. Materials and Methods

2.1 Animals

All protocols for the handling of the animals were approved by the Institute of Animal Care and Use Committee (IAUCUC). Clinically healthy adult female Holstein-Friesian cows in mid- lactation (n=4) from the North Carolina Agricultural and Technical State University dairy farm were used in the study. None of the animals used exhibited any evidence of disease prior to blood sampling.

2.2 Blood Sampling and Treatment of Blood with Nystatin in vitro

Fifty milliliters (50 ml) of whole blood was collected aseptically from the jugular vein of the animals into vacutainer tubes containing 5ml of the anti-coagulant acid citrate dextrose (Macdonald et al., 2006). The tubes were placed on ice immediately after collection and transported to the laboratory. Blood was processed within 2 hours of collection. One milliliter of blood (10^7 cells/ mL of viable cells) was exposed to 100 ng/mL of Nystatin (NYS) (Sigma St Louis, MO). Samples treated with PBS served as control. The cells were incubated at 37°C with 5% CO_2 (Adjei-Fremah et al., 2015), and 85% humidity for 30 minutes. All the reagents used were prepared with endotoxin-free phosphate buffered saline (PBS). Endotoxin assay was performed as described by Adjei-Fremah et al. (2016). All treatments were done in triplicates. At the end of the incubation, period cells were spun down at 1700 x g at 4°C for 5 minutes. Supernatants were collected and stored at -80°C to measure total protein concentration, prostaglandin E2 levels. Trizol was added to cell pellets and stored for RNA isolation.

2.3 Total Cell Count, Viable Cell and White Blood Cell Differential Counts

Cell Viability was assessed using the Trypan blue dye exclusion method on the TC20 cell counting instrument (Bio-rad) as previously described by Asiamah et al. (2016). Samples of whole blood were diluted 1: 100 in saline. Ten (10) μl of diluted whole blood and 10μl of Trypan blue were combined in a 1.5 ml test tube. The mixture was pipetted up and down for five times to mix the cells and the dye. Ten (10) μl of the mixture was loaded in one of two chambers of the counting slide. The slide was then inserted into the TC20 cell counter for automatic cell counting. Cell counting was done in duplicate and an average was taken. Cell viability was expressed as a percentage of [(total viable and non- viable cells/total cells)] on the TC 10 cell counter (Bio-Rad).

2.4 White Blood Cell Differential Counts

White Blood Cell Differential counts were conducted on whole blood treated with all ten treatments or maintained PBS control using Wright's staining procedure. A thin smear of blood was made on a glass slide and left to dry at room temperature overnight. The air-dried slide was dipped in Wright's stain for10 seconds. Excess stain was washed off the stained slide with deionized water. The slide was then air dried before reading under a light microscope (Carolina Biologicals). Smears were read under oil immersion for cell counts. The different cells were counted up to 100 for numerical representation of various cells present in the blood sample.

2.5 Evaluation of the Concentration of Total Plasma Protein

Total protein concentration in plasma from treated and control samples were measured using the Bicinchoninic acid assay (BCA) following the manufacturer's instructions (Thermo Scientific™ Pierce) as described by Worku et al. (2016). Each sample was done in triplicate.

2.6 Evaluation of Prostaglandin (PGE2) Secreted in Plasma

Prostaglandin (PGE2) concentration in plasma from treated and control groups were evaluated using a commercial Enzyme-linked immunosorbent assay (Cayman) following the manufacturer's instructions as described by Ekwemalor et al. (2016). Each sample was done in triplicates.

2.7 Isolation of RNA and cDNA Synthesis

Total RNA was isolated using Trizol (Sigma). The quality and quantity of the RNA was measured with the Nanodrop spectrophotometer. Total RNA was pipetted into an RNA 6000 Nano LabChip® (Agilent) and RNA integrity was determined using Agilent® Bioanalyzer. Manufacturer's protocol was followed. Complimentary DNA (cDNA) synthesis was performed with 500ng/µl RNA using the RT2 first strand synthesis kit per manufacturer's protocol (Qiagen). Each sample was done in triplicates.

2.8 Real-Time PCR

The Cow WNT Signaling Pathway RT² Profiler PCR (Qiagen) was used in profiling the expression of 84 genes related to WNT-mediated signal transduction. The Human Innate & Adaptive Immune Responses RT² Profiler™ PCR Array (Qiagen) was used in profiling the expression of 84 genes involved in the host response to bacterial infection and sepsis.

Gene expression results were analyzed using the Livaks method (Livak and Schmittgen, 2001). Housekeeping gene GAPDH and samples treated with PBS were used to determine the $\Delta\Delta Ct$. Where ΔCt= (Target genes$_{treat}$-GAPDH$_{treat}$) - ΔCt (Target genes $_{PBS}$- GAPDH$_{PBS}$).

Fold change = $2^{(-\Delta\Delta Ct)}$

2.9 Statistical Analysis

Means of White blood cell differential cell counts, Total cell count, concentrations of total plasma protein and PGE2 ELISA of treatment groups were compared with one-way Analysis of variance (ANOVA). SAS 9.2 statistical package was used. Results were considered statistically significant at ($P\leq0.05$). Data for RNA concentration and purity are represented as means.

3. Results

3.1 Total Cell Count, Viable Cell and White Blood Cell Differential Counts

The average cell concentration for cows before treatment was 1.01×10^6 cells/ml(Initial viable cells were considered to be 100%).Cell viability was not affected by treatment at p>0.05 compared to PBS (Nystatin =98% and PBS=97%).White blood cell (WBC) differential count was not affected by Nystatin at p>0.05 (Figure 1).

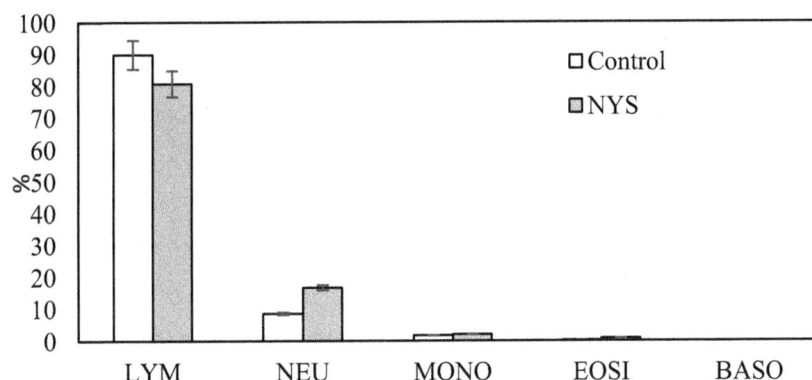

Figure 1. White blood cells differential counts in Nystatin-treated samples compared to control. NYS-Nystatin, LYM-Lymphocytes, NEU-Neutrophils, MONO-Monocytes, EOSI-Eosinophils and BASO-Basophils

3.2 Concentrations and Purity of Total RNA

The concentration and purity of RNA extracted from whole blood was measured to evaluate the treatment effect on mRNA transcription. Concentration and purity levels were not affected by treatment. There were no significant differences in RNA concentrations between the Nystatin treated samples and control samples (p>0.05). All RNA samples averaged 1.7(Figure 2).

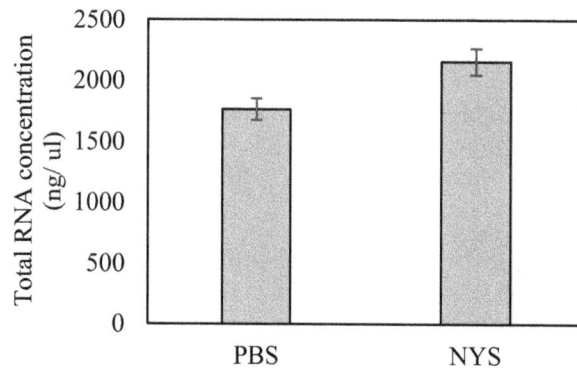

Figure 2. Total RNA concentration in Nystatin treated and PBS (control) samples

3.3 Total Plasma Protein Concentration

All cow blood samples incubated with Nystatin had significantly higher protein concentrations compared to the samples incubated with PBS (control) ($p < 0.05$) (Figure 3).

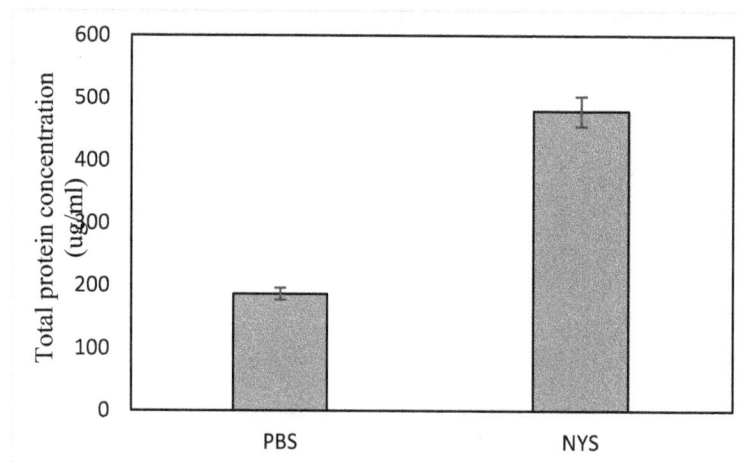

Figure 3. Total protein concentration in Nystatin treated and control group

3.4 Prostaglandin Levels (PGE2)

The results show that prostaglandin levels in blood samples incubated with Nystatin were significantly higher than the ones incubated with PBS (control) ($p < 0.05$) as presented in (Figure 4).

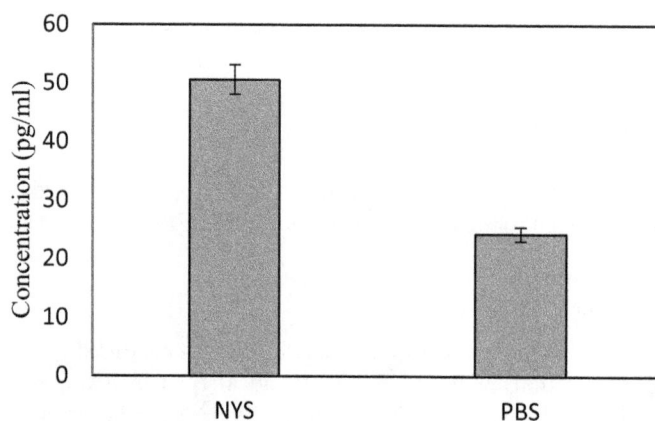

Figure 4. Levels of prostaglandin E2 concentration in blood plasma in NYS-treated and PBS-control group

3.5 Wnt Signaling Genes in Cow Blood

All 84 genes were expressed on the Wnt signaling pathway following Nystatin treatment. Seven of the expressed genes were upregulated (fold change ≥ 2) (Table 1).

Table 1. A list of Up-regulated genes on the Wnt array

Gene name	Fold change	Gene function
DAB2	6.4*	Tumor suppressor
FBXW11	2.0*	WNT Signaling Negative regulation
FZD3	*2.1	Wnt/ Canonical pathway (specific function unknown)
YWHAZ	5.6*	Tyrosine 3-monooxygenase/tryptophan 5- monooxygenase activation protein
WNT9A	3.1*	Cell fate regulator /patterning during embryogenesis
WISP1	2.5*	Wnt-1 Inducible Signaling Pathway/ mediates diverse developmental processes
TBP	2.5*	TATA-box binding factor /coactivator/promoter /modify general transcription factors
		TATA-box binding factor /coactivator/promoter /modify general transcription factors

*= significant, Fold changes ≥ 2 is considered significant.

3.6 Human Innate and Adaptive Pathway Genes in Cow Blood

Genes involved in both innate and adaptive immunity were expressed. Seven (7) out of the 51 new genes that (fold change ≥ 2) after treatment include receptors that respond to both bacteria and viruses (CD 40 and TLR 7, MX1 respectively) (Table 2) (Figure 5). Two genes (APCS and IFNA1) were down-regulated (Table 3).

Table 2. Up-regulated genes in the human innate and adaptive PCR array

Gene name	Fold changes	Gene function
MAPK1	81*	Cell signaling/ innate immunity
MAPK8	31*	Cell signaling/innate immunity
TLR 7	27*	Receptor/defense response to viruses
IL 10	3.7*	Down-regulator /cytokine innate immunity
CD 40	6*	Receptor/defense response to bacteria
MX 1	7.1*	Antiviral response
TBX 21	2.0*	Transcription factor

*= significant, Fold changes ≥2 is considered significant.

Table 3. Down-regulated genes on the human innate and adaptive array

Gene name	Gene function
APCS	Tumor suppressor /innate immunity
IFNA1	Antiviral activity/Adaptive immunity(cytokine)

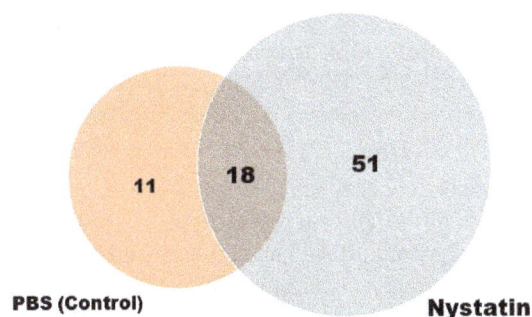

Figure 5 Venn diagram of number of genes expressed in both control and treated samples in Human Innate &Adaptive array

4. Discussion

In this study, we demonstrate that Nystatin modulates both the wnt signaling and the human innate and adaptive immunity pathways in cow blood. Nystatin is an antifungal compound that has been used widely to demonstrate the involvement of lipid rafts in biological processes (Smart et al., 2002). Nystatin disrupts lipid rafts by binding to cholesterol within the plasma membrane it. Lipid rafts are small platforms within the cell membrane, composed of sphingolipids and cholesterol content. It was first discovered in the environment of dairy cows. Scientific research has shown that Nystatin possesses proinflammatory properties (Razonable, 2009). This makes it a potential therapeutic for certain inflammatory diseases in cow e.g. Mastitis.

This study showed that treatment of cow blood with Nystatin modulates wingless signaling pathway. Nystatin activated and modulated the expression of genes associated with Wnt /Ca^{2+} pathway, and inhibited the expression of beta-catenin pathway associated genes. Wnt/Ca^{2+} pathway have been shown to antagonize the beta-catenin pathway (De, 2011). The Wnt/Ca^{2+} signaling pathway regulates cell movement, proliferation, and migration, along with modulation of cell behavior and structural change in a variety of organisms (Angers et al., 2009; Mikel et al., 2006). In this study, cell viability and white blood cell counts were not affected by treatment. Although there was no effect on blood count, cells were live and activated. Sen et al. (2010) in previous studies found no effect of Nystatin on cells at 0.6 µg/ml. In this study, it is evident that 100ng/ml of Nystatin was not enough to have a significant effect on cell counts and viability. Concurrently, genes that were upregulated on both the wnt signaling pathway were mostly involved in cell survival and proliferation inhibition. These include YWHAZ which has attracted interest because of its elevated expression associated with a variety of cancers (Matta et al., 2007). Recently, it is understood that YWHAZ has critical anti- apoptotic functions (Neal et al., 2009). Another gene of interest that increased in fold change was DAB 2. Studies have shown that the over-expression of DAB2 significantly inhibits cell proliferation in cultured lung cancer cells in vitro (Karam et al., 2007).

Nystatin also had effect on activation and expression of immune response genes. Razonable et al. (2009) showed that Nystatin induces the secretion of IL-1, IL-8, and TNF-alpha in TLR2-expressing but not in TLR2-deficient cells. Our results support their findings, that Nystatin is a potential pro-inflammatory agent that uses the TLR-dependent pathway since all TLRs were activated in our experiment (Wiegand et al., 1988; Razonable et al., 2009). Interestingly, TLR 7, which is a receptor for virus detection was upregulated on the innate and adaptive pathway.

Nystatin was also found to increase total protein and prostaglandin E levels in cow. This could be due to the binding of the lipid raft inhibitor to cholesterol in the lipid raft thereby degrading its integrity of the cell through hole formations. Cellular components are then released into the blood stream which then triggers inflammatory response that may lead to increase in prostaglandin formation (Wong et al., 2009). Nystatin increases, e.g., the intracellular concentration of Ca^{2+} (Wiegand et al., 1988). Any cell stimulation is accompanied by an increase in intracellular concentration of calcium. Although prostaglandin released in high amounts may cause tissue damage and other related inflammation problems, it stimulates and regulates a wide array of immune functions, thus playing an important role in the host defense system.

In both Wingless and immune signaling pathways, MAP Kinase activities were activated. These kinases are a chain of proteins in the cell that communicates signals from a receptor on the surface of the cell to the DNA in the nucleus of the cell. Since the receptors in both signaling pathways were upregulated, we can say that MAPK should also be up-regulated in order to transmit signals to the nucleus to ensure homeostasis. Increasing activity of MAPKs and their involvement in the regulation of the synthesis of inflammation mediators at the transcription level, make them potential targets for novel anti-inflammatory therapeutics. This could be a point of cross-talk for both Wnt and TLR pathways. Both wnt and human innate and adaptive pathways rely on membrane-embedded receptors for proper signaling and activation. This study highlights the significance of opportunistic microorganism to animal health. However, in vivo studies in cows are needed to assess the significance of our in vitro findings. Nevertheless, the present study provides new insights into the pharmacological mechanisms of action of Nystatin.

This study demonstrated that in cow blood, Nystatin affects genes involved in both wingless signaling pathway and human innate and adaptive pathway. TLR and Wnt-dependent process may serve as the molecular basis for the proinflammatory properties of Nystatin. It is evident in this study that the lipid raft inhibitor Nystatin, served as an agonist for TLR, whose consequent activation mediates effective antimicrobial properties. An appreciation of the role of TLR and WNT in pharmacologic immunotoxicology highlights their potential as novel therapeutic targets.

5. Conclusion

In this study, PRR associated lipid raft inhibition by Nystatin resulted in modulation of gene activation in cow blood. Exposure of cow blood to Nystatin modulated transcription and translation of genes involved in homeostasis and immunity. This study provides evidence for an association between lipid raft inhibition and alterations in the wingless signaling pathway in ruminant blood. The expression of wingless homolog pathway genes in cow blood in response to Nystatin and other PAMPs needs to be explored since it may be important in homeostasis and inflammation. Furthermore, the results presented may inform antifungal drug design and use in cows.

References

Adjei-Fremah, S., Asiamah, E. K., Ekwemalor, K., Jackai, L., Schimmel, K., & Worku, M. (2016). Modulation of Bovine Wnt Signaling Pathway Genes by Cowpea Phenolic Extract. *Journal of Agricultural Science, 8*(3), 21. http://dx.doi.org/10.5539/jas.v8n3p21

Adjei-Fremah, S., Jackai, L. E., & Worku, M. (2015). Analysis of phenolic content and antioxidant properties of selected cowpea varieties tested in bovine peripheral blood. *American Journal of Animal and Veterinary Sciences, 10*(4), 235-245. http://dx.doi.org/10.3844/ajavsp.2015.235.245

Angers, S., & Moon, R. T. (2009). Proximal events in Wnt signal transduction. Nature reviews Molecular cell biology, 10(7), 468-477. http://dx.doi.org/10.1038/nrm2717

Asiamah, E. K., Adjei-Fremah, S., Osei, B., Ekwemalor, K., & Worku, M. (2016). An Extract of Sericea Lespedeza Modulates Production of Inflammatory Markers in Pathogen Associated Molecular Pattern (PAMP) Activated Ruminant Blood. *Journal of Agricultural Science, 8*(9), 1. http://dx.doi.org/10.5539/jas.v8n9p1

Brown, D. A., & London, E. (2000). Structure and function of sphingolipid-and cholesterol-rich membrane rafts. *Journal of Biological Chemistry, 275*(23), 17221-17224. http://dx.doi.org/10.1074/jbc. R00000520

Cheng, J. T., Witty, R. T., Robinson, R. R., & Yarger, W. E. (1982). Amphotericin B nephrotoxicity: increased renal resistance and tubule permeability. *Kidney international, 22*(6), 626-633. http://dx.doi.org/10.1038/ki. 1982.221

Cong, F., Schweizer, L., & Varmus, H. (2004). Wnt signals across the plasma membrane to activate the β-catenin pathway by forming oligomers containing its receptors, Frizzled and LRP. *Development, 131*(20), 5103-5115. http://dx.doi.org/10.1242/dev.01318

Costa, E. O., Ribeiro, A. R., Watanabe, E. T., & Melville, P. A. (1998). Infectious bovine mastitis caused by environmental organisms. *Journal of Veterinary Medicine, Series B, 45*(1-10), 65-71. http://dx.doi.org/10. 1111/j.1439-0450

De, A. (2011). Wnt/Ca2+ signaling pathway: a brief overview. *Acta biochimica et biophysica Sinica*, gmr079. http://dx.doi.org/10.1093/abbs/gmr079

Ekwemalor, K., Asiamah, E. K., Adjei-Fremah, S., & Worku, M. (2016). Effect of a Mushroom (Coriolus versicolor) Based Probiotic on Goat Health. *American Journal of Animal and Veterinary Sciences, 11*(3), 108-118. http://dx.doi.org/10.3844/ajavsp.2016.108.118

Gori, S., Drouhet, E., Gueho, E., Huerre, M., Lofaro, A., Parenti, M., & Dupont, B. (1998). Cutaneous disseminated mycosis in a patient with AIDS due to a new dimorphic fungus. *Journal de mycologie médicale, 8*(2), 57-63.

Karam, J. A., Shariat, S. F., Huang, H. Y., Pong, R. C., Ashfaq, R., Shapiro, E., .. & Hsieh, J. T. (2007). Decreased DOC-2/DAB2 expression in urothelial carcinoma of the bladder. *Clinical Cancer Research, 13*(15), 4400-4406. http://dx.doi.org/10.1158/1078-0432.CCR-07-0287

Lagneau, P. E., Lebtani, K., & Swinne, D. (1996). Isolation of yeast from bovine milk in Belgium. *Mycopathologia, 135*, 99–102. http://dx.doi.org/10.1007/BF00436458

Logan, C. Y., & Nusse, R. (2004). The Wnt signaling pathway in development and disease. *Annu. Rev. Cell Dev. Biol., 20*, 781-810. http://dx.doi.org/10.1146/annurev.cellbio.20.010403.113126

Masood, A. K., Faisal, S. M., Haque, W., & Owais, M. (2002). Immunomodulator tuftsin augments anti-fungal activity of amphotericin B against experimental murine candidiasis. *Journal of drug targeting, 10*(3), 185-192. http://dx.doi.org/10.1080/10611860290022615

Matta, A., Bahadur, S., Duggal, R., Gupta, S. D., & Ralhan, R. (2007). Over-expression of 14-3-3zeta is an early event in oral cancer. *BMC cancer, 7*(1), 169. http://dx.doi.org/10.1186/1471-2407-7-169

Mikels, A. J., & Nusse, R. (2006). Purified Wnt5a protein activates or inhibits β-catenin–TCF signaling depending on receptor context. *PLoS Biol, 4*(4), e115. http://dx.doi.org/10.1371/journal.pbio.0040115

Neal, C. L., Yao, J., Yang, W., Zhou, X., Nguyen, N. T., Lu, J., ... Hittelman, W. (2009). 14-3-3ζ overexpression defines high risk for breast cancer recurrence and promotes cancer cell survival. *Cancer research, 69*(8), 3425-3432. http://dx.doi.org/10.1158/0008-5472.CAN-08-2765

Netea, M. G., Ferwerda, G., van der Graaf, C. A., Van der Meer, J. W., & Kullberg, B. J. (2006). Recognition of fungal pathogens by toll-like receptors. *Current pharmaceutical design, 12*(32), 4195-4201. https://doi.org/10.2174/138161206778743538

Raja, A., Vignesh, A. R., Mary, B. A., Tirumurugaan, K. G., Raj, G. D., Kataria, R., ... & Kumanan, K. (2011). Sequence analysis of Toll-like receptor genes 1–10 of goat (Capra hircus). *Veterinary immunology and immunopathology, 140*(3), 252-258. http://dx.doi.org/10.1016/j.vetimm.2011.01.007

Razonable, R. R., Henault, M., Lee, L. N., Laethem, C., Johnston, P. A., Watson, H. L., & Paya, C. V. (2005). Secretion of proinflammatory cytokines and chemokines during amphotericin B exposure is mediated by coactivation of toll-like receptors 1 and 2. *Antimicrobial agents and chemotherapy, 49*(4), 1617-1621. http://dx.doi.org/10.1128/AAC.49.4.1617-1621.2005

Reya, T., & Clevers, H. (2005). Wnt signaling in stem cells and cancer. *Nature, 434*(7035), 843-850. http://dx.doi.org/10.1038/nature03319

Sen, M., Lauterbach, K., El-Gabalawy, H., Firestein, G. S., Corr, M., & Carson, D. A. (2000). Expression and function of wingless and frizzled homologs in rheumatoid arthritis. *Proceedings of the National Academy of Sciences, 97*(6), 2791-2796. http://dx.doi.org/10.1073/pnas.050574297

Smart, E. J., & Anderson, R. G. (2001). Alterations in membrane cholesterol that affect structure and function of caveolae. *Methods in enzymology, 353*, 131-139. Choose an option to locate/access this article. http://dx.doi.org/10.1016/S0076-6879(02)53043-3

Wiegand, R., Betz, M., & Hänsch, G. M. (1988). Nystatin stimulates prostaglandin E synthesis and formation of diacylglycerol in human monocytes. *Agents and actions, 24*(3-4), 343-350. http://dx.doi.org/10.1007/BF02028292

Wong, S. L., Leung, F. P., Lau, C. W., Au, C. L., Yung, L. M., Yao, X., ... Huang, Y. (2009). Cyclooxygenase-2–Derived Prostaglandin F2α Mediates Endothelium-Dependent Contractions in the Aortae of Hamsters With Increased Impact During Aging. *Circulation research, 104*(2), 228-235. http://dx.doi.org/10.1161/CIRCRESAHA.108.179770

Worku, M., Abdalla, A., Adjei-Fremah, S., & Ismail, H. (2016). The Impact of Diet on Expression of Genes Involved in Innate Immunity in Goat Blood. *Journal of Agricultural Science, 8*(3), 1.

18

Application of PCR in the Detection of Aflatoxinogenic and Non-aflatoxinogenic Strains of *Aspergillus Flavus* Group of Cattle Feed Isolated in Iran

Sepideh Rahimi[1], Noshin Sohrabi[1], Mohammad Ali Ebrahimi[1], Majid Tebyanian[2], Morteza Taghi Zadeh[2] & Sahar Rahimi[3]

[1] Department of Plant Biotechnology, Tehran Shargh Branch, Faculty of Agriculture Payam Noor University, Iran

[2] Department of Research and development, The Razi Vaccine and Serum Research Institute in the Hessarak district in Karaj, Iran

[3] Department of Food Science and Technology, Pharmaceutical Sciences branch Islamic Azad University, Tehran, Iran

Correspondence: Sepideh Rahimi, Department of Plant Biotechnology, Tehran Shargh Branch, Faculty of Agriculture Payam Noor University, Iran E-mail: srahimi1919@gmail.com

Abstract

Aflatoxins are among the most important Mycotoxins that are mainly produced by various *Aspergillus* species, specially *Aspergillus flavus* and *Aspergillus parasiticus*. Aflatoxins are carcinogenetic and immunosuppressive, so that can lead to acute liver damage, cirrhosis of the liver and hepatocarcinoma induction. Consuming the feed contaminated by *Aspergillus* puts humans and animals under the danger of Aflatoxins that are considered as an important threats for human and animal health. The purpose of the present study was to make distinction between Aflatoxinogenetic and non-Aflatoxinogenetic strains and *Aspergillus Flavus* using PCR and TLC and the expression of five Aflatoxin biosynthesis genes including *aflD (nor-1)*, *aflP(omtA)*, *aflO (omtB)*, *aflQ(ordA)*, *aflR* in 40 strains was investigated using PCR. In this study, a number of 40 *Aspergillus flavus* strains from 67 species of cattle feed from 21 industrial warehouses of various areas of Tehran and Alborz were used. After isolation and culture in exclusive environment of yeast extract of sucrose agar, the isolated *Aspergillus* strains were investigated by microscopic and macroscopic methods. In order to make distinction between Aflatoxinogenetic and non-Aflatoxinogenetic strains, PCR method and TLC techniques were used. The results showed that only 7 strains (1, 3, 5, 14, 22, 34, and 38) were Aflatoxin-producers fungi and the rest 33 samples were non-Afatoxin-producers fungi. Since *Aspergillus flavus* is the main contaminator of cattle feed, there is a need to develop a simple, rapid and sensitive method to identify Aflatoxigenetic fungi, particularly between Aflatoxinogenetic and non-Aflatoxinogenetic strains of AF.

Keywords: Aflatoxinohenetic, non-Aflatoxinogenetic, *Aspergillus flavus* strains, PCR, TLC, Cattel feed

1. Introduction

Fungi growth in stored feed is one of the threatening factors of human health (Richard et al., 2003). The use of unhealthy cattle feed causes distortion in health cycle of cattle, people and diaries (Humans, Organization, & Cancer, 2002). Various studies on cattle feed have shown that the contamination of cattle feed by fungi, particularly *Aspergillus* species, leads to the creation of Aflatoxin (J. W. Bennett & Klich, 2003; Kamei & Watanabe, 2005).

Aflatoxins are considered as secondary fungal metabolites that are easily created along the growth and storage of feed and are mainly produced by *Aspergillus flavus, Aspergillus parasiticus*, and *Aspergillus nomius* (K. C. Ehrlich, Montalbano, & Cotty, 2003; Varga, Frisvad, & Samson, 2009; Yu et al., 2004; Yu, Woloshuk, Bhatnagar, & Cleveland, 2000).

Goldbatt reported that the death of more than one thousands turkeys in England was due to the contamination of Brazilian peanut meal poultry feed to *Aspergillus* and prtoducing Aflatoxin. This disease is caused by toxin produced by *Aspergillus flavus* on birds' feed (Wogan & Pong, 1970). This toxin in 1961 led to more awareness regarding damages resulting from fungi toxins as the contaminator of food and causing disease or even death in humans and animals and years between 1960 and 1975 were the golden years of studies on fungi toxins, since scientists conducted many studies on these toxigenic factors (J. W. Bennett & Klich, 2003).

Aflatoxins, compared to other fungi toxins, due to carcinogenicity and acute toxicity effects, are more important. Among the detrimental effects of Aflatoxins, we can mention weakening and destroying the immune system, genetic mutations and cancer (Chu, 1991; Richard et al., 2003). There are four main types of Aflatoxins including: AFB1, AFG1, AFB2 and AFG2 (J. Bennett, Kale, & Yu, 2007; Murphy, Hendrich, Landgren, & Bryant, 2006). Here, AFB1 is the most toxic one (Van Egmond, Schothorst, & Jonker, 2007; Yu, Bhatnagar, & Ehrlich, 2002). Among effective factors in producing Aflatoxins, we can refer to genetic and environmental factors such as the type of fungi, type of medium, humidity, temperature, growth, storage, ventilation, light, carbon source, and pH. At least 23 enzymatic reactions contribute in Aflatoxin synthesis oath. The synthesis of this toxin is done through a series of oxidation and rehabilitation reactions. The Aflatoxin biosynthetic pathway involves approximately 25 genes clustered in a 70 kb DNA region (Bhatnagar, Cary, Ehrlich, Yu, & Cleveland, 2006; Scherm, Palomba, Serra, Marcello, & Migheli, 2005).

Most of these genes are regulated by the specific path of the protein binding to DNA (AflR) and are produced by *aflR* gene (Chang, Horn, & Dorner, 2005; Yu et al., 1995). Due to the toxic and carcinogenetic characteristics of Aflatoxins, there is an immediate need for a rapid method. There are sensitive and particular methods for detecting Aflatoxigentic matters in feed and the conventional methods to make distinction between toxigenetic and non-toxigenetic isolations in *Aspergillus flavus* including fungi culture in suitable medium, extracting Aflatoxins by organic solvents and monitoring their presence by chromatographic techniques (Abbas et al., 2004; Fente, Ordaz, Vazquez, Franco, & Cepeda, 2001; Sforza, Dall'Asta, & Marchelli, 2006). The current methods used for monitoring the presence of Aflatoxins are time consuming and labor intensive.

Recently, DNA-based detection systems are introduced as powerful tools to detect and identify Aflatoxin (Geisen, 1996). The Polymerase Chain Reaction (PCR) is among the selected methods for this purpose (Criseo, Bagnara, & Bisignano, 2001; Färber, Geisen, & Holzapfel, 1997; Shapira et al., 1996; Sweeney, Pàmies, & Dobson, 2000). Unique DNA sequences of the respective fungus have to be chosen as primer binding sites concluded that genes involved in the Aflatoxin biosynthetic pathway (Geisen, 1996; Scherm et al., 2005). The purpose of the present study was to make distinction between Aflatoxinogenetic and non-Aflatoxinogenetic strains using PCR and TCL and the expression of *aflD (nor-1), aflP(omtA), aflO (omtB), aflQ(ordA), aflR* in 40 strains was investigated using PCR.

2. Materials and Methods

2.1 Fungal Strains and Culture

A number of 40 *Aspergillus flavus* strains from a collection of 67 species from cattle feed including maize (A), domestic barly (B1), imported barly (B2), wheat bran (D) and soybean meal (E) were collected from 21 industrial warehouses of Tehran and Alborz.

Strains were cultured on sabouraud detrose agar (SDA) medium containing chloramphenicol (to prevent the growth of bacteria and yeasts) and plates were incubated at the temperature of 28-30 °C for 3-5 days. After the emergence of colony, the fungal spores were transmitted to yeast extract of sucrose agar (YES) (Samson, Hoekstra, & Frisvad, 2004). All plates were incubated at the temperature of 25 °C for one week. Then, the growing strains in each plate were put under microscopic and macroscopic investigations using valid identification keys (Klich & Pitt, 1988).

2.2 Determination of Aflatoxin Production by Chromatography

Therefore, 500 ul of samples were mixed by 250 ul of chloroform. Then, they were dotted on TLC plate of silica gel, so that the distance of dots with the lower margin of chromatography plate was approximately 2 cm and these dots were dried at the temperature of 27 °C. After that, 50 ml extraction solvent (containing 40 ml toluene, 7/5 ml methanol and 2/5 ml acetic acid) were added to the solvent tank, so that the depth of solvent should be at the bottom of the tank, less than 2 cm. The TLC plate is placed in tank and let the solvent be 4-5 cm higher than the dotted line (about 15 to 20 minutes). Then, bring the TLC plate out of the solvent tank to be dried and observable under ultraviolet light (365 nm). An Aflatoxingenic strain was used a positive control.

2.3 Molecular Characterization

2.3.1 Detection of Aflatoxin Genes aflD (nor-1), aflP(omtA), aflO (omtB), aflQ(ordA) and aflR

In this study, common naming system of Aflaxotin genes was used by (Cary, Ehrlich, Bland, & Montalbano, 2006; K. Ehrlich, Yu, & Cotty, 2005; Yu et al., 2004). The performance of these five genes is indicated in Table 1.

Table 1. Aflatoxin biosynthetic genes and functions[a]

Old name	New name	Enzyme/product	Function in the pathway
nor1	*aflD*	*NOR reductase*	*norsolorinic acid (NOR) →* *averantin (AVN)*
omtA	*aflP*	*O-methyltransferase A*	*sterigmatocystin (ST) →* *O-methylsterigmatocystin (OMST)*
omtB	*aflO*	*O-methyltransferase B*	*DHDMST (dihydrodemethylsterigmatocystin) →* *DHST (dihydrosterigmatocystin)*
ordA	*aflQ*	*Oxydoreductase*	*O-methylsterigmatocystin (OMST) → AFB1 and AFG1,* *dihydro-Omethylsterigmatocystin (DHOMST) → AFB2 and AFG2*
aflR	*aflR*	*Transcription activator* *AflR*	*Pathway regulator*

[a]Clustered pathway genes in aflatoxin biosynthesis (Cary et al., 2006; K. Ehrlich et al., 2005; Yu et al., 2004).

2.4 DNA Extraction

To extract DNA, some microliters of spore suspension of *Aspergillus* isolate were kept, transferred to the plate containing YES, and a one-week colony is employed. 500 microliters lysis buffer (containing 1 mollar Tris-HCI (PH=8), 0.5 mollar EDTA (Ph=8) and 7.45 g KCI), a pile of about 60 mg mycelium was added from *Aspergillus* colony and then crushed by hand and vortex for 45 seconds and finally centrifuged for 10 minutes at 5000 g. The supernatant liquid was transferred to new fresh tube, and 300 microliters cold isopropanol (kept below -20°C) was added and finally cell lysis and isopropanol were mixed through multiple reversal activities of microtube and centrifuged for 10 minutes at 12000 g. The supernatant liquid was discarded and about 0.8 microliter 70 degree alcohol was added to sediment and after 15 minutes was incubated at 37°C. Eventually, 50 microliters deionized distilled water was added to the remaining sediment, and then DNA was mixed with distilled water by gently tapping. The resulting liquid is frozen and stored at -20°C as a pure DNA solution.

2.5 PCR Amplification

In the present study, five pairs of primers were designed based on *Aspergillus flavus* sequences from *aflD (nor-1)*, *aflP(omtA)*, *aflO (omtB)*, *aflQ(ordA)*, *aflR* using OLIGO7 software (Scherm et al., 2005). Oligonucleotides were made by Macro-Gene Company. The primer sequences are presented in Table 2.

Table 2. Sequences of the nucleotide primers used in this study

Primer code	Target gene	Primer sequences	PCR product size (bp)	Accession no
AflD-1for *AflD-2rev*	*aflD(nor-1)*	5'- CTCATCACACGCAGGCATCGG -3' 5'- AGATGCCTGCCACACTGTCT -3'	702	FN398169.1
AflP-1for *AflP-2rev*	*aflP(omtA)*	5'- CCCATCTCGATAGCGCCTG -3' 5'- GCCACCCATACCTAGATCAAAGC -3'	611	FN398191.1
AflO-1for *AflO-2rev*	*aflO(omtB)*	5'- TTACGATTTGATGGAGCAGG -3' 5'- AGGTTCTCTTGGCTACAG -3'	358	HM355030.1
AflQ-1for *AflQ-2rev*	*aflQ(ordA)*	5'- AACATTCTCTGCCTCATCACT -3' 5'- TCGCTCTGGCTTGAACACC -3'	445	Ay510451.1
AflR-1for *AflR-2rev*	*aflR*	5'- AGAGCTACTGAACGTCCCAT -3' 5'- ATCAGGTTGCACGAACTGTCC -3'	1458	AF441430.2

About 5 microliters of extracted DNA, 1 microliters of forward and revers primers, 10 microliters of PCR master mix prepared by Amplicon Company (containing 0.2 U/µl of Taq DNA polymerase, 0.4 milimolar of dATP and dNTP (dTTP, dCTP, dGTP and 3 milimolar of MgCl2) and necessary amount of sterile deionized distilled water (ddH20) were added until achieving the final volume of 20 microliters. The heating program of PCR was done according to 错误!未找到引用源。.

Table 3. Heat program used for PCR

	1 cycle		34 cycle						1 cycle	
PCR steps	Initial denaturation		Denaturation		Annealing		Extention		Final extention	
	Tm	Time	Tm	Time	Tm	Time	Tm	Time	Tm	Time

aflD	95 °C	3 min	95 °C	30 sec	61.4°C	40 sec	72°C	30 sec	72°C	7 min
aflP	95 °C	3 min	95 °C	30 sec	60°C	45 sec	72°C	45 sec	72°C	7 min
aflO	95 °C	3 min	95 °C	30 sec	52.4°C	30 sec	72°C	30 sec	72°C	7 min
aflQ	95 °C	3 min	95 °C	30 sec	57°C	30 sec	72 °C	45 sec	72°C	7 min
aflR	95 °C	2 min	95 °C	30 sec	58 °C	45 sec	72°C	1:30 min	72°C	7 min

3. Results and Discussion

Aspergillus flavus strains were grown on yeast extraction of sucrose agar at the temperature of 25 °C for one week. In the present study, another set of primers exclusive for *aflR* was used. This gene has a very important role in biosynthesis path of Aflatoxin by regulating the activities of other structural genes such as *nor-1, omtA, omtB* and *ordA* (Chang, Yu, Bhatnagar, & Cleveland, 1999; Woloshuk et al., 1994).

PCR was applied for these genes using five sets of primers. Figure 1 (A-E) shows products of PCR obtained from each primer. A series of bands from *aflD (nor-1), aflP(omtA), aflO (omtB), aflQ(ordA)* and *aflR* can be observed in 702, 611, 358, 445, and 1458 bp. Nine strains (1, 2, 3, 5, 13, 14, 22, 34 and 38) show a similar pattern that indicates the presence of all five genes and other strains show different patterns. The obtained results by TLC were investigated. Only 7 strains (1, 3, 5, 14, 22, 34 and 38) were Aflatoxin-producers fungi and other 33 samples were non-Aflatoxin producers fungi. TLC method has shown a clear distinction between Aflatoxin-producing and non-producing *Aspergillus flavus*. The results obtained by PCR and TLC are compared in 错误!未找到引用源。.

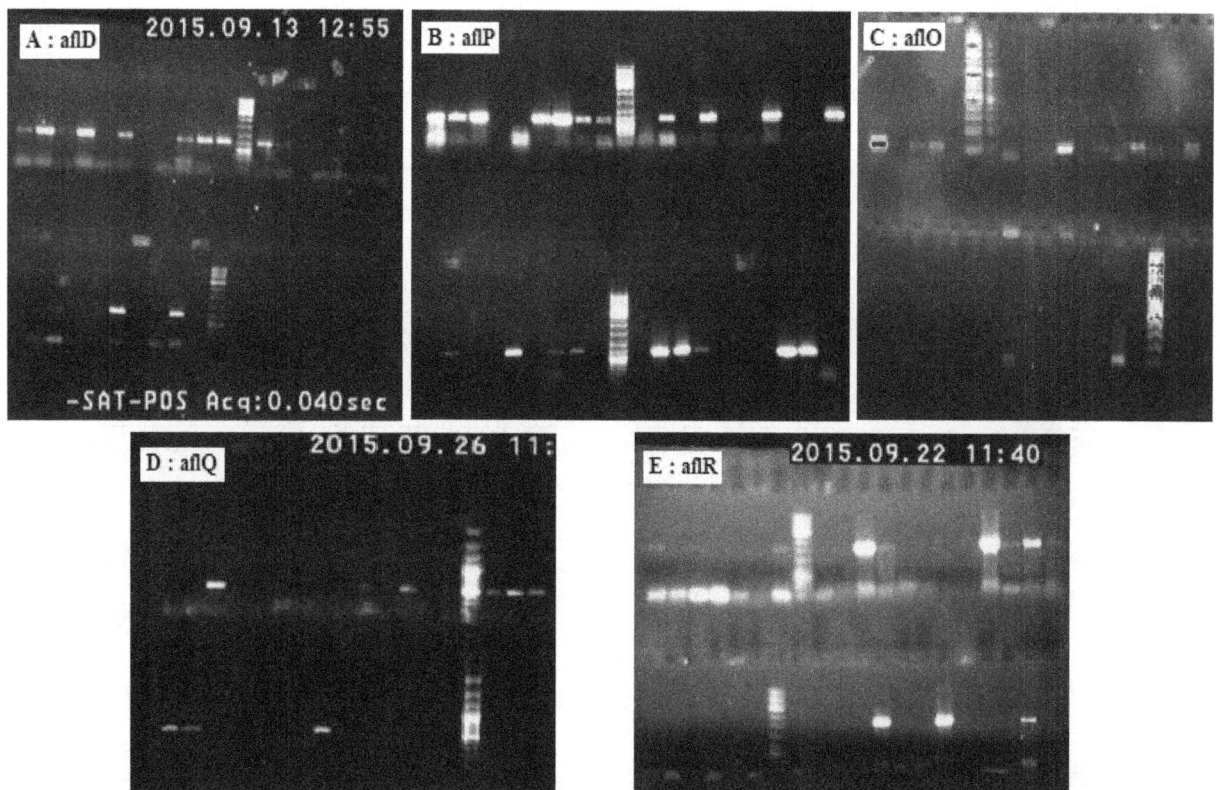

Figure 1. The images of gel electrophoresis of PCR products for expression of genes (aflD, aflP, aflO, aflQ and aflR) of Aspergillus isolated from cattle feed in 1% agarose gel and ladder 1kb

Table 3. This table indicates a comparison between the conventional and molecular methods (TLC and PCR) on aflatoxin production

Sample	PCR results	Aflatoxin production by TLC* method

Sample Type	Strain No.	aflR	aflP	aflO	aflQ	aflD	Alfaotoxin production
1:B2	1	+	+	+	+	+	Positive
1:B2	2	+	+	+	+	+	Negative
1:B2	3	+	+	+	+	+	Positive
1:E	4	-	-	+	+	+	Negative
2:B1	5	+	+	+	+	+	Positive
2:B2	6	-	+	-	+	-	Negative
2:E	7	+	+	+	+	-	Negative
3:B2	8	+	-	+	+	-	Negative
3:E	9	+	+	+	+	-	Negative
4:A	10	-	+	+	+	-	Negative
4:A	11	-	-	+	+	-	Negative
5:B2	12	+	-	+	+	-	Negative
5:D	13	+	+	+	+	+	Negative
5:E	14	+	+	+	+	+	Positive
6:E	15	+	+	+	-	+	Negative
7:A	16	-	+	+	+	+	Negative
7:A	17	-	-	+	+	-	Negative
7:D	18	+	+	+	-	-	Negative
8:B1	19	+	+	+	-	-	Negative
9:A	20	+	+	+	-	-	Negative
9:B2	21	+	+	+	+	-	Negative
9:D	22	+	+	+	+	+	Positive
9:D	23	+	+	-	-	-	Negative
9:D	24	-	+	+	-	+	Negative
10:B1	25	+	-	+	+	+	Negative
10:E	26	+	+	+	-	-	Negative
11:A	27	-	+	+	+	-	Negative
11:E	28	+	-	+	+	-	Negative
12:B1	29	+	+	+	+	-	Negative
13:B1	30	+	+	+	-	-	Negative
13:D	31	+	+	+	+	-	Negative
14:A	32	-	+	+	+	-	Negative
15:A	33	+	-	+	+	-	Negative
15:B2	34	+	+	+	+	+	Positive
15:D	35	-	+	+	+	+	Negative
19:B2	36	-	-	+	+	-	Negative
19:D	37	-	+	+	+	-	Negative
20:A	38	+	+	+	+	+	Positive
20:A	39	-	+	-	-	-	Negative
20:D	40	-	-	+	+	-	Negative
Control		+	+	+	+	+	Positive

*TLC: Thin layer chromatography.

The purpose of the present study was to standardize and optimize PCR method to make distinction between Aflatoxinogenetic and non-aflatoxinogenetic strains of *Aspergillus flavus* that produce Aflatoxins by effective genes in biosynthesis path. For this purpose, we worked on 40 strains of *Aspergillus flavus* selected from 67 species of cattle feed from 21 warehouses of Tehran and Alborz as the samples of this study and an standard strain of *Aspergillus flavus 5004* as a positive control to examine TLC and PCR. The procedure was conducted by *aflD (nor-1)*, *aflP (omtA)*, *aflO (omtB)*, *aflQ (ordA)* and *aflR. Nor-1, omtA, omtB* and *ordA* are four structural gens in gene cluster in Aflatoxin biosynthesis path that code key regulating enzymes in Aflatoxin production. As a result, they are necessary for Aflatoxin production (Yabe & Nakajima, 2004; Yu et al., 2004). After DNA extraction, the heating program of each PCR cycle was optimized within specific time using previous studies (Criseo et al., 2001; Färber et al., 1997; Geisen, 1996; Shapira et al., 1996).

The results showed that 5 primers, constituted sharp and distinct bands in certain area. Another studies show regulating Aflatoxin biosynthesis in *Aspergillus* containing a complex pattern of positive and negative transcription regulating factors that are under the influence of nutritional and environmental parameters (Chang, Yu, Bhatnagar, & Cleveland, 2000; K. C. Ehrlich et al., 2003; Flaherty & Payne, 1997; Takahashi et al., 2002). The results showed that seven samples of 40 *Aspergillus flavus* species were positive using TLC and in general, 40 samples with the five mentioned primers were positive using PCR technique and positive control in both methods was positive, as well. The same results were obtained using PCR and multiplex PCR (Criseo et al., 2001; Shapira et al., 1996). The results showed that PCR is a sensitive, rapid and specific method in Aflatoxinogenetic molds detection, but it cannot make distinction between toxigenic and non-toxigenic fungi. According to the results of this study regarding non-Aflatoxin producing strains, there is not any relationship between the obtained results by PCR and conventional methods. Lack of Aflatoxin production is related to an incomplete pattern in PCR. This proposes that various types of mutations can deactivate Aflatoxin biosynthesis path of these strains. Geisen (1996) stated that the lack of Aflatoxin production can be the result of substitution of some bases (Geisen, 1996). Also, Liu and Chu (1998) showed that various physiologic conditions can be effective in Aflatoxin biosynthesis (Liu & Chu, 1998). In this study, it was observed that we can refer to PCR as a screening test for initial isolation regarding high sensitivity and speed (100%). The positive samples should be more investigated such as chromatography and RT-PCR. RT-PCR is a complementary measurement of PCR and the presence of gene. Mayer et al. (2003) and Sweeney et al. (2000) suggested that the presence and absence of mRNA could allow direct distinction between them (Mayer, Färber, & Geisen, 2003; Sweeney et al., 2000).

In this regard, multiplex RT-PCR by having the advantage of unique response to expressing several genes enclosed in the biosynthesis of Aflaxotin and real experimental time of RT-PCR can be used for fungi growth kinetic and the presence of designed AFB simultaneously. However, none of these methods has yet been applicable to make distinction between non-taxogenic and taxogenic strains of *Aspergillus flavus*.

4. Conclusion

Since investigations show that the highest pollution of cattle feed is related to toxigenic *Aspergillus*, therefore, there is need for developing a simple, rapid and sensitive method to detect Aflatoxigentic fungi, particularly, making distinction between Aflatoxin-producing and non-producing strains of AF. As a result, we can prevent the entrance of Aflatoxin to the health cycle of human and cattle.

References

Abbas, H. K., Zablotowicz, R., Weaver, M., Horn, B., Xie, W., & Shier, W. (2004). Comparison of cultural and analytical methods for determination of aflatoxin production by Mississippi Delta Aspergillus isolates. *Canadian Journal of Microbiology, 50*(3), 193-199.

Bennett, J., Kale, S., & Yu, J. (2007). Aflatoxins: background, toxicology, and molecular biology *Foodborne diseases* (pp. 355-373): Springer.

Bennett, J. W., & Klich, M. (2003). Mycotoxins. *Clinical Microbiology Reviews, 16*(3), 497-516. doi: 10.1128/CMR.16.3.497-516.2003

Bhatnagar, D., Cary, J. W., Ehrlich, K., Yu, J., & Cleveland, T. E. (2006). Understanding the genetics of regulation of aflatoxin production and Aspergillus flavus development. *Mycopathologia, 162*(3), 155-166.

Cary, J. W., Ehrlich, K. C., Bland, J. M., & Montalbano, B. G. (2006). The aflatoxin biosynthesis cluster gene, aflX, encodes an oxidoreductase involved in conversion of versicolorin A to demethylsterigmatocystin. *Applied and environmental microbiology, 72*(2), 1096-1101.

Chang, P.-K., Horn, B. W., & Dorner, J. W. (2005). Sequence breakpoints in the aflatoxin biosynthesis gene cluster and flanking regions in nonaflatoxigenic Aspergillus flavus isolates. *Fungal Genetics and Biology, 42*(11), 914-923.

Chang, P.-K., Yu, J., Bhatnagar, D., & Cleveland, T. E. (1999). The Carboxy-Terminal Portion of the Aflatoxin Pathway Regulatory Protein AFLR of Aspergillus parasiticus ActivatesGAL1:: lacZ Gene Expression inSaccharomyces cerevisiae. *Applied and environmental microbiology, 65*(6), 2508-2512.

Chang, P.-K., Yu, J., Bhatnagar, D., & Cleveland, T. E. (2000). Characterization of the Aspergillus parasiticus major nitrogen regulatory gene, areA. *Biochimica et Biophysica Acta (BBA)-Gene Structure and Expression, 1491*(1), 263-266.

Chu, F. S. (1991). Mycotoxins: food contamination, mechanism, carcinogenic potential and preventive measures. *Mutation Research/Genetic Toxicology, 259*(3), 291-306.

Criseo, G., Bagnara, A., & Bisignano, G. (2001). Differentiation of aflatoxin-producing and non-producing strains of Aspergillus flavus group. *Letters in applied microbiology, 33*(4), 291-295.

Ehrlich, K., Yu, J., & Cotty, P. (2005). Aflatoxin biosynthesis gene clusters and flanking regions. *Journal of Applied Microbiology, 99*(3), 518-527.

Ehrlich, K. C., Montalbano, B. G., & Cotty, P. J. (2003). Sequence comparison of aflR from different Aspergillus species provides evidence for variability in regulation of aflatoxin production. *Fungal Genetics and Biology, 38*(1), 63-74.

Färber, P., Geisen, R., & Holzapfel, W. (1997). Detection of aflatoxinogenic fungi in figs by a PCR reaction. *International journal of food microbiology, 36*(2), 215-220.

Fente, C., Ordaz, J. J., Vazquez, B., Franco, C., & Cepeda, A. (2001). New additive for culture media for rapid identification of aflatoxin-producingAspergillus strains. *Applied and environmental microbiology, 67*(10), 4858-4862.

Flaherty, J. E., & Payne, G. A. (1997). Overexpression of aflR leads to upregulation of pathway gene transcription and increased aflatoxin production in Aspergillus flavus. *Applied and environmental microbiology, 63*(10), 3995-4000.

Geisen, R. (1996). Multiplex polymerase chain reaction for the detection of potential aflatoxin and sterigmatocystin producing fungi. *Systematic and Applied Microbiology, 19*(3), 388-392.

Humans, I. W. G. o. t. E. o. C. R. t., Organization, W. H., & Cancer, I. A. f. R. O. (2002). *Some traditional herbal medicines, some mycotoxins, naphthalene and styrene*: World Health Organization.

Kamei, K., & Watanabe, A. (2005). Aspergillus mycotoxins and their effect on the host. *Medical mycology, 43*(sup1), 95-99.

Klich, M., & Pitt, J. (1988). Differentiation of Aspergillus flavus from A. parasiticus and other closely related species. *Transactions of the British Mycological Society, 91*(1), 99-108.

Liu, B.-H., & Chu, F. S. (1998). Regulation of aflR and its product, AflR, associated with aflatoxin biosynthesis. *Applied and environmental microbiology, 64*(10), 3718-3723.

Mayer, Z., Färber, P., & Geisen, R. (2003). Monitoring the production of aflatoxin B1 in wheat by measuring the concentration of nor-1 mRNA. *Applied and environmental microbiology, 69*(2), 1154-1158.

Murphy, P. A., Hendrich, S., Landgren, C., & Bryant, C. M. (2006). Food mycotoxins: an update. *Journal of food science, 71*(5), R51-R65.

Richard, J., Payne, G., Desjardins, A., Maragos, C., Norred, W., & Pestka, J. (2003). Mycotoxins: risks in plant, animal and human systems. *CAST Task Force Report, 139*, 101-103.

Samson, R. A., Hoekstra, E. S., & Frisvad, J. C. (2004). *Introduction to food-and airborne fungi*: Centraalbureau voor Schimmelcultures (CBS).

Scherm, B., Palomba, M., Serra, D., Marcello, A., & Migheli, Q. (2005). Detection of transcripts of the aflatoxin genes aflD, aflO, and aflP by reverse transcription–polymerase chain reaction allows differentiation of aflatoxin-producing and non-producing isolates of Aspergillus flavus and Aspergillus parasiticus. *International Journal of Food Microbiology, 98*(2), 201-210.

Sforza, S., Dall'Asta, C., & Marchelli, R. (2006). Recent advances in mycotoxin determination in food and feed by hyphenated chromatographic techniques/mass spectrometry. *Mass Spectrometry Reviews, 25*(1), 54-76.

Shapira, R., Paster, N., Eyal, O., Menasherov, M., Mett, A., & Salomon, R. (1996). Detection of aflatoxigenic molds in grains by PCR. *Applied and environmental microbiology, 62*(9), 3270-3273.

Sweeney, M. J., Pàmies, P., & Dobson, A. D. (2000). The use of reverse transcription-polymerase chain reaction (RT-PCR) for monitoring aflatoxin production in Aspergillus parasiticus 439. *International journal of food microbiology, 56*(1), 97-103.

Takahashi, T., Chang, P.-K., Matsushima, K., Yu, J., Abe, K., Bhatnagar, D., ... Koyama, Y. (2002). Nonfunctionality of Aspergillus sojae aflR in a strain of Aspergillus parasiticus with a disrupted aflR gene. *Applied and environmental microbiology, 68*(8), 3737-3743.

Van Egmond, H. P., Schothorst, R. C., & Jonker, M. A. (2007). Regulations relating to mycotoxins in food. *Analytical and bioanalytical chemistry, 389*(1), 147-157.

Varga, J., Frisvad, J., & Samson, R. (2009). A reappraisal of fungi producing aflatoxins. *World Mycotoxin Journal,* *2*(3), 263-277.

Wogan, G. N., & Pong, R. S. (1970). AFLATOXINS*. *Annals of the New York Academy of Sciences, 174*(2), 623-635.

Woloshuk, C., Foutz, K., Brewer, J., Bhatnagar, D., Cleveland, T., & Payne, G. A. (1994). Molecular characterization of aflR, a regulatory locus for aflatoxin biosynthesis. *Applied and environmental microbiology, 60*(7), 2408-2414.

Yabe, K., & Nakajima, H. (2004). Enzyme reactions and genes in aflatoxin biosynthesis. *Applied Microbiology and Biotechnology, 64*(6), 745-755.

Yu, J., Bhatnagar, D., & Ehrlich, K. C. (2002). Aflatoxin biosynthesis. *Revista iberoamericana de micología, 19*(4), 191-200.

Yu, J., Chang, P.-K., Cary, J. W., Wright, M., Bhatnagar, D., Cleveland, T. E., ... Linz, J. E. (1995). Comparative mapping of aflatoxin pathway gene clusters in Aspergillus parasiticus and Aspergillus flavus. *Applied and environmental microbiology, 61*(6), 2365-2371.

Yu, J., Chang, P.-K., Ehrlich, K. C., Cary, J. W., Bhatnagar, D., Cleveland, T. E., ... Bennett, J. W. (2004). Clustered pathway genes in aflatoxin biosynthesis. *Applied and environmental microbiology, 70*(3), 1253-1262.

Yu, J., Woloshuk, C. P., Bhatnagar, D., & Cleveland, T. E. (2000). Cloning and characterization of avfA and omtB genes involved in aflatoxin biosynthesis in three Aspergillus species. *Gene, 248*(1), 157-167.

19

The Synthesis & Characterizes of Nano-Metallic Particles Against Antibiotic Resistant Bacteria, Isolated from Rasoul-e-Akram Hospital's Patients, Tehran, Iran

Alireza Jafari[1], Ali Majidpour[1,5], Roya Safarkar[2], Seyyedeh Masumeh Mirnurollahi[3] & Shahrdad Arastoo[4]

[1]Antimicrobial Resistance Research Center, Rasoul-e-Akram Hospital, Iran University of Medical Sciences, Tehran, Iran

[2]Department of Microbiology, Islamic Azad University, Ardabil Branch, Ardabil, Iran

[3] Department of Biology, Science and Research, Islamic Azad University, Tehran, Iran

[4]Department of Microbiology, Islamic Azad University Qom Branch, Qom, Iran

[5] Department of Infection Disease, School of Medicine

Correspondence: Ali Majidpour, Antimicrobial Resistance Research Center, Rasoul-e-Akram Hospital, Iran University of Medical Sciences, Tehran, Iran. E-mail: alimajidpour@yahoo.com

Abstract

The emergence of antimicrobial resistance of microorganisms to antibiotics, Also, an increase in nosocomial infections, particularly by *Methicillin Resistant Staphylococcus aureus (MRSA), Pseudomonas aeruginosa*, the need to discover new antibacterial agents with a mechanism of action different from killing bacteria were more than ever before. The Ag nanoparticles (NPs), ZnO (NPs) and Ag/ZnO (NPs) were synthesized through the thermal decomposition of the precursor of oxalate. Gram-negative antibiotic resistant bacteria and Gram-positive antibiotic resistant bacteria were prepared from the Central laboratory of Rasoul-e-Akram hospital. All of isolates were confirmed by biochemical tests. For determine of antibiotic resistance patterns of isolated, disk diffusion method in accordance with the standard CLSI were used, again. Antibacterial effects of (NPs) against antibiotic resistance bacteria were conducted by MIC and MBC tests. The particles size was less of 50 nm, approximately. Curiously, the silver (NPs) was not exposed the antibacterial properties against all of isolated bacteria. Also, *klebsiella pneumonia* and *MRSA* had greatest sensitivity to the ZnO (NPs). Also, Gram-positive antibiotic resistant bacteria showed high sensitivity to Ag/ZnO (NPs), compared to other bacteria. Interestingly, The MBC for ZnO (NPs) against *Pseudomonas aeruginosa* ≥8192 was observed. The Ag (NPs) had not the ability to inhibit the nosocomial infection. *Klebsiella pneumonia* and *MRSA* had greatest sensitivity to the ZnO (NPs). The Ag/ZnO (NPs) was ability to kill antibiotics resistant bacteria. The antibacterial agents can open a new leaf in our life in the treatment of nosocomial infections.

Keywords: Nano-metallic particles, Antibiotic Resistant Bacteria, Rasoul-e-Akram Hospital

1. Introduction

Antibiotic resistance is a worldwide problem (Roberts et al., 2009). Majority shape of resistance spread with remarkable speed. World health guidance have described antibiotic-resistant microorganisms as "nightmare bacteria" that "pose a catastrophic threat" to people in every country in the world (Roberts et al., 2009). Each year in the United States, at least 2 million people acquire serious infections with bacteria that are resistant to more of the antibiotics designed to treat those infections. Approximately, 23000 people die each year of these antibiotic-resistant infections. Many more die from other conditions that were intricate by an antibiotic-resistant infection. Antibiotic-resistant infections add avoidable costs to the already overburdened U.S. health care system. In another cases, antibiotic-resistant infections require prolonged treatments, extend hospital stays, necessitate additional doctor visits and result in greater disability and death compared with infections that are easily treatable with antibiotics (Roberts et al., 2009).

Nowadays, researchers have suggested the use of nano-metal oxides, specially Silver and Zinc oxide (NPs) as superior disinfectants and antimicrobial agent for nosocomial Infections microorganisms (Blanc, Carrara, Zanetti,

& Francioli, 2005; Reddy et al., 2007; Yu-sen, Vidic, Stout, McCartney, & Victor, 1998). Investigates shown that residual these metal ions may adversely affect human health (Sondi & Salopek-Sondi, 2004). Another word, they report on the toxicity of ZnO (NPs) to gram-negative and gram-positive bacterial systems, *Escherichia coli* (*E. coli*) and *Staphylococcus aureus* (*S. aureus*) and primary human immune cells. Those results shows that ZnO (NPs)may potentially prove useful as antimicrobial agents at selective therapeutic dosing regimens (Reddy et al., 2007). Also, they believed that silver (NPs) is incorporated in the cell membrane, which causes leakage of intracellular substances and eventually causes cell death (Kim et al., 2007; Ruparelia, Chatterjee, Duttagupta, & Mukherji, 2008). Some of the silver (NPs) also penetrate into the cells (Morones et al., 2005).

Jayesh assumed that combination of metal oxide (NPs) may give rise to more complete bactericidal effect against mixed bacterial population (Ghosh, Das, Jena, & Pradhan, 2015). We know that the bactericidal effect of metal (NPs) has been attributed to their small size, photo-catalytic of activity and high surface to volume ratio, which allows them to interact closely with microbial membranes and is not merely due to the release of metal ions in solution (Jafari, Ghane, Sarabi, & Siyavoshifar, 2011). The aim of this work was to synthesize nano-metallic particles with potent antibacterial activity, also with simple and cost-effective method that is capable kill of hospital resistant bacteria. Continue the process of the investigation to the In-vitro, In-vivo and Ex-vivo condition, may be lead to the discovery of nano-drugs with potent antibacterial activity, in the future.

2. Material and Method

2.1 Synthetizes of Nano-Metallic Particles via Oxalate Decomposition

The general reaction for decomposition basic compounds of metal oxalate following:

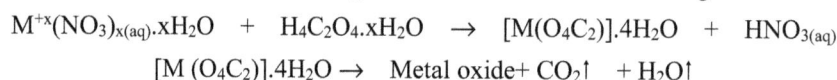

$$M^{+x}(NO_3)_{x(aq)}.xH_2O \quad + \quad H_4C_2O_4.xH_2O \quad \rightarrow \quad [M(O_4C_2)].4H_2O \quad + \quad HNO_{3(aq)}$$

$$[M(O_4C_2)].4H_2O \rightarrow \quad Metal\ oxide + CO_2\uparrow \quad + H_2O\uparrow$$

End products of metal oxides depended to the stability of the crystalline oxide and the metal cautions desired stability which can be M^+, M^{2+}, $M^{+8/3}$ and the M^{3+}.

2.2 Synthesis of Ag, ZnO, Ag/ZnO Nanoparticles and Characterization

The Ag, ZnO, Ag/ZnO(NPs) were synthesized in Antimicrobial Resistance Research Center (ARRC) of Iran University of Medical Sciences (IUMS), according to Jafari and their Colleague's protocol. To study of the crystal structure of nano-metallic particles, X-ray diffractometer set (XRD, Bruker D8-Advance diffract meter using Cu Kα radiation) in X-Ray Laboratory School of Mining Engineering, University of Tehran, were used. The FT-IR spectrum was record don a Bruker spectrophotometer in KBr pellets in Institute of Materials and Energy (MERC) of Tehran. Surface morphology of product was characterized by using a Scanning Electronic Microscopy (SEM, Cam Scan MV 2300, nano-electronics laboratory, Tehran University) with an accelerating voltage of 30 KV (Dabbagh, Moghimipour, Ameri, & Sayfoddin, 2010; Gan, Liu, Zhong, Liu, & Li, 2004).

2.3 Sampling and Collection of Isolated Bacteria

Klebsiella pneumoniae, Staphylococcus epidermidis, Pseudomonas aeruginosa, Escherichia coli, Acinetobacter baumannii, Methicillin resistant Staphylococcus aureus (MRSA), had been delivered with respect for the ethical considerations and appropriate licenses to the Central Laboratory of Rasoul-e-Akram Hospital and then was transferred to the Antimicrobial Resistance Research Center laboratory. Sampling and collection of isolated antibiotic-resistant bacteria, since early May 2015 until late June 2015 had been conducted. According to the reports of central laboratory of Rasoul-e-Akram Hospital, Klebsiella pneumonia and Escherichia coli had been isolated from urine samples of patients. Pseudomonas aeruginosa, Staphylococcus epidermidis and Acinetobacter baumannii were isolated from burn wounds and MRSA that was obtained from blood samples.

2.4 Identification & Determination of Antibiotic Resistance Patterns of Bacteria

To confirm the identification of bacteria resistant to antibiotics, biochemical tests were used. For determine of antibiotic resistance patterns of isolated, disk diffusion method (Kirby-Bauer) in accordance with the standard CLSI (Clinical and Laboratory Standards Institute) and the National Committee for clinical laboratory Standards (NCCLS) were used, again. Antibiotic discs for each bacteria isolated were listed in Tables 1, 2, 3, 4.

Table 1. The lists of antibiotic discs that used in *Klebsiella pneumoniae, Escherichia coli & Pseudomonas aeruginosa*

Antimicrobial Agent	Symbol & Count.	Made in
Cefazolin	CEF10	PADTAN TEB Co. Iran
Gentamycin	GM10	PADTAN TEB Co. Iran
Amikacin	An	PADTAN TEB Co. Iran
Cefepime	FEP30	PADTAN TEB Co. Iran
Cefotaxime	CTX30	PADTAN TEB Co. Iran
Ciprofloxacin	CP10	PADTAN TEB Co. Iran
Trimethoprim sulfa methoxazole	TMP5	PADTAN TEB Co. Iran
Meropenem	MEN10	PADTAN TEB Co. Iran
Ceftazidime	CAZ30	PADTAN TEB Co. Iran
Nitrofurantoin	FM300	PADTAN TEB Co. Iran
Piperacillin Tazobactam	PTZ100/10	PADTAN TEB Co. Iran
Ampicillin	AM30	PADTAN TEB Co. Iran
Imipenem	IPM10	MAST Co. UK
Colistine	CL10	MAST Co. UK
Aztreonam	AZ15	PADTAN TEB Co. Iran

Table 2. The lists of the antibiotic discs that used in *Methicillin Resistance of Staphylococcus aureus (MRSA)*

Antimicrobial Agent	Symbol & Count.	Made in
Nitrofurantoin	FM300	PADTAN TEB Co. Iran
Trimethoprim sulfa meth oxazole	TMP5	PADTAN TEB Co. Iran
Erythromycin	E15	PADTAN TEB Co. Iran
Methicillin	ME5	PADTAN TEB Co. Iran

Table 3. The lists of the antibiotic discs that used in *Acinetobacter baumannii.*

Antimicrobial Agent	Symbol & Count.	Made in
Imipenem	IPM10	PADTAN TEB Co. Iran
Ceftazidime	CT30	PADTAN TEB Co. Iran
Ticarcillin	TIC75	PADTAN TEB Co. Iran
Tobramycin	TOB10	PADTAN TEB Co. Iran
Gentamycin	GM10	PADTAN TEB Co. Iran
Cefotaxime	CTX30	PADTAN TEB Co. Iran
Ciprofloxacin	CP10	PADTAN TEB Co. Iran
Co-trimoxazole	SXT25	PADTAN TEB Co. Iran
Colistine	CL10	PADTAN TEB Co. Iran

Table 4. The lists of the antibiotic discs that used in *Staphylococcus epidermidis*

Antimicrobial Agent	Symbol & Count.	Made in
Cefazolin	CEF10	PADTAN TEB Co. Iran
Rifampicin	RA5	PADTAN TEB Co. Iran
Vancomycin	V30	PADTAN TEB Co. Iran
Clindamycin	CC2	PADTAN TEB Co. Iran
Co-trimoxazole	SXT25	PADTAN TEB Co. Iran
Minocycline	MI30	PADTAN TEB Co. Iran
Linezolid	LZ30	PADTAN TEB Co. Iran
Azithromycin	AZM15	PADTAN TEB Co. Iran
Clarithromycin	CLR15	PADTAN TEB Co. Iran
Oxacillin	OX1	PADTAN TEB Co. Iran

2.5 Supplying of Standard McFarland

In order to providing of 0.5 McFarland concentration, 0.5 ml of pure sulfuric acid and 9.95 ml of barium chloride to the clean test tube were stirred, slowly. The test tube was kept in a dark location away from light and heat.

2.6 Determining the Sensitivity of Bacteria to Ag, ZnO and Ag/ZnO NPs via Disk Diffusion and Cavity Method

First at all, we were poured 0.327 gr of Ag, ZnO and Ag/ZnO (NPs) into the sterile test tubes, containing 20 ml of liquid medium Mueller Hinton broth (MHB) (Merck, Germany). In each of the test tubes, sterile blank discs were placed. Then for 30 min were sonicated at room temperature by ultrasonic waves at room temperature and frequency of 28 KHz (PULSE Co. Germany). Next, all of discs in order to lose of moisture were placed in desiccators at room temperature. Immediately, several colonies of freshly bacteria were injected into test tubes containing 10 ml of sterile saline and equivalent to 0.5 McFarland. Then 100 λ of bacterial suspension was culture don MHB. All of discs impregnated with (NPs) was polluted on MHB, also was incubated at 37 °C for at least 18 hours. In order to performance of cavity test, we were drilled several cavity on MHB, also 100 λ of (NPs) is poured into it.

2.7 MIC and MBC Tests

The serial dilution method was used for determine the Minimum Inhibitory Concentration (MIC) of the Ag, ZnO and Ag/ZnO (NPs) (Dabbagh et al., 2010; Jafari, Ghane, Sarabi, et al., 2011). In this way, all of test tubes were filled with 1 ml of the liquid Muller Hinton broth (MHB) medium. Then, all of (NPs) had been sonicated with the culture medium were added and mixed. Subsequently, one ml of the content of test tube number two was added to test tube number 3 and mixed completely and then this process was performed serially to last test tube. Totally, microbial suspensions of *Klebsiella pneumoniae*, *Staphylococcus epidermidis*, *Pseudomonas aeruginosa*, *Escherichia coli*, *Acinetobacter baumannii*, MRSA, containing 1.5×10^8 CFUml^{-1} were added to test tubes and were incubated at 37 °C for 24 h. All the experiments were carried out in triplicate (Dabbagh et al., 2010; Jafari, Ghane, Sarabi, et al., 2011).

The minimum bactericidal concentration (MBC), i.e., the lowest concentration of nanoparticles that kills 99.9% of the bacteria was also determined from the batch culture studies. To experiments for bactericidal effect, a loop-full from each test tubes (Specially, negative & positive test tubes) was inoculated on Muller Hinton agar and incubated at 37 °C for 24 h. The nanoparticles concentration illustrating bactericidal effect was picked out based on absence of colonies on the agar plate (Gan et al., 2004; Jafari, Ghane, & Arastoo, 2011; Kim et al., 2007).

3. Result

3.1 The FT-IR Spectra Analysis of Ag, ZnO, Ag/ZnO Nanoparticles

The nanoparticles obtained at a temperature of 550°C, FT-IR spectra were taken. The interpretation is as follows; (Figures 1a, 1b, 1c) shows FT-IR spectra of Ag, ZnO and Ag/ZnO (NPs), respectively. Figures 1a shows that the shoulder at 1428.89 cm^{-1} is present in the spectrum evidence of (N-O) tremble and the closely spaced bands at 875.31 cm^{-1} and 577.35 cm^{-1} are presents in the spectrum evidence of (O-C-O) tensional tremble and (M-O) tremble respectively.

Also, figure 1b that depended to ZnO (NPs) FT-IR spectrum was demonstrated the band at 1428.89 cm^{-1} is present in the spectrum evidence of (N-O) tremble and the closely spaced bands at 876.14 cm^{-1} and 551.12 cm^{-1} are presents in the spectrum evidence of (O-C-O) tensional tremble and (Zn-O) tensional tremble respectively[10].

At last, Figure 1c was related on Ag/ZnO (NPs) FT-IR spectrum. Consistent with the results obtained by Jafari et al. the shoulder at 1458.22 cm^{-1} is present in the spectrum evidence of (N-O) tremble and the closely spaced bands 625.36 cm^{-1} are presents in the spectrum evidence of (Ag/ZnO) (NPs) tensional tremble respectively (Dabbagh et al., 2010).

3.2 The XRD spectra, SEM Images Analysis

The XRD patterns of Ag, ZnO and Ag/ZnO (NPs) (Figures 2a, 2b, 2c) were compared and interpreted with standard data of International Centre of Diffraction Data (ICDD). Results of XRD spectra and SEM of Ag, ZnO and Ag/ZnO (NPs) (Figures 3a, 3b, 3c) consistent with the results of Jafari and their colleagues.

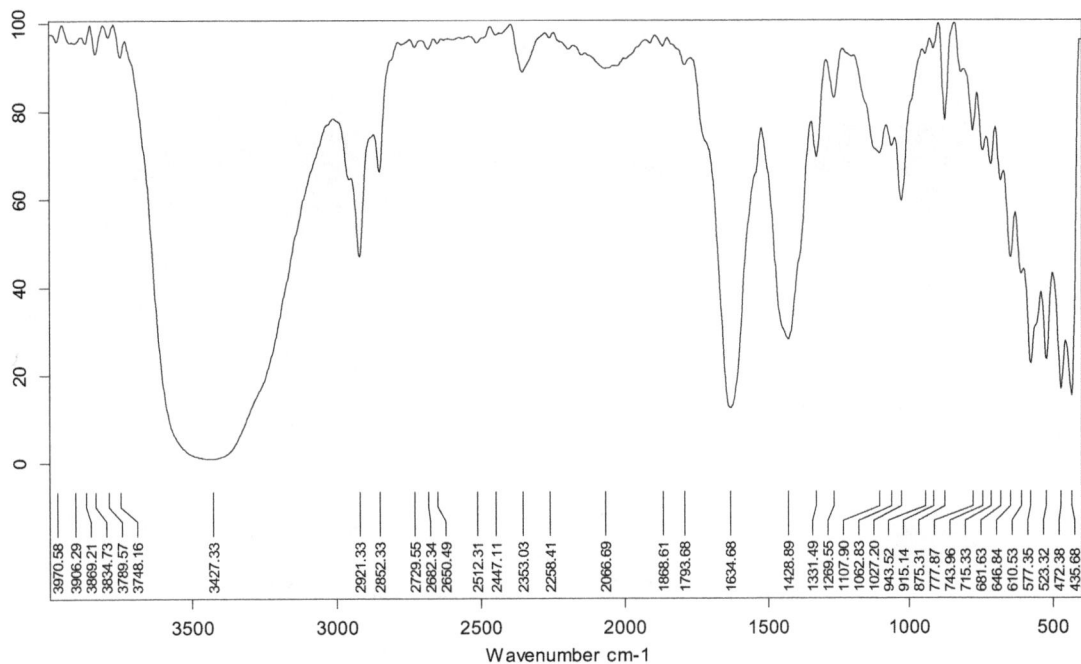

Figure 1a. The FT-IR spectra analysis of Ag NPs

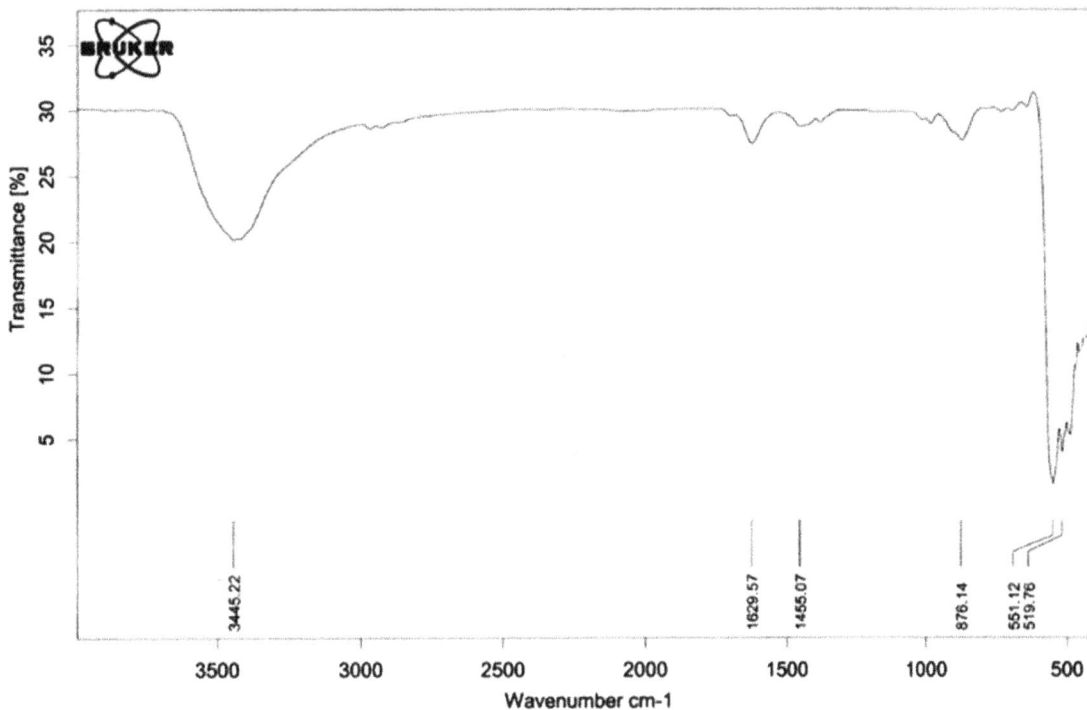

Figure 1b. The FT-IR spectra analysis of ZnO NPs

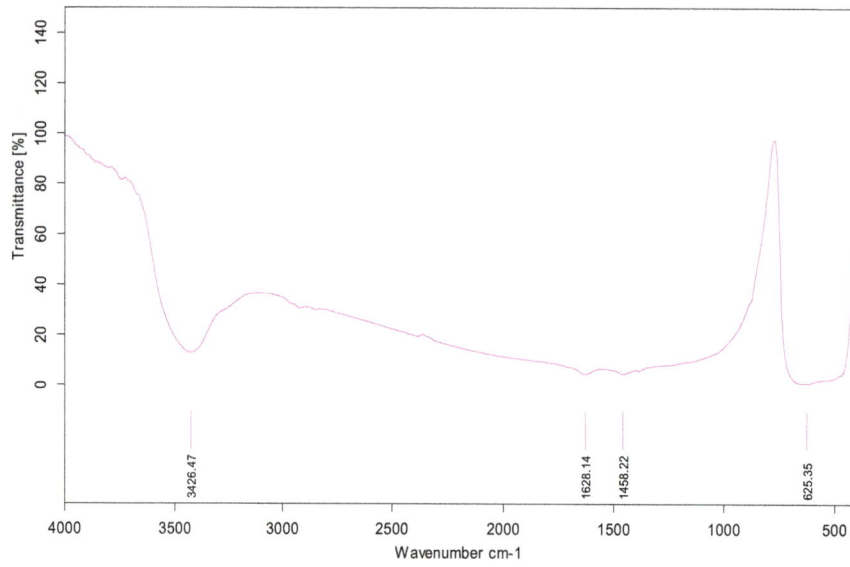

Figure 1c. The FT-IR spectra analysis of Ag/ZnO NPs

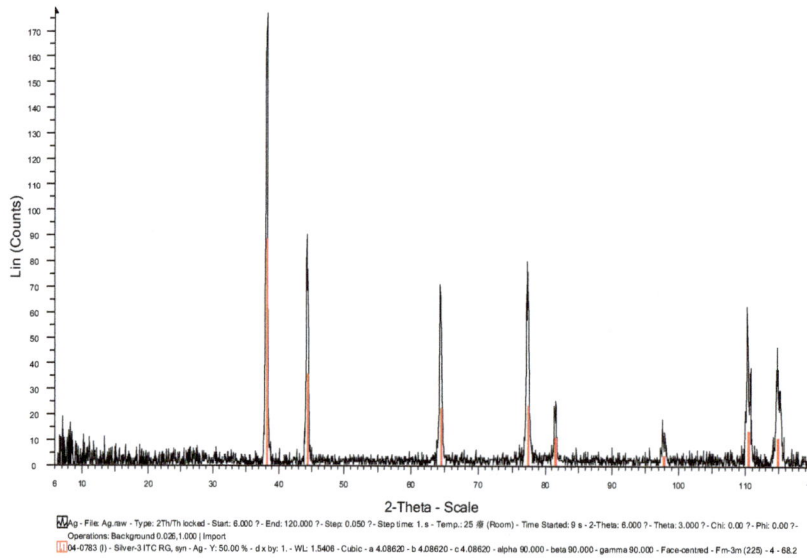

Figure 2a. The XRD patterns of Ag (NPs)

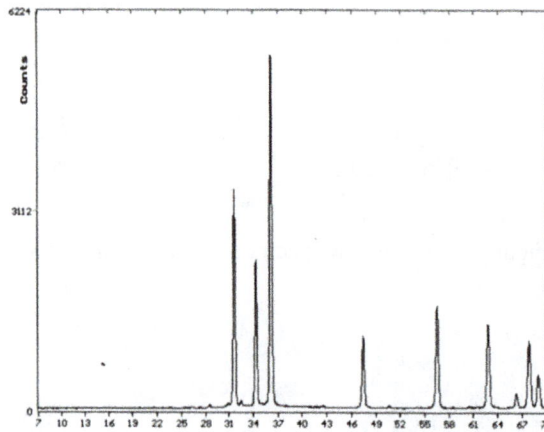

Figure 2b. The XRD patterns of ZnO (NPs)

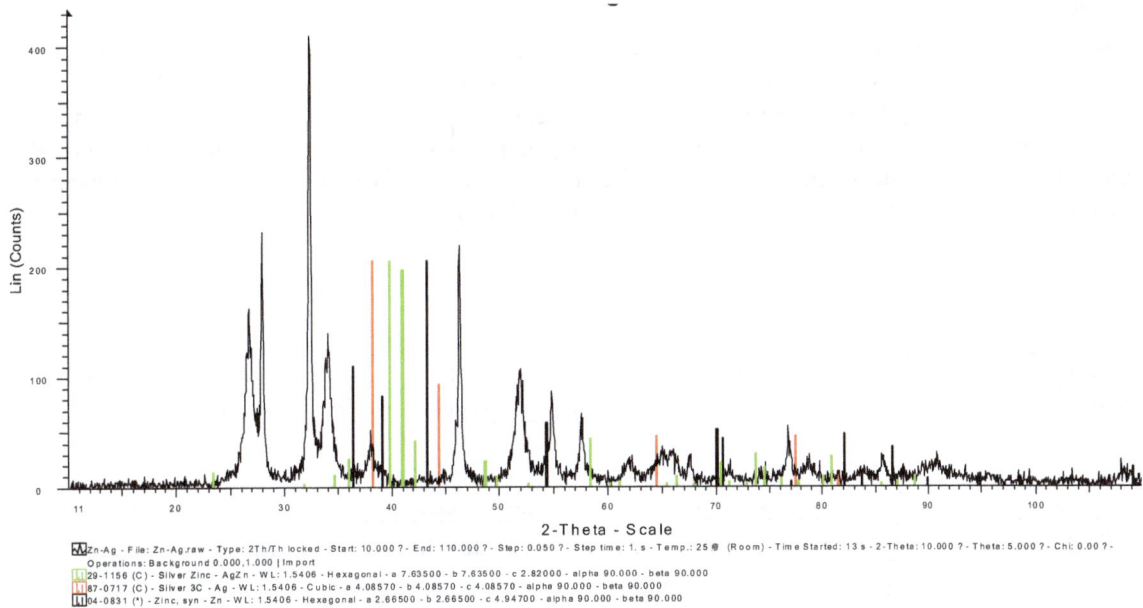

Figure 2c. The XRD patterns of Ag/ZnO (NPs)

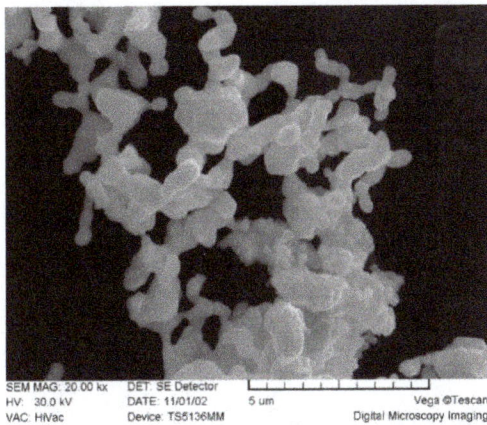

Figure 3a. SEM images analysis of Ag (NPs)

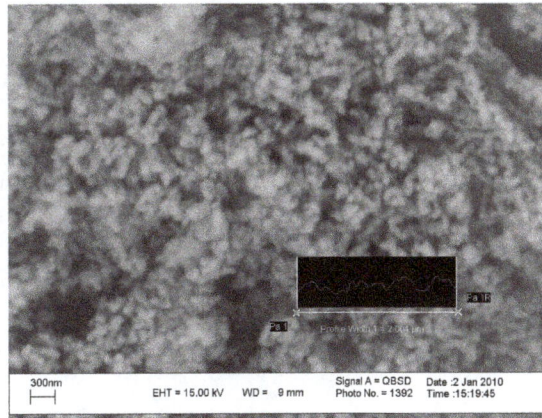

Figure 3b. SEM images analysis of ZnO (NPs)

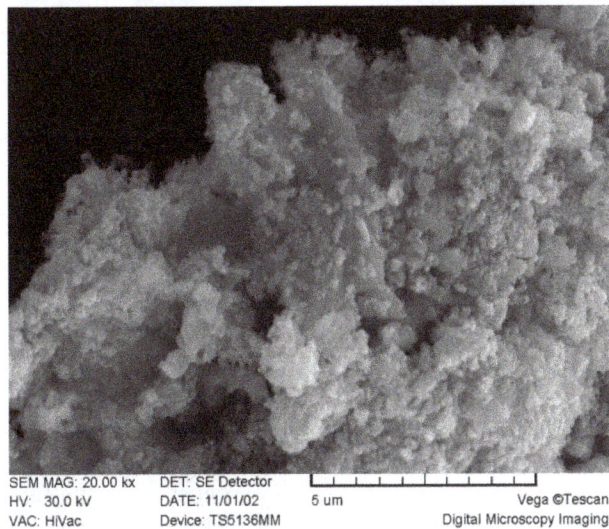

Figure 3c. SEM images analysis of Ag/ZnO (NPs)

3.3 Identification & Determination of Antibiotic Resistance Patterns of Bacteria

The results of biochemical tests for each of the bacteria shown in Table 5. Klebsiella pneumoniae, Staphylococcus epidermidis, Pseudomonas aeruginosa, Escherichia coli, Acinetobacter baumannii, Methicillin resistant Staphylococcus aureus (MRSA) were identified by biochemical tests and were confirmed Genus and species. Based on the results, Klebsiella pneumonia was resistant to Cefixime, Ceftriaxoneand Aztreonam. While Nitrofurantoin, Ceftazidime, Imipenem, Amikacin, Cefepime, Ciprofloxacin, Piperacillin, Colistin, Trimethoprim, Cefazolin, Gentamycin and Ampicillin showed sensitivity (Table 6).

Table 5. The results of biochemical tests for each of the bacteria

Escherichia coli

Gram stain	Catalase test	Oxidasetest	TSI	Motile	Urease test	SH₂	Citrate test	MR	VP	Indole test
Negative	Positive	Negative	A/AG	Positive	Negative	Negative	Negative	Positive	Negative	Positive

MRSA

Gram stain	β-hemolysis	Catalase	Coagulase	Motile	Urease	Acetoin	Mannitol	OF Aerobic	OF Anaerobic	Baird Parker agar
Positive	Positive	Positive	Positive	Negative	Negative	Positive	Positive	Positive	Positive	Positive

Klebsiella pneumoniae

Gram stain	Lysine	Citrate	Indole	TSI	motile	Ornithine
Negative	Positive	Positive	Negative	Acid/ Acid (with gas)	Negative	Negative

Staphylococcus epidermidis

Gram stain	Coagulase Test	Bacitracin Test	Novobiocin test	Catalase Test	β-hemolysis	Glucose	Urease
Positive	Negative	Resistance	sensitive	Positive	Negative	Positive	Positive

Acinetobacter baumannii

Gram stain	lactose	TSI	Oxidase	OF Glucose	Motile	Catalase	Polymyxin B
Negative	Negative	ALK/ALK	Negative	Positive	Negative	Positive	sensitive

Pseudomonas aeruginosa

Gram stain	Motile	Oxidase	lactose	TSI	SH₂	Catalase	Citrate	OF Aerobic	OF Anaerobic	β-hemolysis
Negative	Positive	Positive	Negative	ALK/ALK	Negative	Positive	Positive	Positive	Negative	Positive

Table 6. Determination of antibiotic resistance patterns of *Klebsiella pneumonia*.

Antimicrobial Agent	Symbol & Count.	
Cefazolin	CEF	Sensitive
Gentamycin	GM10	Sensitive
Amikacin	An	Sensitive
Cefepime	FEP30	Sensitive
Cefotaxime	CTX30	Sensitive
Ciprofloxacin	CP10	Sensitive
Trimethoprim sulfa methoxazole	TMP5	Resistance
Meropenem	MEN10	Sensitive
Ceftazidime	CAZ30	Sensitive
Nitrofurantoin	FM300	Sensitive
Piperacillin Tazobactam	PTZ100/10	Sensitive
Ampicillin	AM30	Resistance
Imipenem	IPM10	Sensitive
Colistine	CL10	Sensitive
Aztreonam	AZ	Sensitive
Ceftriaxone	CRO30	Sensitive
Cefixime	CFM5	Sensitive
Nitrofurantoin	FM300	Resistance

Pseudomonas aeruginosa to Gentamicin, Imipenem, Colistin, Ciprofloxacin, Meropenem, Amikacin, Ceftazidime, Cefepime and Cefotaxime was sensitivity. However, Trimethoprim, Ampicillin, Nitrofurantoin, showed resistance (Table 7).

Table 7. Determination of antibiotic resistance patterns of *Pseudomonas aeruginosa*

Antimicrobial Agent	Symbol & Count.	
Cefazolin	CEF	Sensitive
Gentamycin	GM10	Sensitive
Amikacin	An	Sensitive
Cefepime	FEP30	Sensitive
Cefotaxime	CTX30	Sensitive
Ciprofloxacin	CP10	Sensitive
Trimethoprim sulfa methoxazole	TMP5	Sensitive
Meropenem	MEN10	Sensitive
Ceftazidime	CAZ30	Sensitive
Nitrofurantoin	FM300	Sensitive
Piperacillin Tazobactam	PTZ100/10	Sensitive
Ampicillin	AM30	Sensitive
Imipenem	IPM10	Sensitive
Colistine	CL10	Sensitive
Aztreonam	AZ	Resistance
Ceftriaxone	CRO30	Resistance
Cefixime	CFM5	Resistance
Nitrofurantoin	FM300	Sensitive

Escherichia coli to Piperacillin, Nitrofurantoin, Imipenem, Gentamicin, Ampicillin, Amikacin, Ceftazidime, Cefepime and Cefotaxime were sensitive. However, *Trimethoprim* and *Cefazolin*, showed resistance. (Table 8).

Table 8. Determination of antibiotic resistance patterns of *E.coli*

Antimicrobial Agent	Symbol & Count.	
Cefazolin	CEF	Resistance
Gentamycin	GM10	Sensitive
Amikacin	An	Sensitive
Cefepime	FEP30	Sensitive
Cefotaxime	CTX30	Sensitive
Ciprofloxacin	CP10	Sensitive
Trimethoprim sulfa methoxazole	TMP5	Resistance
Meropenem	MEN10	Sensitive
Ceftazidime	CAZ30	Sensitive
Nitrofurantoin	FM300	Sensitive
Piperacillin Tazobactam	PTZ100/10	Sensitive
Ampicillin	AM30	Sensitive
Imipenem	IPM10	Sensitive
Colistine	CL10	Sensitive
Aztreonam	AZ	Sensitive
Ceftriaxone	CRO30	Sensitive
Cefixime	CFM5	Sensitive
Nitrofurantoin	FM300	Sensitive

MRSA to *Nitrofurantoin, Trimethoprim* showed sensitivity. But it showed resistance to *Erythromycin* and *Methicillin* (Table IX). *Staphylococcus epidermidis* was resistance to *Cefazolin, Clindamycin, Clarithromycin, Minocycline, Azithromycin, and Oxacillin.* However, *Staphylococcus epidermidis* was sensitive to *Rifampicin, Vancomycin, co-trimoxazole* and *Linezolid* (Table 9).

Table 9. Determination of antibiotic resistance patterns of *Staphylococcus epidermidis*

Antimicrobial Agent	Symbol & Count.	
cefazolin	CEF	Resistance
Rifampicin	RA5	Sensitive
Vancomycin	V30	Sensitive
Clindamycin	CC2	Resistance
Co-trimoxazole	SXT25	Sensitive
Minocycline	MI30	Resistance
Linezolid	LZ30	Sensitive
Azithromycin	AZM15	Resistance
Clarithromycin	CLR15	Resistance
Oxacillin	OX1	Resistance

Acinetobacter baumannii, isolated from burn wounds was resistance to *Imipenem, Ceftazidime, Cefepime, Ticarcillin, Tobramycin, Gentamycin, Cefotaxime, Ciprofloxacin, Co-trimoxazole* and also it was sensitive to *Colistin* (Table 10).

Table 10. Determination of antibiotic resistance patterns of *Acinetobacter baumannii*

Antimicrobial Agent	Symbol & Count.	
Imipenem	IPM10	Resistance
Ceftazidime	CT30	Resistance
Ticarcillin	TIC75	Resistance
Tobramycin	TOB10	Resistance
Gentamycin	GM10	Resistance
Cefotaxime	CTX30	Resistance
Ciprofloxacin	CP10	Resistance
Co-trimoxazole	SXT25	Resistance
Colistin	CL10	Sensitive

3.4 Disk Diffusion and Cavity Method

According to the results in the Table 12, silver (NPs) after being exposed to ultrasonic waves has no effect on the resistant antibiotics bacteria. *Staphylococcus epidermidis* and *Methicillin resistance Staphylococcus aureus* had most sensitive to ZnO and Ag/ZnO (NPs). Meanwhile, Gram-negative antibiotic-resistant bacteria such as *Klebsiella pneumoniae, Pseudomonas aeruginosa* and *Escherichia coli* were weaker.

3.5 MIC and MBC Tests

According to the results recorded in the Table 13, no one of the bacteria showed any sensitivity to silver (NPs). Compared with Gram-negative bacteria, *Staphylococcus epidermidis* and *Methicillin Resistance Staphylococcus aureus* growth, at concentration slower of ZnO and Ag/ZnO (NPs) had stopped. The greatest resistances to zinc oxide (NPs) were observed in *Pseudomonas aeruginosa* (≥ 8192 µg.ml^{-1}).

Meanwhile, with combination of silver and zinc oxide (NPs), the dose of MIC and MBC were decreased. It means that 256 µgml^{-1} and 4096 µgml^{-1}, respectively. The results were evident in the case of other gram-negative bacteria.

Table 12. Disk diffusion and Cavity method for Ag, ZnO, Ag/ZnO (NPs) against *Klebsiella pneumoniae, Staphylococcus epidermidis, Pseudomonas aeruginosa, Escherichia coli, Acinetobacter baumannii, Methicillin Resistant Staphylococcus aureus* (*MRSA*)

		Ag (NPs)	ZnO (NPs)	Ag/ZnO (NPs)
Klebsiella pneumoniae	Disc Diffusion Test	Negative	8mm	12mm
	Cavity Test	Negative	10mm	14mm
Staphylococcus epidermidis	Disc Diffusion Test	Negative	12mm	18mm
	Cavity Test	Negative	10mm	14mm
Pseudomonas aeruginosa	Disc Diffusion Test	Negative	8mm	10mm
	Cavity Test	Negative	5mm	8mm
Escherichia coli	Disc Diffusion Test	Negative	8mm	12mm
	Cavity Test	Negative	8mm	10mm
Acinetobacter baumannii	Disc Diffusion Test	Negative	8mm	12mm
	Cavity Test	Negative	8mm	10mm
MRSA	Disc Diffusion Test	Negative	14mm	18mm
	Cavity Test	Negative	12mm	20mm

Table 13. MIC and MBC tests for Ag, ZnO, Ag/ZnO (NPs) against *Klebsiella pneumoniae, Staphylococcus epidermidis, Pseudomonas aeruginosa, Escherichia coli, Acinetobacter baumannii, Methicillin Resistant Staphylococcus aureus* (*MRSA*)

		Ag(NPs)	ZnO(NPs)	Ag/ZnO(NPs)
Klebsiella pneumoniae	MIC	$\geq$8192 µg/ml	512 µg/ml	256 µg/ml
	MBC	$\geq$8192 µg/ml	1024 µg/ml	512 µg/ml
Staphylococcus epidermidis	MIC	$\geq$8192 µg/ml	64 µg/ml	128 µg/ml
	MBC	$\geq$8192 µg/ml	512 µg/ml	512 µg/ml
Pseudomon asaeruginosa	MIC	$\geq$8192 µg/ml	512 µg/ml	256 µg/ml
	MBC	$\geq$8192 µg/ml	$\geq$8192 µg/ml	4096 µg/ml
Escherichia coli	MIC	$\geq$8192 µg/ml	256 µg/ml	256 µg/ml
	MBC	$\geq$8192 µg/ml	2048 µg/ml	512 µg/ml
Acinetobacter baumannii	MIC	$\geq$8192 µg/ml	256 µg/ml	128 µg/ml
	MBC	$\geq$8192 µg/ml	1024 µg/ml	512 µg/ml
MRSA	MIC	$\geq$8192 µg/ml	128 µg/ml	128 µg/ml
	MBC	$\geq$8192 µg/ml	256 µg/ml	256 µg/ml

4. Discussion

The MIC and MBC results obtained by Gan and his colleagues, in 2004, showed that the metal oxide (NPs) were able to inhibiting or destroying many pathogenic bacteria (Sondi & Salopek-Sondi, 2004). Regarding this theory, Guogang and Jayesh dispersed suspension of metal oxide (NPs) with the ultrasonic waves. They believed that the antibacterial properties of (NPs) will increase (Avadi et al., 2004; Lok et al., 2006). So far, many studies by researchers around the world in the field of antibacterial properties of silver (NPs) have been carried out (Avadi et al., 2004; Batarseh, 2004; Lok et al., 2006; Sondi & Salopek-Sondi, 2004; Thirumurugan, Shaheedha, & Dhanaraju, 2009). Though, studies of several authors in recent years, confirmed the antibacterial effects of Ag (NPs) (Batarseh, 2004; Lok et al., 2006; Sondi & Salopek-Sondi, 2004).

In the current study, zinc oxide, silver and zinc oxide/silver (NPs), by thermal decomposition of a precursor oxalate, were synthesized. FTIR spectroscopy, X-ray powder diffraction (XRD), scanning electron microscopy (SEM) and Transmission Electron Microscopy (TEM) for identification, structure, and surface morphology were used. Analysis of the results obtained in this study was interesting. Regarding that the diameter of inhibition zone (DIZ), reflects the sensitivity of the organism, strains of sensitive, show larger DIZ and the resistant strains show smaller DIZ (Thirumurugan et al., 2009). Results of disc diffusion test of antibiotic resistant *Pseudomonas aeruginosa* showed the least sensitivity to ZnO and Ag/ZnO (NPs). While DIZ obtained from the disc diffusion test, showed the greatest sensitivity to *Staphylococcus epidermidis* and MRSA.

Silver (NPs) have not antibacterial effect against six strains of antibiotic resistant bacteria. Additionally, Silver (NPs) had not bacteriostatic effect on *Klebsiella pneumoniae, Staphylococcus epidermidis, Pseudomonas aeruginosa, Escherichia coli, Acinetobacter baumannii,* and *Methicillin resistant Staphylococcus aureus* at concentrations of 8192 µg.ml⁻¹ to 0.2µg.ml⁻¹. Whereas, results of the disc diffusion with Ag µg.ml⁻¹, by Thirumurugan against strains of pathogens *E. coli, S. typhi, B. subtilis, S. aureus*, indicated higher sensitivity to silver µg.ml⁻¹ which is in contrast with the results of our study (Thirumurugan et al., 2009).

In 2005, Cho studies the MIC of Ag nanoparticles against *Pseudomonas aeruginosa* bacteria in which its growth in concentration of 7.5 µg.ml⁻¹ completely inhibited (Cho, Park, Osaka, & Park, 2005). However, antibiotic resistance of *Pseudomonas aeruginosa* used in the current study, was resistance to Ag (NPs). Lowest MIC in *Pseudomonas aeruginosa* observed Ag/ZnO with 256 µgml⁻¹ it has the highest inhibitory effect on *Pseudomonas aeruginosa*. In fact, the greatest resistance to zinc oxide (NPs) was seen in *Pseudomonas aeruginosa*(≥8192 µgml⁻¹). In another study, Cho reported the MIC rate of silver (NPs) for *S. aureus*as 12.6 µgml⁻¹, but interestingly, *MRSA* was resistance to the Ag (NPs) completely (Cho et al., 2005). The *MRSA* used in the current study, showed the least and the most sensitivity to silver and Ag/ZnO (NPs), respectively. Actually, the least degree of MIC in *MRSA* was related to combine (NPs) of silver and zinc oxide with concentration 128 µgml⁻¹, this (NPs) had the most growth inhibitory effect in *MRSA*.

In fact, the results of our tests, MIC and MBC, was also interesting. In our study, Silver/Zinc oxide (NPs) had a highest inhibitory effect against, all of the antibiotics resistant bacteria. However, our results had been consistent with Jafari and their colleagues in 2009 (Jafari, Ghane, Sarabi, et al., 2011). Jayesh recorded the MIC rate ranged 40-180 µgml⁻¹ using tests of determining the sensitivity of silver (NPs) against different strains of *Escherichia coli* (Jafari, Ghane, Sarabi, et al., 2011). Kim studied the gram negative bacteria *E. coli* and gram positive *S. aureus,* also reported that the antibacterial silver (NPs) mostly affects the *E. coli,* which is due to the difference between cell wall of gram negative & positive microorganisms (Kim et al., 2007). Reddy were worked on the toxicity of the ZnO (NPs) in gram negative & positive bacteria (Reddy et al., 2007). They found that this (NPs) are able to completely inhibit the growth of *E. coli*. Also, Reddy and colleagues found ZnO (NPs) do not have any toxicity against eukaryotic cells. Actually, Ling Yang believed that photo catalytic ability of ZnO (NPs) plus silver (NPs) improves and also increases its oxidation and reduction abilities, while suppressing bacteria growth (Yang et al., 2006).

However, silver ions, eventually release during sterilization and kill bacteria due to their high antibacterial activation. They theorized that silver ions release following bacteria death and colloid with other bacteria and repeat their sterilization behavior. It was also mentioned that silver covered in the surface of ZnO (NPs) has the ability to involve the electrons produced through photo catalytic reactions of ZnO (NPs) which increases electron isolation and makes gaps in cell membrane, so increase its antimicrobial activity. Regarding studies of these authors, antibacterial property of silver and zinc oxide (NPs) improves with their combination. Based on the results obtained in this study, it was found that the composition of the metallic nanoparticle(NPs), silver and zinc oxide (NPs), it increases the antibacterial properties.

5. Conclusion

Silver (NPs) had not the ability to inhibit the antibiotic resistant bacteria. Also, we shows that Gram-positive antibiotic resistant bacteria such as MRSA, *Staphylococcus epidermidis* and also, *klebsiella pneumonia*, showed high sensitivity to Ag/ZnO and ZnO (NPs), compared to other bacteria.

Acknowledgements

We are indebted to research Vice Chancellor of Iran University of Medical Sciences, for supporting this research. Also, we gratefully acknowledge to Institute of Materials & Energy (MERC), for the XRD and SEM analysis. The authors would like to acknowledge to Central Laboratory of University of Tehran for FTIR analysis. We gratefully acknowledge Executive Director of Iran-Nanotechnology Organization (Govt. of Iran). The anonymous reviewers are acknowledged for providing valuable comments and insights for improving the manuscript.

Refferences

Avadi, M., Sadeghi, A., Tahzibi, A., Bayati, K., Pouladzadeh, M., Zohuriaan-Mehr, M., & Rafiee-Tehrani, M. (2004). Diethylmethyl chitosan as an antimicrobial agent: Synthesis, characterization and antibacterial effects. *European Polymer Journal, 40*(7), 1355-1361.

Batarseh, K. I. (2004). Anomaly and correlation of killing in the therapeutic properties of silver (I) chelation with glutamic and tartaric acids. *Journal of Antimicrobial Chemotherapy, 54*(2), 546-548.

Blanc, D., Carrara, P., Zanetti, G., & Francioli, P. (2005). Water disinfection with ozone, copper and silver ions, and temperature increase to control Legionella: seven years of experience in a university teaching hospital. *Journal of Hospital Infection, 60*(1), 69-72.

Cho, K.-H., Park, J.-E., Osaka, T., & Park, S.-G. (2005). The study of antimicrobial activity and preservative effects of nanosilver ingredient. *Electrochimica Acta, 51*(5), 956-960.

Dabbagh, M. A., Moghimipour, E., Ameri, A., & Sayfoddin, N. (2010). Physicochemical characterization and antimicrobial activity of nanosilver containing hydrogels. *Iranian Journal of Pharmaceutical Research*, 21-28.

Gan, X., Liu, T., Zhong, J., Liu, X., & Li, G. (2004). Effect of silver nanoparticles on the electron transfer reactivity and the catalytic activity of myoglobin. *ChemBioChem, 5*(12), 1686-1691.

Ghosh, T., Das, A. B., Jena, B., & Pradhan, C. (2015). Antimicrobial effect of silver zinc oxide (Ag-ZnO) nanocomposite particles. *Frontiers in Life Science, 8*(1), 47-54.

Jafari, A., Ghane, M., & Arastoo, S. (2011). Synergistic antibacterial effects of nano zinc oxide combined with silver nanocrystales. *African Journal of Microbiology Research, 5*(30), 5465-5473.

Jafari, A., Ghane, M., Sarabi, M., & Siyavoshifar, F. (2011). Synthesis and antibacterial properties of zinc oxide combined with copper oxide nanocrystals. *Oriental Journal of Chemistry, 27*(3), 811.

Kim, J. S., Kuk, E., Yu, K. N., Kim, J.-H., Park, S. J., Lee, H. J., ... Hwang, C.-Y. (2007). Antimicrobial effects of silver nanoparticles. *Nanomedicine: Nanotechnology, Biology and Medicine, 3*(1), 95-101.

Lok, C.-N., Ho, C.-M., Chen, R., He, Q.-Y., Yu, W.-Y., Sun, H., ... Che, C.-M. (2006). Proteomic analysis of the mode of antibacterial action of silver nanoparticles. *Journal of Proteome research, 5*(4), 916-924.

Morones, J. R., Elechiguerra, J. L., Camacho, A., Holt, K., Kouri, J. B., Ramírez, J. T., & Yacaman, M. J. (2005). The bactericidal effect of silver nanoparticles. *Nanotechnology, 16*(10), 2346.

Reddy, K. M., Feris, K., Bell, J., Wingett, D. G., Hanley, C., & Punnoose, A. (2007). Selective toxicity of zinc oxide nanoparticles to prokaryotic and eukaryotic systems. *Applied physics letters, 90*(21), 213902.

Roberts, R. R., Hota, B., Ahmad, I., Scott, R. D., Foster, S. D., Abbasi, F., . . . Supino, M. (2009). Hospital and societal costs of antimicrobial-resistant infections in a Chicago teaching hospital: implications for antibiotic stewardship. *Clinical Infectious Diseases, 49*(8), 1175-1184.

Ruparelia, J. P., Chatterjee, A. K., Duttagupta, S. P., & Mukherji, S. (2008). Strain specificity in antimicrobial activity of silver and copper nanoparticles. *Acta Biomaterialia, 4*(3), 707-716.

Sondi, I., & Salopek-Sondi, B. (2004). Silver nanoparticles as antimicrobial agent: a case study on E. coli as a model for Gram-negative bacteria. *Journal of colloid and interface science, 275*(1), 177-182.

Thirumurugan, G., Shaheedha, S., & Dhanaraju, M. (2009). In vitro evaluation of antibacterial activity of silver nanoparticles synthesised by using Phytophthora infestans. *Int J Chem Tech Res, 1*, 714-716.

Yang, L., Mao, J., Zhang, X., Xue, T., Hou, T., Wang, L., & Tu, M. (2006). Preparation and characteristics of Ag/nano-ZnO composite antimicrobial agent. *Nanoscience, 11*(1), 44-48.

Yu-sen, E. L., Vidic, R. D., Stout, J. E., McCartney, C. A., & Victor, L. Y. (1998). Inactivation of Mycobacterium avium by copper and silver ions. *Water Research, 32*(7), 1997-2000.

20

Biofilm of *Pseudomonas aeruginosa* in Nosocomial Infection

Z. Mahmmudi[1] & A. A. Gorzin[2]

[1] Kazeroun Branch, Islamic Azad University, Kazeroon, Iran

[2] School of Medicine, Shiraz University of Medical Sciences, Shiraz, Iran

Correspondence: A. A. Gorzin, School of Medicine, Shiraz University of Medical Sciences, Shiraz, Iran.
E-mail: z.mahmmudi.792@gmail.com

Abstract

Bacteria in natural, industrial and clinical settings predominantly live in biofilms, i.e., sessile structured microbial communities encased in self-produced extracellular matrix material. One of the most important characteristics of microbial biofilms is that the resident bacteria display a remarkable increased tolerance toward antimicrobial attack. Biofilms formed by opportunistic pathogenic bacteria are involved in devastating persistent medical device-associated infections, and chronic infections in individuals who are immune-compromised or otherwise impaired in the host defense. Because the use of conventional antimicrobial compounds in many cases cannot eradicate biofilms, there is an urgent need to develop alternative measures to combat biofilm infections. The present review is focussed on the important opportunistic pathogen and biofilm model organism *Pseudomonas aeruginosa*. Initially, biofilm infections where P. aeruginosa plays an important role are described. Subsequently, current insights into the molecular mechanisms involved in P. aeruginosa biofilm formation and the associated antimicrobial tolerance are reviewed. And finally, based on our knowledge about molecular biofilm biology, a number of therapeutic strategies for combat of P. aeruginosa biofilm infections are presented.

Keywords: *Pseudomonas aeruginosa*; biofilm matrix; exopolysaccharides; gene regulation; anti-biofilm

1. Introduction

Biofilms are microbial communities encased in extracellular polymeric substances (EPS) (Epps & Walker, 2006). Biofilm formation represents a protective mode of growth that allows microorganisms to survive in hostile environments and disperse seeding cells to colonize new niches under desirable conditions. Biofilms can form on a variety of surfaces and are prevalent in natural, industrial, and hospital niches. These sessile OPEN ACCESS Int. J. Mol. Sci. 2013, 14 20984 microbial communities are physiologically distinct from free-living planktonic counterparts (English & Gaur, 2010; Guarner & Malagelada, 2003). Clinically, biofilms are responsible for many persistent and chronic infections due to their inherent resistance to antimicrobial agents and the selection for phenotypic variants. A better understanding of the genetic and molecular mechanisms of biofilm formation may provide strategies for the control of chronic infections and problems related to biofilm formation. The EPS of biofilm is a mixture of polysaccharides, extracellular DNA (eDNA), and proteins, which function as matrix, or glue, holding microbial cells together. The biofilm matrix contributes to the overall architecture and the resistance phenotype of biofilms (Beaugerie & Petit, 2004; Høiby, Ciofu, & Bjarnsholt, 2010). Uncovering roles played by EPS matrices in biofilm formation will be beneficial for the design of targeted molecules to control biofilm formation. In this review, advances in biofilm formation and regulation are presented with a focus on the biofilm matrix in P. aeruginosa, a model organism for biofilm research.

2. The Mechanism of Biofilm Formation

Formation of a biofilm begins with the attachment of free-floating microorganisms to a surface. While still not fully understood, it is thought that the first colonists of a biofilm adhere to the surface initially through weak, reversible adhesion via van der Waals forces and hydrophobic effects (Kalia & Purohit, 2011; Blackledge, Worthington, & Melander, 2013). If the colonists are not immediately separated from the surface, they can anchor themselves more permanently using cell adhesion structures such as pili. Hydrophobicity also plays an important role in determining the ability of bacteria to form biofilms, as those with increased hydrophobicity have reduced repulsion between the extracellular matrix and the bacterium (Sharma et al., 2014).

Some species are not able to attach to a surface on their own but are instead able to anchor themselves to the matrix or directly to earlier colonists. It is during this colonization that the cells are able to communicate via quorum sensing (QS) using products such as N-acyl homoserine lactone (AHL). Some bacteria are unable to form biofilms as successfully due to their limited motility. Non-motile bacteria cannot recognize the surface or aggregate together as easily as motile bacteria (Sharma et al., 2014). Once colonization has begun, the biofilm grows through a combination of cell division and recruitment. Polysaccharide matrices typically enclose bacterial biofilms. In addition to the polysaccharides, these matrices may also contain material from the surrounding environment, including but not limited to minerals, soil particles, and blood components, such as erythrocytes and fibrin (Sharma et al., 2014). The final stage of biofilm formation is known as dispersion, and is the stage in which the biofilm is established and may only change in shape and size.

The development of a biofilm may allow for an aggregate cell colony (or colonies) to be increasingly resistant to antibiotics. Cell-cell communication or quorum sensing has been shown to be involved in the formation of biofilm in several bacterial species (Deep, Chaudhary, & Gupta, 2011).

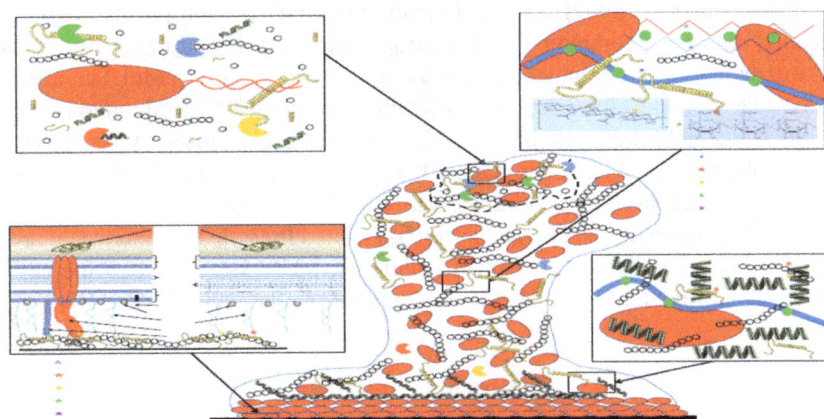

Figure 1. The biofilm matrix is comprised of entangled polymers (polysaccharides, DNA, proteins) that affect the permeability and mechanical properties of the entire biofilm. To understand the biophysical properties of the biofilm several questions need to be addressed. For example, what is the pore size of the matrix? Does a specific substrate interact with the matrix components? Which structural components of the matrix regulate the permeability properties? Is the matrix a static arrangement or do the individual components engage in dynamic rearrangements?

2.1 Biofilm associated infections and their implications in nosocomial infection

According to a recent public announcement from the National Institutes of Health, more than 60% of all the infections are caused by biofilms (Karatan & Watnick, 2009). As described by Prasanna et al, about 40-50% of adults had biofilm related gingival infections. Among 4000 infants with cerebrospinal- fluid shunts, 15-20% had biofilm related infections. 95% of the urinary tract infections were associated with urinary catheters. 86% pneumonias were associated with mechanical ventilation and 85% of the blood stream infections were closely related to intravascular devices (Guarner & Malagelada, 2003).

2.2 The Detection of Biofilm Producing Microorganisms

Early biofilm formation detection might result in a greater success in the treatment, because in long standing cases, they may be very damaging and may produce immune complex sequelae (Yang et al., 2012). There are two methods for the detection of biofilms – 1). The Phenotypic method a. The tissue culture plate (TCP) method – The wells of the tissue culture plates are inoculated with a bacterial suspension along with positive and negative controls and these are incubated for 24 to 48 hours. Planktonic cells are removed by washing with phosphate buffered saline. Biofilms are fixed with 2% sodium acetate and are stained with 0.1% crystal violet. The excess dye is washed away with deionised water. The plates are dried properly and the optical densities of the stained biofilms are obtained spectrophotometrically. b. The tube method(TM) – 10 ml of Tripticase soy broth with 1% glucose is inoculated with a loopful of test organisms, along with positive and negative controls. The broths are incubated at for 24 – 48 hours. The culture supernatants are decanted and the tubes are washed with phosphate buffered saline. The tubes are dried and are stained with 0.1% crystal violet. The excess stain is washed away with deionised water. The tubes are dried in an inverted position. c. The Congo red agar (CRA) method – The Congo red stain is prepared as a concentrated aqueous solution and is autoclaved at 1210c for 15 minutes. This is added to autoclaved Brain heart infusion agar with sucrose at 550c. The plates are inoculated with the test organisms along with positive and negative controls and are incubated at 370C for 24 to 48 hours aerobically. Black colonies with a dry crystalline consistency indicate biofilm production. Various studies have established that TCP is a better screening test for biofilm production than the TM and the CRA methods. The test is easy to perform and to assess biofilms, both qualitatively and quantitatively (Nikolaev & Plakunov, 2007; Flemming, Neu, & Wozniak, 2007). 2). The Genotypic method Sonications and PCR amplification methods have been shown to improve the detection of biofilms. Biofilm non producers are negative for ica A and ica D and lack the entire ica ADBC operon. But this requires specialized equipments and techniques (Sutherland, 2001; Branda, Vik, Friedman, & Kolter, 2005).

2.3 Regulation of Biofilm Matrix in P. aeruginosa

Gene regulation is important for our understanding of biofilm formation. Generally, organisms form a biofilm in response to several factors including nutritional cues, secondary messengers, host-derived signals or, in some cases, to sub-inhibitory concentrations of antibiotics (Hentzer, Eberl, & Givskov, 2005; Shrout et al., 2006). When a cell switches to the biofilm mode of growth, it undergoes a phenotypic shift in behavior whereby a large array of genes is differentially regulated (Ma et al., 2009). Biofilm formation is a multicellular process involving environmental signals and a concerted regulation combining both environmental signals and regulatory networks. Due to the major roles of EPS matrix in biofilm formation, its regulation is discussed. Int. J. Mol. Sci. 2013, 14 20990 3.1. c-di-GMP Bis-(3'-5')-cyclic dimeric guanosine monophosphate (c-di-GMP), a ubiquitous intracellular second messenger widely distributed in bacteria, was discovered in 1987 as an allosteric activator of the cellulose synthase complex in Gluconacetobacter xylinus (Shrout et al., 2006). In general, c-di-GMP stimulates the biosynthesis of adhesins and exopolysaccharide mediated biofilm formation and inhibits bacterial motilities, which controls the switch between the motile planktonic and sessile biofilm-associated lifestyle of bacteria (Figure 2). Moreover, c-di-GMP controls the virulence of animal and plant pathogens, progression through the cell cycle, antibiotic production and other cellular functions (Branda, Vik, Friedman, & Kolter, 2005; Ryder, Byrd, & Wozniak, 2007).

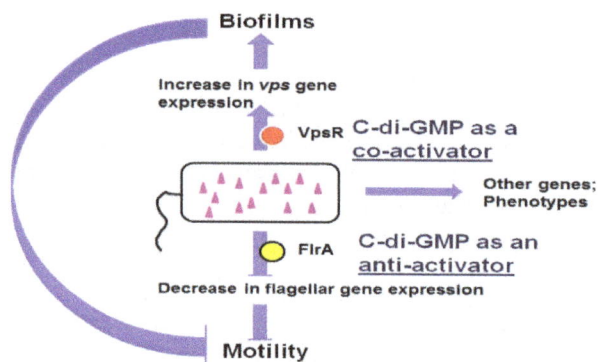

Figure 2. Schematic presentation of physiological functions of c-di-GMP. In bacterial cells, c-di-GMP is generated by diguanylate cyclases (DGC) and broken down by specific phosphodiesterases (PDE). As a second messenger, low levels of c-di-GMP can promote motility by upregulating flagellar expression, assembly or interfering with flagellar motor function and are required for the expression of acute virulence genes. High levels of c-di-GMP however favor sessility and stimulate the synthesis of various matrix exopolysaccharides, such as Pel (mediated by PelD) and alginate (mediated by Alg44) (Masák, Čejková, Schreiberová, & Řezanka, 2014; Donlan, 2002; Karatan & Watnick, 2009)

2.4 Matrix-Driven Strategies against Biofilms

Once biofilms develop into a mature stage, they become extremely difficult to eradicate from infections sites with traditional antimicrobial agents (Cegelski, Marshall, Eldridge, & Hultgren, 2008). Agents that inhibit biofilm formation or transform bacteria from biofilm life style to free-living individuals are ideal to eradicate biofilm. The strategies used for anti-biofilm mainly stem from two basic ways: matrix synthesis and its regulatory mechanisms. For example, disruption of the initial attachment that is dependent on a large array of adhesins would contribute to inhibition of the establishment of biofilms, while the digestion of the EPS matrix may be another method to interfere with biofilm formation. As we mentioned before, DNase I treatment has already shown efficacy in the inhibition of the early development of biofilm. It was also reported that alginate lyase could enhance antibiotic killing of mucoid P. aeruginosa in biofilms (Bjarnsholt, 2013). In addition, the macrolide antibiotic azithromycin was shown to block alginate formation and quorum sensing signaling (Nikolaev & Plakunov, 2007) and was further reported to improve lung function of CF patients, especially in the subgroup chronically colonized by Pseudomonas (Flemming, Neu, & Wozniak, 2007). Antagonizing the intracellular signaling molecules to control biofilm formation has also been investigated. One example is the identification of furanones, which have shown their ability to inhibit Int. J. Mol. Sci. 2013, 14 20996 the biofilm formation of P. aeruginosa in vitro (Sutherland, 2001; Branda, Vik, Friedman, & Kolter, 2005). Molecules of this type have been reported to function through inhibiting the AHL-dependent QS systems in P. aeruginosa. Iron has also been employed in distinct aspects to control the formation of biofilms. Singh and his colleagues have identified an innate immunity component, lactoferrin, which prevents P. aeruginosa biofilm formation by chelating iron and stimulating the type IV pili-mediated twitching motility (Epps & Walker, 2006; Yang et al., 2012). Furthermore, iron salts such as ferric ammonium citrate were found to not only perturb biofilm formation but also disrupt existing biofilms by P. aeruginosa (Shrout et al., 2006). In a screen of co-therapy of antibiotics against P. aeruginosa, 14-alpha-lipoyl and rographolide, a diterpenoid lactone derivative from the herb Andrographis paniculata appeared to inhibit biofilm formation by decreasing EPS production and to sensitize the bacterium to a variety of antibiotics (Yang et al., 2012). Recently, it was found that Gram-positive bacterium Bacillus subtilis produced a factor that prevented biofilm formation and could break down existing biofilms. The factor was identified to be a mixture of D-leucine, D-methionine, D-tyrosine, and D-tryptophan that could disassemble at nanomolar concentrations. D-amino acid treatment subserved the release of amyloid fibers that linked cells together in the biofilm. In addition, D-amino acids also prevented biofilm formation by Staphylococcus aureus and P. aeruginosa, indicating it may be a widespread signal for biofilm disassembly (Chicurel, 2000). Furthermore, the same group identified another biofilm disassembly compound, norspermidine, which targets directly and specifically with the exopolysaccharide matrix and this biofilm inhibition effect could be enhanced together with D-amino acids and is effective in other bacterial species (Masák, Čejková, Schreiberová, & Řezanka, 2014).

2.5 Perspectives

Accumulating data presented in the recent literature provides valuable insights into the novel roles of the biofilm matrix and its regulatory mechanism in P. aeruginosa biofilm formation. A deep understanding of the mechanisms involved in biofilm formation will ultimately shed light on the generation of alternative treatments for P. aeruginosa infections. There is no doubt that future studies will reveal additional biofilm matrix components and identify more elaborate regulatory circuits for biofilm formation. Finally, the interaction of the biofilm matrix and the synergistic effects of different anti-biofilm strategies should also be regarded as major concerns.

3. Conclusion

Many biofilm infections develop slowly, producing very few symptoms initially, but in the long run, they may produce immune complex sequelae and may act as reservoirs of infection (English & Gaur, 2010). Standard, in vitro antibiotic susceptibility tests are not predictive of the therapeutic outcome of biofilm associated infections (Ghafoor, Hay, & Rehm, 2011). The overall healthcare costs which are attributed to the treatment of biofilm associated infections are much higher due to their persistence. Besides, a longer hospital stay is another factor for higher costs. Early detection of biofilm associated infections and newer treatment options for the management of the same are needed.

References

Beaugerie, L., & Petit, J. C. (2004). Antibiotic-associated diarrhoea. *Best practice & research Clinical gastroenterology, 18*(2), 337-352.

Bjarnsholt, T. (2013). The role of bacterial biofilms in chronic infections. *Apmis, 121*(s136), 1-58.

Blackledge, M. S., Worthington, R. J., & Melander, C. (2013). Biologically inspired strategies for combating bacterial biofilms. *Current opinion in pharmacology, 13*(5), 699-706.

Branda, S. S., Vik, Å., Friedman, L., & Kolter, R. (2005). Biofilms: the matrix revisited. *Trends in microbiology, 13*(1), 20-26.

Cegelski, L., Marshall, G. R., Eldridge, G. R., & Hultgren, S. J. (2008). The biology and future prospects of antivirulence therapies. *Nature Reviews Microbiology, 6*(1), 17-27.

Chicurel, M. (2000). Bacterial biofilms and infections. Slimebusters. *Nature, 408*(6810), 284–286.

Costerton, J. W., Stewart, P. S., & Greenberg, E. P. (1999). Bacterial biofilms: a common cause of persistent infections. *Science, 284*(5418), 1318-1322.

Davey, M. E., Caiazza, N. C., & O'Toole, G. A. (2003). Rhamnolipid surfactant production affects biofilm architecture in *Pseudomonas aeruginosa* PAO1. *Journal of bacteriology*, *185*(3), 1027-1036.

Deep, A., Chaudhary, U., & Gupta, V. (2011). Quorum sensing and bacterial pathogenicity: from molecules to disease. *Journal of laboratory physicians*, *3*(1), 4-11.

Donlan, R. M. (2002). Biofilms: microbial life on surfaces. *Emerg Infect Dis*, *8*(9), 881–890.

English, B. K., & Gaur, A. H. (2010). The use and abuse of antibiotics and the development of antibiotic resistance. In *Hot Topics in Infection and Immunity in Children VI* (pp. 73-82). Springer New York.

Epps, L. C., & Walker, P. D. (2006). Fluoroquinolone consumption and emerging resistance. *US Pharm*, *10*, 47-54.

Flemming, H. C., & Wingender, J. (2010). The biofilm matrix. *Nature Reviews Microbiology*, *8*(9), 623-633.

Flemming, H. C., Neu, T. R., & Wozniak, D. J. (2007). The EPS matrix: the "house of biofilm cells". *Journal of bacteriology*, *189*(22), 7945-7947.

Ghafoor, A., Hay, I. D., & Rehm, B. H. (2011). Role of exopolysaccharides in *Pseudomonas aeruginosa* biofilm formation and architecture. *Applied and environmental microbiology*, *77*(15), 5238-5246.

Guarner, F., & Malagelada, J. R. (2003). Gut flora in health and disease. *The Lancet*, *361*(9356), 512-519.

Hentzer, Á., Eberl, Á., & Givskov, Á. (2005). Transcriptome analysis of *Pseudomonas aeruginosa* biofilm development: anaerobic respiration and iron limitation. *Biofilms*, *2*(01), 37-61.

Høiby, N., Ciofu, O., & Bjarnsholt, T. (2010). *Pseudomonas aeruginosa* biofilms in cystic fibrosis. *Future microbiology*, *5*(11), 1663-1674.

Kalia, V. C., & Purohit, H. J. (2011). Quenching the quorum sensing system: potential antibacterial drug targets. *Critical reviews in microbiology*, *37*(2), 121-140.

Karatan, E., & Watnick, P. (2009). Signals, regulatory networks, and materials that build and break bacterial biofilms. *Microbiology and Molecular Biology Reviews*, *73*(2), 310-347.

Ma, L., Conover, M., Lu, H., Parsek, M. R., Bayles, K., & Wozniak, D. J. (2009). Assembly and development of the *Pseudomonas aeruginosa* biofilm matrix. *PLoS Pathog*, *5*(3), e1000354.

Ma, L., Wang, J., Wang, S., Anderson, E. M., Lam, J. S., Parsek, M. R., & Wozniak, D. J. (2012). Synthesis of multiple *Pseudomonas aeruginosa* biofilm matrix exopolysaccharides is post-transcriptionally regulated. *Environmental microbiology*, *14*(8), 1995-2005.

Masák, J., Čejková, A., Schreiberová, O., & Řezanka, T. (2014). Pseudomonas biofilms: possibilities of their control. *FEMS microbiology ecology*, *89*(1), 1-14.

Nikolaev, Y. A., & Plakunov, V. K. (2007). Biofilm—"City of microbes" or an analogue of multicellular organisms?. *Microbiology*, *76*(2), 149–163.

Ryder, C., Byrd, M., & Wozniak, D. J. (2007). Role of polysaccharides in *Pseudomonas aeruginosa* biofilm development. *Current opinion in microbiology*, *10*(6), 644-648.

Sharma, G., Rao, S., Bansal, A., Dang, S., Gupta, S., & Gabrani, R. (2014). *Pseudomonas aeruginosa* biofilm: potential therapeutic targets. *Biologicals*, *42*(1), 1-7.

Shrout, J. D., Chopp, D. L., Just, C. L., Hentzer, M., Givskov, M., & Parsek, M. R. (2006). The impact of quorum sensing and swarming motility on *Pseudomonas aeruginosa* biofilm formation is nutritionally conditional. *Molecular microbiology*, *62*(5), 1264-1277.

Sutherland, I. W. (2001). The biofilm matrix–an immobilized but dynamic microbial environment. *Trends in microbiology*, *9*(5), 222-227.

Watnick, P., & Kolter, R. (2000). Biofilm, city of microbes. *Journal of bacteriology*, *182*(10), 2675-2679.

Yang, L., Liu, Y., Wu, H., Song, Z., Høiby, N., Molin, S., & Givskov, M. (2012). Combating biofilms. *FEMS Immunology & Medical Microbiology*, *65*(2), 146-157.

21

Study of Resistance to 82 Clinical Cases Enterobacteriaceae to Beta-lactam Antibiotics

Mahnaz Milani[1]

[1] Department of Microbiology, Shahid Beheshti University, Tehran, Iran

Correspondence: Mahnaz Milani, Department of Microbiology, Shahid Beheshti University, Tehran, Iran.
E-mail: mahnazmilani@yahoo.com

Abstract

Knowledge of antimicrobial resistance patterns in *E. coli*, the predominant pathogen associated with urinary tract infections (UTI) is important as a guide in selecting empirical antimicrobial therapy. To describe the antimicrobial susceptibility of *E. coli* associated with UTI in a major university hospital in Tehran (Iran), seventy-six clinical isolates of *E. coli* were studied for susceptibility to β-lactam antibiotics by the disc diffusion method and Minimal Inhibitory Concentrations determination. All isolates were resistant to ampicillin, amoxicillin and oxacillin. Resistance to the other tested antibiotics was shown to be 93.4% to cefradine, 76.3% to carbenicillin, 47.3% to cefazoline, 50% to cefalexin and 32.8% to cephalothin while 1.3% expressed resistance to cefoxitime, and 2.6% were resistant to ceftizoxime and ceftriaxone. Substrate hydrolysis by ultra violet spectroscopy showed that 87.4% harbored penicillinases, 9% produced cephlosporinases and 3.6% degraded both substrates. Clavulanic acid inhibited enzyme activity in 82.9%, of which 78.95% was penicillinases (group IIa) and 3.95% was cephalosporinases (group IIb) of the Bush classification system. These results indicate that *E. coli* can posses a variety of β-lactamases that are responsible for β-lactam resistance. Members of the family Enterobacteriaceae, particularly Escherichia coli is the most common causes of urinary tract infections in hospitals and societies. Beta-lactam antibiotics, particularly the third and fourth generation of cephalosporins are effective in treating these infections.

Keywords: Beta-lactams, Enterobacteriaceae, *Escherichia coli*

1. Introduction

Urinary Tract Infections (UTI) are the second most common infections present in community practice (Gonzalez & Schaeffer, 1999). Members of Enterobacteriaceae, specifically, *E. coli* are the main causes of urinary infections (Gupta, 2003). Extensive use of β-lactams in veterinary medicine and human practice is believed to be associated with selection of resistance in both pathogenic and nonpathogenic isolates of *E. coli* (Livermore, 1995). More than two hundred βlactamase enzymes are recognized which are classified into 4 main groups and 8 subgroups (Bush & Jacoby, 1996; Bush, 1989). The resistance of Enterobacter spp. to β-lactam antibiotics is most frequently mediated by production of TEM, SHV and AmpC β-lactamase (Barnaud et al., 2001). In the last decade, production of plasmidmediated ESBL which hydrolyzes a wide range of the most recently developed cephalosporins, has been recognized as an additional important emerging mechanism of resistance among members of the family Enterobacteriaceae including clinical isolates of *E. coli* (Bradford, 2001; Pitout et al., 1998). The first plasmid mediated βlactamases (TEM-1) was described in *E. coli* in 1960 and within a few years, it was found in many different genera of Gram-negative bacteria (Bradford, 2001). The AmpC family of β-lactamases occurs both chromosomally and plasmid-mediated in *E. coli* and plasmid encoded AmpC β-lactamases are found to be responsible for global outbreaks (Cudron, Moland, & Sanders, 1997; Eftekhar, 2005). We studied 76 urinary isolates of *E. coli* for their susceptibility to 12 β-lactam antibiotics. Preferred substrate hydrolysis was performed to determine the class of β-lactamases. DNA amplification of βlactamase types TEM, SHV and AmpC genes was carried out by PCR using type specific primers of blaTEM, ampC and SHV genes for all of the isolates.

2. Materials and Methods

Bacteria. Seventy-six clinical isolates of *E. coli* were selected from a collection of urinary Enterobacteriaceae from the Bacteriology Laboratory of Vali-E-Asr Hospital in Tehran (Iran) (Hosseini-Mazinani, Jafar-Nejad, & Ghandili, 2003). *E. coli* ATCC 25922 was used as a control for antibiotic susceptibility tests. K. pneumoniae 57-1 carrying plasmid mediated SHV gene, *E. coli* MK148 carrying the ampC gene and *E. coli* harboring pTEM were used as positive controls for DNA amplification by PCR (Hosseini-Mazinani, 2996). Antibiotic susceptibility. The antibiotic susceptibility of bacteria was initially carried out by the disc diffusion method according to the NCCLS recommendations (National Committee for Clinical Laboratory Standards [NCCLS], 2000). The antibiotic discs were ampicillin (10 μg), amoxicillin (25 μg), carbenicillin (100 μg), cefalexin (30 μg), cephalothin (30 μg), cefazoline (30 μg), cefradine (30 μg), oxacillin (1μg), ceftazidime (30 μg), ceftriaxone (30 μg), ceftizoxime (30 μg) (Padtan Teb, Tehran, Iran) and amoxicillin-clavulonic acid (20/10 μg, Difco, USA). Minimum Inhibitory Concentrations (MIC) of the isolates was determined for ampicillin, ceftazidime, cefotaxime, ceftriaxone, cefepime and imipenem by the microdilution broth method using the NCCLS standard procedure ([NCCLS], 2000). Screening for ESBL production. The double disc synergy test was used to screen for ESBL production (Gonzalez and Schaeffer, 1999). Cefotaxime (30 μg), ceftriaxone (30 μg) and ceftizoxime (30 μg) were placed on Mueller Hinton agar plates adjacent to amoxicillin-clavulanic acid discs (20/10 μg). ESBL production was inferred when cephalosporin inhibition zones expanded by the clavulanate.

Substrate hydrolysis. Relative hydrolysis rates of benzylpenicillin and cephaloridine were evaluated by UV spectroscopy. β-lactamase activity was determined by measuring the decrease in optical density of a 0.1 mM solution of cephaloridine (255 nm) or benzylpenicillin (240 nm). Enzymes were called penicillinase if the relative rate of benzylpenicillin hydrolysis was approximately 30% higher than that of observed for cephaloridine, or cephalosporinase if cephaloridine was hydrolyzed at least 30 % faster than penicillin (Livermore,1 995; Ross & O'Callaghan, 1975).

DNA amplification. Plasmid DNA extraction was carried out using a rapid alkaline lysis method (Winokur, 2001). The oligonucleotide primers used for the PCR assays were; 5'-ATAAAATTCTTGAAGACGAAA3' and 5'-GTCAGTTACCAATGCTTAATC-3' for TEM, 5'-TGGTTATGCGTTATATTCGCC-3' and 5'GGTTAGCGTTGCCAGTGCT-3' for SHV and 5'ATGCAACAACGACAATCCATC-3' and 5'GTTGGGGTAGTTGCGATTGG-3' for AmpC βlactamases (Sutcliffe,1978 ; Bret, 1998). blaTEM and SHV primers were synthesized at the National Research Center for Genetic Engineering and Biotechnology, Iran and ampC primer was synthesized at Faza Pajooh (Tehran, Iran). Reactions were carried out in a Techne DNA thermocycler (Germany) in 25 μl mixtures containing 10 mM Tris-HCl (pH 8.3), 1 mM EDTA, 1.5 mM MgCl2, 200 μM of each deoxyribonucleoside triphosphate, 2-10 pM of oligonucleotide primers and 1 u of Taq DNA polymerase (Fermentas, Lithuania). Following a 4min incubation time at 94°C, 35 cycles were run with the following temperature profile for each cycle: 94°C for 1 min, the proper annealing temperature for each primer (58°C for blaTEM, 59°C for ampC and 52o C for SHV) for 1 min and 72°C for 1 min. An additional 5-10 min incubation time was also carried out at 72°C. PCR experiments for amplification of the SHV gene failed to produce a single DNA product regardless of numerous standardization strategies. Therefore, presence or absence of the desired fragment was determined on the basis of comparing the resulting bands with a positive control as well as DNA size markers.

3. Findings

3.1 Detection and Identification of Bacteria

Identifying bacteria by biochemical version and was confirmed by API 20E test.

The results showed that among 82 cases of *Escherichia coli* cases, 76 cases were *Citrobacter freundii*, a case was *Enterobacter*, a case was *Klebsiella pneumoniae*, a case was *Klebsiella oxytoca*, and a case was *Hafnia*.

3.1.1 Antibiotic Susceptibility Results

A) by disk diffusion method:

All tested bacteria were resistant to antibiotics like amoxicillin, ampicillin and Oxacillin and only 9% of cases were resistant to coamoxiclave. As it is shown in Figure 1 and Figure 2, resistance to Cefradine and carbenicillin was 97.5% and 79.20%; and resistance to other antibiotics was 20.7% Cephalothin, 26.8% cefalexin and 31.7% cefazoline, 1.2% to Cefoxitime, 1.2% Ceftizoxime 1.2%, and Ceftiaxime.

B) The method MIC:

The results of the MIC are shown in Figure 3. MIC test showed susceptibility to the samples.

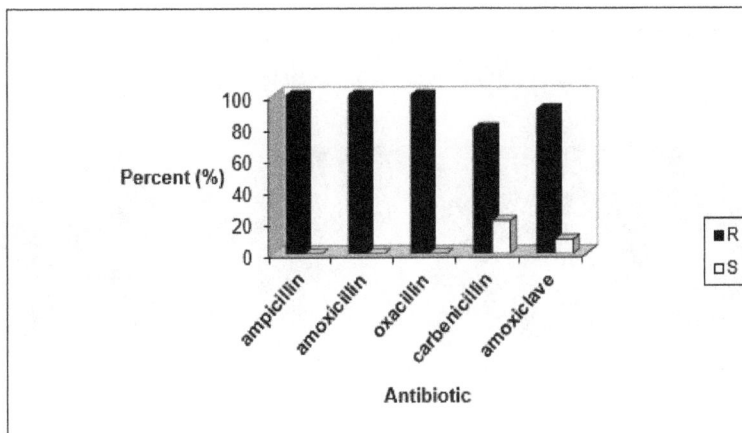

Figure 1. Results of Penicillin antibiotics

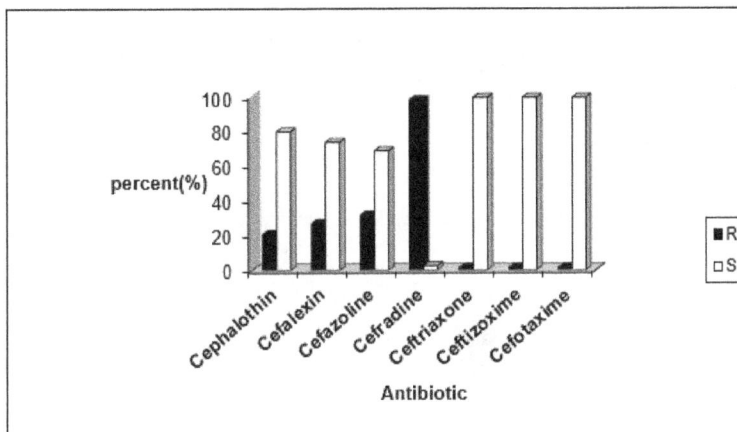

Figure 2. Results of Cephalosporin antibiotics

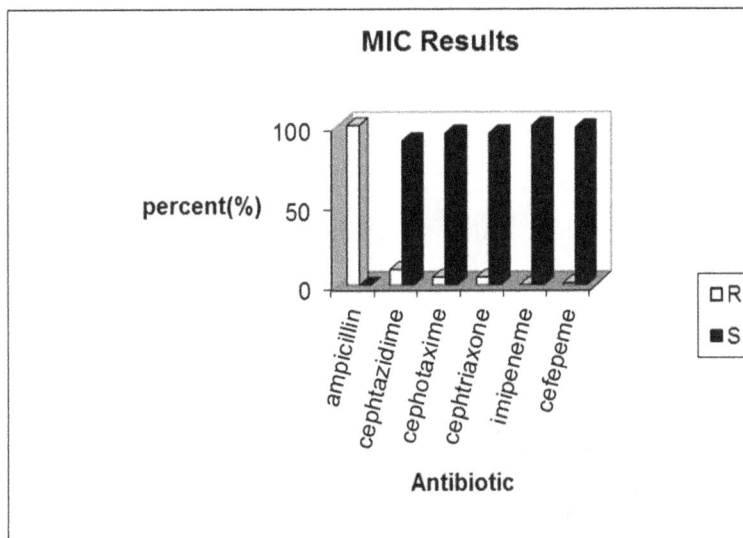

Figure 3. The results of the MIC

3.1.2 Results of colonies Test with iodometer

Iodometer test showed the presence of enzyme in all bacteria.

As it is seen in Figure 4, colonies of bacteria with beta-lactamase enzyme have a white halo.

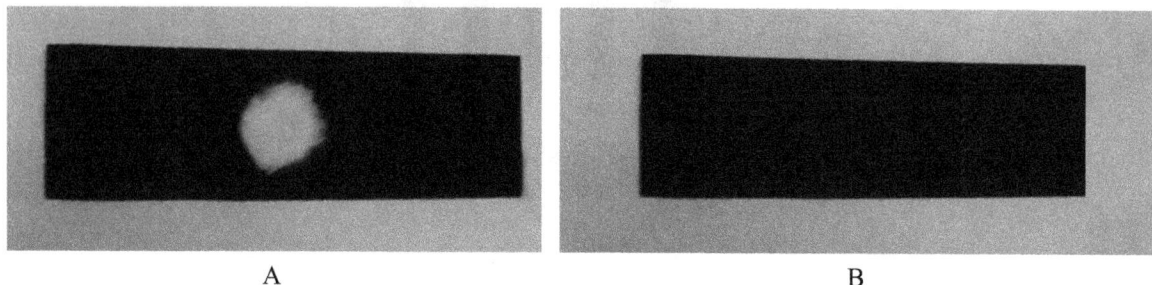

A   B

Figure 4. The results of the colony by Iodometer. A) Bacteria containing beta-lactamase   B) bacteria lacking the enzyme beta-lactamase

3.1.3 The Results of Substrate Hydrolysis

As it is shown in Figure 5, depending on the rate of hydrolysis of benzyl penicillin and Cefaloridine to each other (spectrophotometry) 87.4% organisms contain enzyme penicillinase, 9% organisms contain enzymes Cephalosprinase and 3.6 percent contain both enzymes equally.

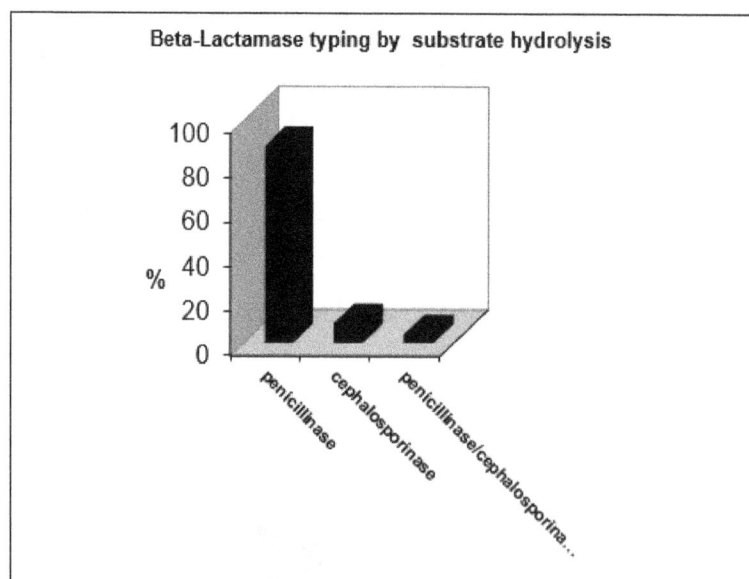

Figure 5. The results of hydrolysis of the substrate by Spectrophotometer

3.1.4 Lactamase Classification Results

None of the beta-lactamase enzymes tested showed no inhibition with EDTA. This does not mean that Group III was among the beta-lactamase. However, 91% enzymes were inhibited by clavulanic acid, and taking into account the results of hydrolysis of the substrate (Figure 5) showed that 87.4% organisms were penicillinase that were resistant to clavulanic acid, and this class of beta-lactamase belonged to Group 2a, 3.6% organisms equally showed penicillinase and cephalosporinase and can be inhibited by clavulanic acid and belonged to the Group 2b, and 9% were penicillinase that cannot be inhibited by clavulanic acid which belonged to Group 4. Among the studied organisms, Group I, i.e. cephalosporinases were not resistant to clavulanic acid (Figure 6).

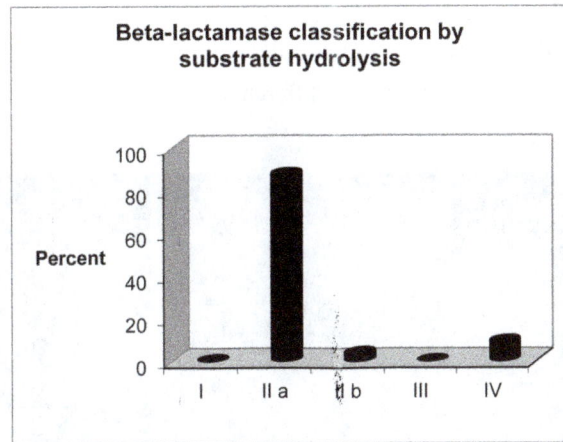

Figure 6. The results of the classification of beta-lactamases

3.1.5 Results of E. Test

The results of 82 clinical cases revealed a broad spectrum beta-lactamase enzymes (ESBL) in three samples of bacteria (17,23a, b29). As it is seen in Figure 7, clover leaf form indicates the presence of ESBL, and creating a clover leaf form in these cases because of resistant to third generation cephalosporins and enzyme containing beta-lactamase (ESBL) are inhibited with clavulanic acid. Lack of beta-lactamase enzyme (ESBL) is shown in figure A.

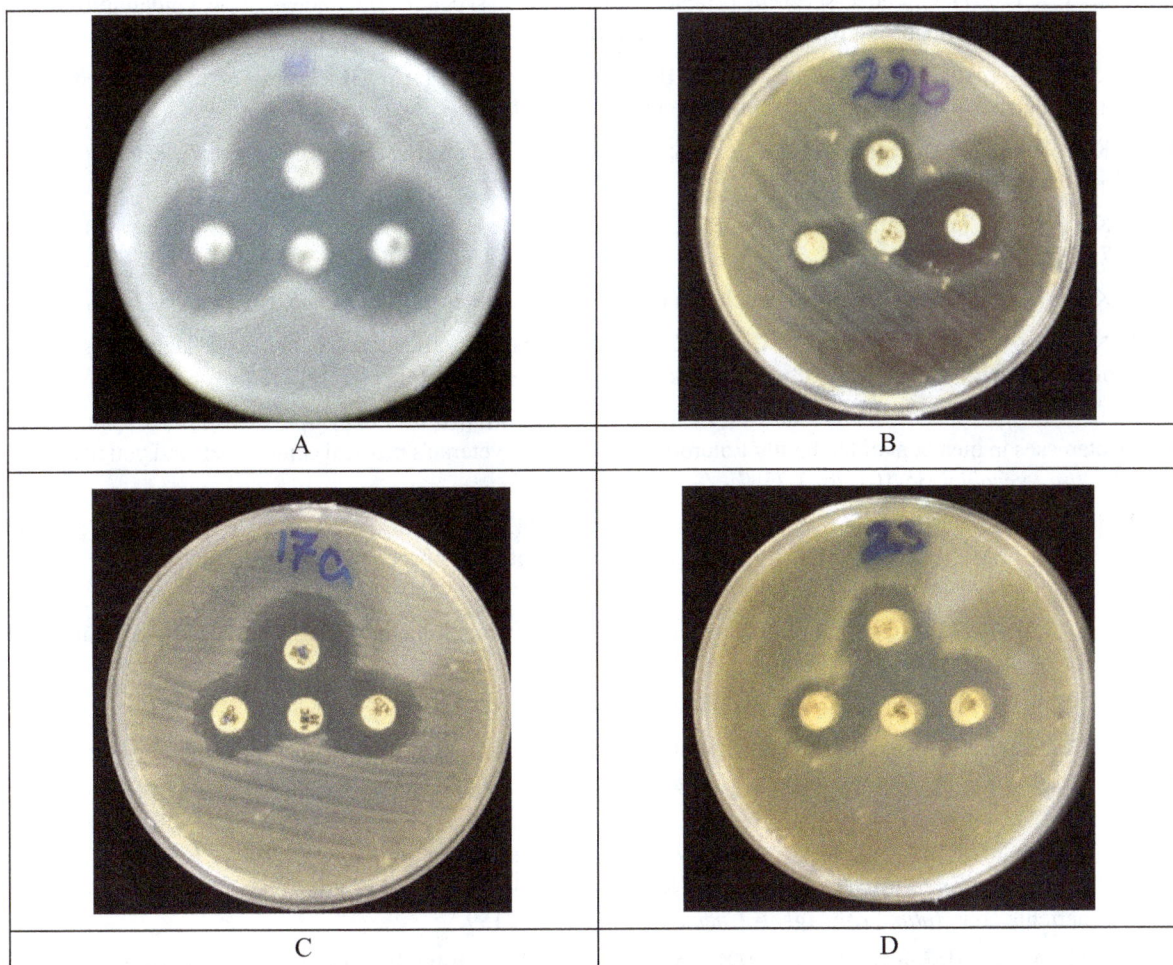

Figure 7. The results of A: ESBLS= bacterial non-ESBL, C, B and D= bacteria with ESBL

3.2 The Plasmid DNA Extraction

Plasmid DNA extraction on 20% of the samples (21 samples) by agarose gel 0.8% is shown in Figure 8. As it can be seen in the figure, plasmid samples have different bands.

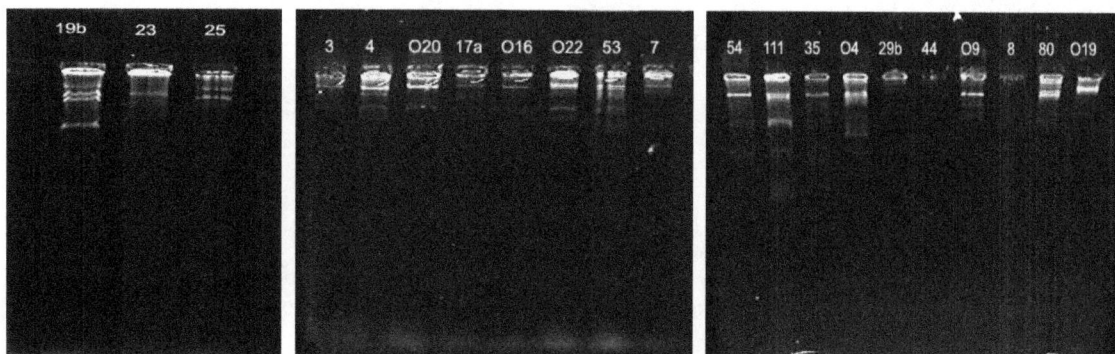

Figure 8. Results of the extraction of plasmid DNA (agarose gel 0.8%)

References

Barnaud, G., Labia, R., Raskine, L., Sanson-Le Pors, M. J., Philippon, A., & Arlet, G. (2001). Extension of resistance to cefepime and cefpirome associated to a six amino acid deletion in the H-10 helix of the cephalosporinase of an Enterobacter cloacae clinical isolate. *FEMS Microbiol. Lett, 195*, 185-190.

Bradford, P.A. (2001). Extended-spectrum betalactamases in the 21st century: characterization, epidemiology, and detection of this important resistance threat. *Clin. Microbiol. Rev, 14*, 933-951.

Bret, L., Chanal-Claris, C., Sirot, D., Chaibi, E. B., Labia, R., & Sirot, J. (1998). chromosomally encoded ampC type β-lactamase in a clinical isolate of Proteus mirabilis. *Antimicrob. Agents Chemother, 42*(5), 1110-4.

Bush, K. (1989a). Classification of β-lactamases: groups 1, 2a, 2b and 2b'. *Antimirob. Agents Chemother, 33*(3), 264-270.

Bush, K. (1989b). Classification of β-lactamases: groups 2c, 2d, 2e, 3 and 4. *Antimirob. Agents Chemother, 33*(3), 271-276.

Bush, K., & Jacoby, G. A. (1997). Nomenclature of TEM β-lactamases. *J. Antimicrob. Chemother, 39*, 1-3.

Bush, K., Jacoby, G. A., & Medeiros, A. A. (1995). A functional classification scheme for β- lactamases and its correlation with molecular structure. *Antimicrob. Agents Chemother, 39*, 1211-1233.

Cudron, P. E., Moland, E. S., & Sanders, C. C. (1997). Occurrence and detection of extended spectrum β-lactamases in members of the family Enterobacteriaceae at a veteran's medical center: seek and you may find. *J. Clin. Microbiol, 35*(10), 2593-2597.

Eftekhar, F., Hosseini-Mazinani, S. M., Ghandili, S., Hamraz, M., & Zamani, S. (2005). PCR detection of plasmid mediated TEM, SHV and AmpC β-lactamases in community and nosocomial urinary isolates of Escherichia coli. Iran. *J. Biotechnol, 3*(1), 48-54.

Gonzalez, C. M., & Schaeffer, A. J. (1999). Treatment of urinary tract infection: what's old, what's new and what works? *World J. Urol, 17*, 372-382.

Gupta, K. (2003). Addressing antibiotic resistance. *Dis. Mon, 49*, 99-110.

Hosseini-Mazinani, S. M., Jafar-Nejad, H., & Ghandili, S. (2003). Beta-lactam resistance patterns of *E. coli* isolated from urinary tract infections at a major university hospital in Iran. *J. Sci. Islam. Rep. Iran, 16*, 209-212.

Hosseini-Mazinani, S. M., Nakajima, E., Ihara, Y., Kameyama, K. Z., & Sugimoto, K. (1996). Recovery of active β-lactamases from Proteus vulgaris and RTEM-1 hybrid by random mutagenesis by using dnaQ of Escherichia coli. *Antimicrob. Agents Chemother, 40*, 2152-2159.

Jacoby, G. A., & Medeiros, A. A. (1991). More extended-spectrum beta-lactamases. Antimicrob. *Agents Chemother, 35*, 1697-1704.

Kim, J., Kwon, Y., Pai, H., Kim, J. W., & Cho, D. T. (1998). Survey of Klebsiella pneumoniae strains producing extended spectrum β-lactamases: prevalence of SHV-12 and SHV-2a in Korea. *J. Clin. Microbiol, 36,* 1446-1449.

Livermore, D. M. (1995). Beta-lactamases in laboratory and clinical resistance. *Clin. Microbiol, Rev., 8,* 557-584.

National Committee for Clinical Laboratory Standards. (2000a). Methods for dilution antimicrobial susceptibility tests for bacteria that grow aerobically, fifth edition: approved standard M7-A5. NCCLS, Wayne, Pennsylvania, USA.

National Committee for Clinical Laboratory Standards. (2000b). Performance standards for antimicrobial disc susceptibility tests, seventh edition: approved standard M2-A7. CCLS, Wayne, Pennsylvania, USA.

Pitout, J. D., Thomson, K. S., Hanson, N. D., Ehrhardt, A. F., Coudron, P., & Sanders, C. C. (1998). Plasmidmediated resistance to expanded-spectrum cephalosporins among Enterobacter aerogenes strains. *Antimicrob. Agents Chemother, 42,* 596-600.

Ross, G. W., & O'Callaghan, C. H. (1975). β-lactamase assays. *Methods Enzymol, 43,* 68-85.

Sutcliffe, J. G. (1978). Nucleotide sequence of the ampicillin resistance gene of Escherichia coli plasmid pBR322. *Proc. Natl. Acad. Sci. USA, 75,* 3737-3741.

Winokur, P. L., Vonstein, D. L., Hoffman, L. G., Uhlenhopp, E. K., & Doren, G. (2001). Evidence for transfer of CMY-2 AmpC β-lactamase plasmids between Escherichia coli and Salmonella isolates from food, animals and humans. *Antimicrob. Agents Chemother, 45,* 2716-2722.

PERMISSIONS

LIST OF CONTRIBUTORS

Mohamed El-Mogy and Yousef Haj-Ahmad
Department of Biological Sciences, Brock University, ON, Canada

Leila Arbabi, Mina Boustanshenas, Mastaneh Afshar, Maryam Adabi and Parwiz Owlia
Antimicrobial Resistance Research Center, Rasoul-e-Akram Hospital, Iran University of Medical Sciences, Tehran, Iran

Mohammad Rahbar
Antimicrobial Resistance Research Center, Rasoul-e-Akram Hospital, Iran University of Medical Sciences, Tehran, Iran
Department of Microbiology, Reference Health Laboratories Research Center, Deputy of Health, Ministry of Health and Medical Education, Tehran, Iran

Ali Majidpour and Mahshid Talebi-Taher
Antimicrobial Resistance Research Center, Rasoul-e-Akram Hospital, Iran University of Medical Sciences, Tehran, Iran
Department of Infectious Diseases, Rasoul-e-Akram General Teaching Hospital, Iran University of Medical Sciences, Tehran, Iran

Nasrin Shayanfar
Rasoul-e-Akram Hospital, Iran University of Medical Sciences, Tehran, Iran

Gérard Lucotte
Institute of Molecular Anthropology, 44 Monge Street, Paris 75 005, France

Peter Hrechdakian
Unifert Group S.A., 54 Louise Avenue, Immeuble Stéphanie, Bruxelles 1050, Belgium

Ibraimov A. I.
Institute of Balneology and Physiotherapy, Bishkek and Laboratory of Human Genetics, National Center of Cardiology and Internal Medicine, Bishkek, Kyrgyzstan

Jennifer Turner Waldo, Tsering Dolma and Emily Rouse
State University of New York at New Paltz, USA

Tinh T. Nguyen, Ricardo Martí-Arbona, Richard S. Hall, Tuhin Maity, Yolanda E. Valdez,
John M. Dunbar, Clifford J. Unkefer and Pat J. Unkefer
Bioscience Division, Los Alamos National Laboratory, Los Alamos, NM, USA

Roya Zand
Department of biology, Islamic Azad University, Tehran north brach, Tehran, Iran

Rana Azeez Hameed
Biology Department, Al-Mustansiriyah University, Baghdad, Iraq

Nidhal Neema Hussain
Biology Department, Baghdad University, Baghdad, Iraq

Abd aljasim Muhisen Aljibouri
Plant Biotechnology Department, Biotechnology Research Center, Al-Nahrain University, Baghdad, Iraq

Kingsley Ekwemalor and Emmanuel Asiamah
Department of Energy and Environmental Systems, North Carolina Agricultural and Technical State University, USA

Mulumebet Worku
Department of Animal Sciences, North Carolina Agricultural and Technical State University, USA

Johann Gross, Heidi Olze and Birgit Mazurek
Molecular Biology Research Laboratory, Department of Otorhinolaryngology, Charité Universitätsmedizin Berlin, Campus Charité Mitte, Berlin, Germany

Yingchun Liu, Jiaming Yin, Meili Xiao, Caihua Gao, Honglei Liu and Jiana Li
Engineering Research Center of South Upland Agriculture of Ministry of Education, College of Agronomy and Biotechnology, Southwest University, Chongqing, China

Donghui Fu
Key Laboratory of Crop Physiology, Ecology and Genetic Breeding, Ministry of Education, Jiangxi Agricultural University, Nanchang, China

Annaliese S. Mason
School of Agriculture and Food Sciences and ARC Centre for Integrative Legume Research, The University of Queensland, Brisbane, Australia

Ahmad Ebrahimi
Faculty of Agriculture, Zabol University, Iran
Department of Agriculture, Islamic Azad University,
Iranshahr Branch, Iranshahr, Iran

Mohammad Galavi and Mahmood Ramroudi
Faculty of Agriculture, Zabol University, Zabol, Iran

Payam Moaveni
Department of Agriculture, Islamic Azad University,
Shahr-e-Qods Branch, Tehran, Iran

**Antônio J. Rocha, José E.C. Freire and Antônio J.S.
Sousa**
Departamento de Bioquímica e Biologia Molecular,
Avenida Humberto Monte, s/n - Pici - CEP 60440-
900, Fortaleza - CE, Brasil

José E. Monteiro-Júnior
Departamento de Biologia, Avenida Humberto Monte,
s/n - Pici - CEP 60440-900, Fortaleza - CE, Brasil

Cristiane S.R. Fonteles
Departmento de Clínica Odontológica, Universidade
Federal do Ceará, Rua Monsenhor Furtado, s/n –
Rodolfo Teófilo - CEP 60430-350, Fortaleza - CE, Brasil

M. M. Corley and J. Ward
Agriculture Research Station, Virginia State University,
Virginia, USA

**Meysam Khosravifarsani, Ali Shabestani-Monfared,
Mahdi Pouramir and Ebrahim Zabihi**
Cellular and Molecular Biology Research Center, Babol
University of Medical Sciences, Babol, Iran

**Lijun Yang, Qifan Kuang, Yanping Jiang, Ling Ye,
Yiming Wu, Menglong Li and Yizhou Li**
College of Chemistry, Sichuan University, Chengdu,
China

**Emmanuel K Asiamah, Sarah Adjei-Fremah, Kingsley
Ekwemalor and Mulumebet Worku**
North Carolina A&T State University, USA

**Sepideh Rahimi, Noshin Sohrabi and Mohammad
Ali Ebrahimi**
Department of Plant Biotechnology, Tehran Shargh
Branch, Faculty of Agriculture Payam Noor University,
Iran

Majid Tebyanian and Morteza Taghi zadeh
Department of Research and development, The Razi
Vaccine and Serum Research Institute in the Hessarak
district in Karaj, Iran

Sahar Rahimi
Department of Food Science and Technology,
Pharmaceutical Sciences branch Islamic Azad
University, Tehran, Iran

Alireza Jafari
Antimicrobial Resistance Research Center, Rasoul-e-
Akram Hospital, Iran University of Medical Sciences,
Tehran, Iran

Ali Majidpour
Antimicrobial Resistance Research Center, Rasoul-e-
Akram Hospital, Iran University of Medical Sciences,
Tehran, Iran
Department of Infection Disease, School of Medicine

Roya Safarkar
Department of Microbiology, Islamic Azad University,
Ardabil Branch, Ardabil, Iran

Seyyedeh Masumeh Mirnurollahi
Department of Biology, Science and Research, Islamic
Azad University, Tehran, Iran

Shahrdad Arastoo
Department of Microbiology, Islamic Azad University
Qom Branch, Qom, Iran

Z. Mahmmudi
Kazeroun Branch, Islamic Azad University, Kazeroon,
Iran

A. A. Gorzin
School of Medicine, Shiraz University of Medical
Sciences, Shiraz, Iran

Mahnaz Milani
Department of Microbiology, Shahid Beheshti
University, Tehran, Iran

Index

www.ingramcontent.com/pod-product-compliance
Lightning Source LLC
Chambersburg PA
CBHW080655200326

41458CB00013B/4866